AF704641

Microbiology Monographs

Volume 25

Series Editor: Alexander Steinbüchel

Münster, Germany

For further volumes:
http://www.springer.com/series/7171

Gerold Barth
Editor

Yarrowia lipolytica

Biotechnological Applications

Springer

Editor
Gerold Barth
Inst. Mikrobiologie
TU Dresden
Dresden
Germany

Series Editor
Alexander Steinbüchel
Institut für Molekulare Mikrobiologie und Biotechnologie
Westfälische Wilhelms-Universität
Münster
Germany

ISSN 1862-5576 ISSN 1862-5584 (electronic)
ISBN 978-3-642-38582-7 ISBN 978-3-642-38583-4 (eBook)
DOI 10.1007/978-3-642-38583-4
Springer Heidelberg New York Dordrecht London

Library of Congress Control Number: 2013944661

© Springer-Verlag Berlin Heidelberg 2013
This work is subject to copyright. All rights are reserved by the Publisher, whether the whole or part of the material is concerned, specifically the rights of translation, reprinting, reuse of illustrations, recitation, broadcasting, reproduction on microfilms or in any other physical way, and transmission or information storage and retrieval, electronic adaptation, computer software, or by similar or dissimilar methodology now known or hereafter developed. Exempted from this legal reservation are brief excerpts in connection with reviews or scholarly analysis or material supplied specifically for the purpose of being entered and executed on a computer system, for exclusive use by the purchaser of the work. Duplication of this publication or parts thereof is permitted only under the provisions of the Copyright Law of the Publisher's location, in its current version, and permission for use must always be obtained from Springer. Permissions for use may be obtained through RightsLink at the Copyright Clearance Center. Violations are liable to prosecution under the respective Copyright Law.
The use of general descriptive names, registered names, trademarks, service marks, etc. in this publication does not imply, even in the absence of a specific statement, that such names are exempt from the relevant protective laws and regulations and therefore free for general use.
While the advice and information in this book are believed to be true and accurate at the date of publication, neither the authors nor the editors nor the publisher can accept any legal responsibility for any errors or omissions that may be made. The publisher makes no warranty, express or implied, with respect to the material contained herein.

Printed on acid-free paper

Springer is part of Springer Science+Business Media (www.springer.com)

Preface

Yarrowia lipolytica is an outstanding yeast, phylogenetically distant from *Saccharomyces cerevisiae* or other well-studied yeast species and standing alone at the bottom of the tree of hemiascomycetous yeasts. It was formerly called *Candida lipolytica*, *Endomycopsis lipolytica*, and *Saccharomycopsis lipolytica* and finally reclassified by van der Walt and von Arx (1980) as the type species of the genus *Yarrowia*. This genus was newly created and dedicated to David Yarrow, who has identified this genus. To differentiate this and other yeasts from the two well-studied yeast species *S. cerevisiae* and *Schizosaccharomyces pombe*, the artificial group of "nonconventional yeasts" has been created and was first used in the title of the conference on "Genetics of non-conventional yeasts" held in Weimar (Germany) in 1987. In the meantime, *Y. lipolytica* is the best-studied organism of this group. The detection of sexual reproduction in this yeast by Wickerham and colleagues in 1970 has forced genetic investigations and development of genetic tools as well as construction of laboratory strain lineages in American, French, and German groups for comparative studies. The sequencing of the complete genome of this yeast to which groups from Canada, France, Germany, Japan, Korea, and Spain have contributed has further encouraged studies with this yeast since 2004.

Y. lipolytica is used now as model organism for several studies of academic interest like degradation of *n*-alkanes, utilization as well as accumulation of lipids, biosynthesis and degradation of peroxisomes, secretion of proteins as well as metabolites, stress response, dimorphism, alternative intron splicing, genome evolution, and analysis of mitochondrial respiratory chain complex I.

The special physiological features of *Y. lipolytica* and the acceptance of its generally recognized as safe (GRAS) status make this dimorphic yeast significant for biotechnological applications. Properties like intracellular accumulation of oil, production of hydroxyl or dicarboxylic acids, as well as secretion of large amounts of organic acids focus the interest on this yeast as a potential producer of basic commodities, fine chemicals, or building blocks for chemical industry in the post-alkane area. Furthermore, the high capacity for secretion of proteases and lipases favors this yeast as a producer of these enzymes as well as of heterologous proteins.

The aim of the two *Microbiology Monograph* volumes is to summarize the huge knowledge of the biology and applications of this fascinating yeast. The first volume covers the genetics, genomics, and physiology of *Y. lipolytica*. The second volume provides a broad survey of biotechnological applications of this yeast and discusses new developments.

The editor wants to thank all authors for their excellent contributions, their substantial efforts, and especially for their lots of patience. Many thanks are also given to the series editor, Alexander Steinbüchel, for his help and encouragement and to Jutta Lindenborn, Springer, for her help during the publishing process. Finally, the editor wants to thank the people behind the scenes, especially Falk Matthäus, who supported the project in putting into effect.

Dresden Gerold Barth
March 2013

Contents

Heterologous Protein Expression and Secretion in *Yarrowia lipolytica* 1
Catherine Madzak and Jean-Marie Beckerich

Acid and Alkaline Extracellular Proteases of *Yarrowia lipolytica* 77
David Ogrydziak

The Lipases from *Y. lipolytica*: Genetics, Production, Regulation, and Biochemical Characterization 99
Patrick Fickers, Vincent Sauveplane, and Jean-Marc Nicaud

Biotechnological Applications of *Yarrowia lipolytica* Lipases: An Overview 121
Patrick Fickers and Jean-Marc Nicaud

Production of Organic Acids by *Yarrowia lipolytica* 137
Christina Otto, Martina Holz, and Gerold Barth

Production of Dicarboxylic Acids and Flagrances by *Yarrowia lipolytica* 151
Yves Waché

Cytochrome P450 Expression in *Yarrowia lipolytica* and Its Use in Steroid Biotransformation 171
Stephan Mauersberger, Ludmila A. Novikova, and Vladimir M. Shkumatov

Cytochromes P450 of the Alkane-Utilising Yeast *Yarrowia lipolytica* . . . 227
Stephan Mauersberger

Index 263

Heterologous Protein Expression and Secretion in *Yarrowia lipolytica*

Catherine Madzak and Jean-Marie Beckerich

Abstract Heterologous protein production constitutes an important research field, having both academic and commercial applications. The use of yeasts as host systems took advantage of the combination of their easy manipulation and high growth capacity, with their eukaryotic subcellular organisation allowing post-translational processing. Among available yeast host systems, *Yarrowia lipolytica* (Yarrowia) appears as one of the most attractive. This non-conventional dimorphic yeast has been distinguished due to its remarkable regularity of performance in the efficient secretion of various heterologous proteins. We will present in this chapter Yarrowia's characteristics regarding heterologous protein production and describe the genetic and molecular tools available. Recent developments of the Yarrowia expression system, such as surface display vectors, engineered strains and high-throughput screening processes, will be reviewed. A comprehensive survey of the literature allowed us to list more than 100 heterologous proteins, from more than 60 species, successfully produced until now in this yeast. This amount of data sometimes offers a comparison with other yeast host systems and globally demonstrates the reliability and versatility of Yarrowia as host for heterologous production.

Contents

1 Introduction 2
 1.1 Heterologous Expression in Yeast Host Systems 2
 1.2 Yarrowia as a Host for Heterologous Production 4
 1.3 Chronology of Heterologous Production in Yarrowia 7
2 Genetic and Molecular Tools for Heterologous Production 7
 2.1 Expression/Secretion Vectors 8
 2.2 Recipient Strains 12
 2.3 Selection Marker Genes 13
 2.4 Expression Cassettes 14

C. Madzak (✉) • J.-M. Beckerich
INRA, UMR1319 Micalis, Domaine de Vilvert, 78352 Jouy-en-Josas, France
e-mail: Catherine.Madzak@grignon.inra.fr; Jean-Marie.Beckerich@grignon.inra.fr

G. Barth (ed.), *Yarrowia lipolytica*, Microbiology Monographs 25,
DOI 10.1007/978-3-642-38583-4_1, © Springer-Verlag Berlin Heidelberg 2013

2.5 Development of a Commercial Yarrowia Expression Kit ... 37
2.6 Amplification of Expression Cassette Copy Number ... 37
2.7 High-Throughput Screening Processes ... 41
3 Review of Heterologous Protein Production ... 42
3.1 Promoters and Transcription Efficiency ... 43
3.2 Codon Usage Bias and Translation Efficiency ... 45
3.3 Secretion Signals, Maturation and Secretion Efficiency ... 46
3.4 Glycosylation and Overglycosylation ... 48
3.5 Cultivation Conditions and Scale-Up ... 49
3.6 Comparison with Competitor Expression Systems ... 53
4 Conclusions and Prospects ... 55
4.1 Use of Yarrowia for Genetic Engineering ... 56
4.2 Use of Yarrowia as Whole-Cell Biocatalyst ... 57
4.3 Use of Yarrowia as Biosensor ... 62
4.4 Projects for "Universal" vectors ... 63
4.5 Projects for Engineered and "Humanised" Strains ... 64
References ... 66

1 Introduction

Heterologous protein production consists in using an easy-to-handle organism (the host) to produce a protein of interest from another (generally more complex) organism. Since several decades, this research area has rapidly developed and been applied to both academic and commercial purposes. Recombinant proteins are used in a variety of biomedical applications (e.g. interferon, insulin, interleukins or viruses antigens) or of industrial processes (e.g. amylases for food processing and paper industry, cellulases for biofuel production, laccases for bleaching or chymosin for cheesemaking). Historically, the first host system used for the industrial production of heterologous proteins has been the well-studied eubacterium *Escherichia coli*. But prokaryotic hosts present some drawbacks: unability to perform post-translational processing, folding problems with formation of insoluble complexes and possible contamination with toxins. Thus, as more complex proteins were tentatively expressed, the development of new transformation systems increased greatly the range of interesting hosts. Among those now currently used are prokaryotes (mostly *E. coli*), several yeasts (e.g. *Saccharomyces cerevisiae*, *Pichia pastoris* and *Kluyveromyces lactis*), filamentous fungi (e.g. *Aspergillus niger*), moss protoplasts, insect cells (baculoviral expression system), mammalian cells (e.g. episomal adenoviral and integrated retroviral expression systems) and even transgenic animals or plants (e.g. production in goat milk or in tobacco leaves).

1.1 Heterologous Expression in Yeast Host Systems

Yeasts offer a number of advantages as expression systems for complex proteins. Being unicellular organisms, they present an ease of manipulation and a growth capacity similar to those of bacteria. Additionally, in contrast to prokaryotes, they possess a eukaryotic subcellular organisation able to perform the post-translational processing necessary for expression of complex proteins from "higher organisms". In yeasts, secretion occurs via a complex multi-component apparatus, encompassing two major pathways: a post-translational and a co-translational one. This organisation allows proteolytic maturation, formation of disulfide bonds, N-linked and O-linked glycosylation of defined sites and other minor post-translational modifications. The first yeast used for heterologous production has been the well-studied ascomycete *S. cerevisiae*. The choice of this model organism allowed to beneficiate from the high amount of knowledge accumulated on its genetics and physiology. *S. cerevisiae* also presented a safety guaranty, considering its millennial use in baking and brewing, and, more recently, the Generally Regarded As Safe (GRAS) classification of industrial processes involving this yeast (FDA, Food and Drug Administration, USA). However, its first uses revealed some limitations: low product yield, poor plasmid stability, difficulties in scaling-up production, hyperglycosylation of recombinant proteins and low secretion capacity. Since then, heterologous production using *S. cerevisiae* has much progressed, especially by the design of improved strains overexpressing chaperones and foldases (Shusta et al. 1998). Meanwhile, some laboratories chose in contrast to develop alternative yeast expression systems.

These competitor systems include the well-studied fission yeast *Schizosaccharomyces pombe* and the ill-assorted group of so-called "non-conventional" yeasts: mainly the methylotrophic *P. pastoris* and *Hansenula polymorpha*, the dairy yeast *K. lactis*, the amylolytic yeast *Schwanniomyces occidentalis* and the dimorphic *Arxula adeninivorans* and *Yarrowia lipolytica*. These alternative expression systems benefited from the technical know-how developed on *S. cerevisiae*. Their performances frequently surpassed those of *S. cerevisiae* in terms of product yield, reduced hyperglycosylation and secretion efficiency, especially for large complex proteins (reviewed by Buckholz and Gleeson 1991; Dominguez et al. 1998; Gellissen et al. 2005). The prevalent expression system among "non-conventional" yeasts was developed in *P. pastoris* and commercialised since the 1990s as a kit (available from Invitrogen, USA). It has already allowed the production of several hundreds of heterologous proteins (reviewed by Cereghino and Cregg 2000; Cregg et al. 2000; Cereghino et al. 2002). However, even this successful system encounters some obstacles in some cases (lower secretion level, hyperglycosylation) (Cereghino et al. 2002), giving room to improvements or to competitors. More recently, expression kits have also been developed in *K. lactis* (commercialised by New England Biolabs, UK) and in Yarrowia (commercialised by Yeastern Biotech Co., Taiwan—cf. Sect. 2.5).

It remains difficult to evaluate comparatively the performances of all the different yeast expression systems, due to differences in the array of heterologous proteins tested and to disparities between the characteristics of the strains/vectors used and of the methodologies employed. For these reasons, the prevalent comparative test up to now still remains the work by Muller et al. (1998): *S. cerevisiae*, *H. polymorpha*, *K. lactis*, *S. pombe* and Yarrowia were evaluated for their capacity to secrete active forms of six fungal enzymes. All of the examined alternative hosts performed better than *S. cerevisiae*, but their relative efficiency varied significantly with each heterologous protein. The most attractive host, especially in terms of performance reproducibility, was Yarrowia.

1.2 Yarrowia as a Host for Heterologous Production

Since more than 40 years, Yarrowia has promoted interest for industrial applications. Firstly oriented towards the production of metabolites with commercial value, industrial processes were secondly aimed at producing heterologous proteins, as soon as genetic and molecular tools were developed in Yarrowia.

1.2.1 Main Characteristics

Yarrowia is a dimorphic organism: either yeast cells or hyphae and pseudohyphae are formed, depending on growth conditions (Dominguez et al. 2000; Ruiz-Herrera and Sentandreu 2002). This yeast was noticed at first for its capacity to grow on *n*-paraffins and to produce high amounts of organic acids. Large-scale industrial use of Yarrowia for production of citric acid, or of single-cell proteins, permitted the accumulation of extensive data on its cultivation. These processes have been classified as GRAS by the FDA. Wild-type strains can use glucose (but not sucrose, in contrast to some engineered strains—cf. Sect. 2.2), alcohols, acetate and hydrophobic substrates (such as alkanes, fatty acids and oils) as carbon sources (reviewed in Barth and Gaillardin 1997). In accordance with its species name (lipolytica), this yeast is characterised by its capacity to metabolise very efficiently lipids (Fickers et al. 2005). The very effective uptake of hydrophobic substrates can be attributed to protrusions on the surface of the cell that connect it through a channel to the inner side (Fickers et al. 2005). This peculiar feature is particularly interesting for the use of Yarrowia as an effective whole-cell biocatalyst for hydrophobic substrates bioconversion (cf. Sect. 4.2). Yarrowia has now become a model organism in several research areas, among which are the study of fatty acid metabolism, which leads to the selection of interesting promoters (Juretzek et al. 2000a), and the study of general secretion pathway (Beckerich et al. 1998; Swennen and Beckerich 2007). At last, the sequencing of Yarrowia genome by the Genolevures Consortium (Dujon et al. 2004) provided new tools for Yarrowia research (http://www.genolevures.org/), which could benefit to the heterologous production domain. A review on environmental and industrial applications of Yarrowia has recently been made by Bankar et al. (2009). They highlighted the applications of this yeast in bioremediation of polluted environments and also presented industrially important processes recently developed. These various applications make use of both inherent enzymes (such as inulinases, α-mannosidases, esterases and lipases) and heterologous proteins expressed in Yarrowia. The examples cited in this review demonstrate the use of wild-type, mutant or genetically manipulated strains for developing various industrial processes and technologies.

1.2.2 Secretion of Proteins

Since a long time, wild-type Yarrowia strains have been distinguished for the high secretion yield of numerous proteins, among which are two proteases (the alkaline extracellular protease, AEP, and the acid extracellular protease, AXP), several lipases and phosphatases, an RNase and an esterase (reviewed in Barth and Gaillardin 1997). Under convenient inducing conditions, some Yarrowia strains are able to secrete very large amounts (1–2 g/l—Tobe et al. 1976; Ogrydziak and Scharf 1982) of AEP. Consequently, this protease became a model in secretion research, and the corresponding gene, *XPR2*, a useful molecular tool for both secretion studies and heterologous production development (Matoba et al. 1988; Fabre et al. 1991, 1992). Namely, the *XPR2* promoter, the *XPR2* prepro region (secretion signal + folding-helping prodomain) and the *XPR2* terminator have been the first elements used at the beginnings of heterologous production in Yarrowia.

At the early steps of their synthesis, eukaryotic proteins intended to be secreted can be translocated from the cytoplasm to the endoplasmic reticulum (ER) by following either a co-translational pathway or a post-translational one. In mammalian cells, as soon as the signal peptide emerges from the ribosome, it is recognised by a cytoplasmic signal recognition particle (SRP), in which binding causes a translational pause and directs the complex to the ER membrane through interaction with a SRP receptor (Walter and Blobel 1980, 1983). Then translation resumes: the nascent protein is directly released in the ER lumen, where its folding takes place in presence of chaperones. In contrast, in *S. cerevisiae*, the components involved in the SRP-dependent pathway are not essential, and several secreted proteins have been shown to cross the ER membrane following their complete synthesis in the cytoplasm (Brown et al. 1994; Rothblatt and Meyer 1986). Thus, the post-translational pathway is predominant during protein secretion in *S. cerevisiae*. Otherwise, Yarrowia is remarkable in this domain since, although being a yeast, its protein secretion mainly follows the co-translational pathway (Boisramé et al. 1998), as it is the case in "higher" eukaryotes (e.g. in mammalian cells). This salient feature constitutes an important advantage of Yarrowia over *S. cerevisiae* for the efficient production of complex heterologous proteins. In particular, it can be supposed that the folding of proteins of mammalian origin could be optimised in the very special environment of the ER, in the presence of a large array of chaperones, with redox conditions very different from those encountered in the cytoplasm. An extensive study of the homology between 165 proteins implied in vesicular secretion in Yarrowia and their counterparts in other organisms has shown that 72 % of them were closest to their fungal homologues and that 40 % were closer to their mammalian homologues than to *S. cerevisiae* ones (Swennen and Beckerich 2007). Thus, as it has been previously observed for ER translocation pathway (Boisramé et al. 1998), vesicular secretion in Yarrowia also appears to be closer in many aspects to that in fungi or in mammals than to that in *S. cerevisiae* (Swennen and Beckerich 2007). A recent study of the quality control of membrane proteins in Yarrowia has been performed in our laboratory, by making use of mutated forms of uracil permease that were retained in the cell, in which immunodetection allowed mass spectrometry analysis of the associated proteins. This work contributed to identify the components of Yarrowia folding proteome, from translocation to folding and vesicular transport, or degradation via the proteasome (Swennen et al. 2010).

1.2.3 Glycosylation of Proteins

N-linked glycosylation constitutes a major post-translational modification of proteins, occurring during their passage in the secretory pathway. The characteristics of glycosylation in a given host can affect activity and stability of heterologous proteins. In addition, as many proteins of therapeutic interest are glycosylated, the precise type of the modifications could affect their clearance rate and their immunogenicity. Unfortunately, there are major differences in glycosylation pathways from yeasts and mammals (reviewed in Cereghino and Cregg 2000). Briefly, N-linked glycosylation in all eukaryotic cells begins with the addition of a core oligosaccharide to Asp residue from consensus sequence Asp-X-Ser/Thr (where X could be any amino acid except Pro). Following trimming of this core, glycosylation patterns of "lower" (yeasts) and "higher" eukaryotes (mammals) differ (Goochee et al. 1991): yeasts add only mannose outer chains, when mammals generate three different types of further modifications (high-mannose, complex and hybrid types). Although yeast glycosylation resembles the high-mannose type of mammalian glycosylation, the length of mannose outer chains can vary greatly for heterologous proteins and can be much longer than in mammals. In *S. cerevisiae*, outer chains contain a long α-1,6-linked backbone with short α-1,2 and α-1,3 side chains; backbone mannose chains are typically of 50–200 residues, a condition referred to as hyperglycosylation (Dean 1999). Hyperglycosylated heterologous proteins can lose activity, and problems become tougher for therapeutic proteins, which can exhibit a rapid clearance or be exceedingly antigenic (reviewed in Hamilton and Gerngross 2007).

Fortunately, some "non-conventional" yeasts present less hyperglycosylation problems, which renders them more adapted to the production of therapeutic proteins. The length of mannose chains added to heterologous proteins by *P. pastoris* and *H. polymorpha* was shown to be in the range of 8–14 residues (Grinna and Tschopp 1989; Gellisen and Hollenberg 1997). Until recently, the only detailed information available on the length of mannose chains in Yarrowia was an unpublished study which showed that human tissue plasminogen activator secreted from this yeast contained only short oligosaccharide chains, of 8–10 residues (A. Franke, personal communication). These results have been confirmed recently by an extensive study on the glycosylation of a homologous lipase (Jolivet et al. 2007): the major YlLip2p isoform exhibited 17 mannoses distributed between the two glycosylation sites (Man8GlcNAc2 and Man9GlcNAc2 glycoforms). However, a minor heavier YlLip2p isoform was shown to exhibit a total of 27 mannoses for these two sites. Although N-glycosylation has been reported to be required for folding, secretion and activity of proteins, especially lipases, Jolivet et al. (2007) provided evidence that glycosylation of YlLip2p was not necessary, neither for its secretion nor for its activity. Concerning heterologous proteins, maize cytokinin oxidase I (*Zea mays* CKO1p) secreted in Yarrowia was shown to exhibit an average chain length of 17 mannose units, and *Trichoderma reesei* endoglucanase I was found to have oligosaccharide chains similar to those of homologous YlLip2p, ranging from 8 to 12 mannoses (Kopecny et al. 2005; Song et al. 2007—cf. Sect. 3.4).

Another problem linked to glycosylation in *S. cerevisiae* is that terminal mannose residues are linked by a α-1,3-bond, which is immunogenic in humans (Nakajima and Ballou 1975). In contrast, only non-immunogenic terminal α-1,2-linked mannoses were found in *P. pastoris* and *H. polymorpha* (Trimble et al. 1991; Kim et al. 2004). Similarly, a recent study by Song et al. (2007) concluded to the probable absence of immunogenic terminal α-1,3-mannose linkage in Yarrowia, confirming the interest of this yeast host system for the production of recombinant proteins with human-compatible glycosylation.

In conclusion, the glycosylation pattern of some "non-conventional yeasts", including Yarrowia, appears closer to the mammalian high-mannose type of N-glycosylation than to that of *S. cerevisiae*.

1.3 Chronology of Heterologous Production in Yarrowia

Since the development of a transformation method in 1985 (Davidow and Dezeeuw 1985; Gaillardin et al. 1985), and the first examples of recombinant proteins in 1987 (Davidow et al. 1987a, b; Gaillardin and Ribet 1987), heterologous production in Yarrowia has known a rapid development, as exemplified in Fig. 1. This figure presents a chronological report of the evolution of this research domain. When, nearly 20 years ago, Buckholz and Gleeson (1991) wrote one of the first review of heterologous production in yeasts, they listed five heterologous proteins produced using Yarrowia. In a more recent review from our laboratory (Madzak et al. 2004), they were 42, and they are now more than 100, from more than 60 different organisms of various phylogenetic origins (cf. Sect. 3). The landmarks from Fig. 1 refer to main methodological innovations (patents, kit development) which will be described in details in Sect. 2.

2 Genetic and Molecular Tools for Heterologous Production

In addition to efficient transformation processes, the design of expression systems for heterologous proteins requires the development of numerous genetic and molecular tools. The former correspond to recipient strains, carrying auxotrophy

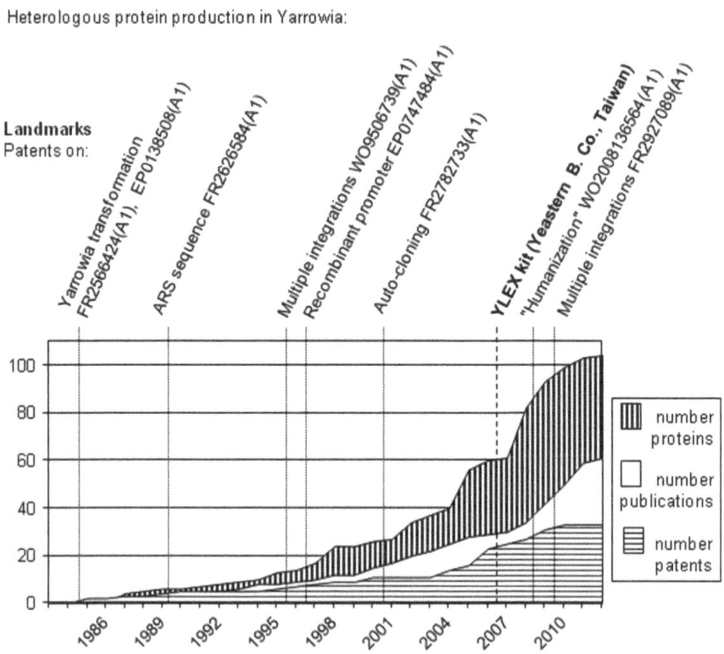

Fig. 1 Chronology of heterologous protein production in Yarrowia. The numbers of proteins produced, of corresponding scientific publications, and of patents related to either methodology or examples of heterologous production in Yarrowia are shown as a function of time (years). Landmarks corresponding to the main patents describing methodological progresses (description of the use of native promoters has been omitted) and to the commercialisation of a kit for heterologous expression in Yarrowia (in *bold*) are also indicated

marker genes, or compatible with the use of heterologous dominant marker genes. The latter correspond to selectable vectors able to replicate in the recipient strains or to integrate into their genome. The different possibilities will be exemplified hereafter.

2.1 Expression/Secretion Vectors

For reasons of convenience, the vectors for transforming Yarrowia, like for other yeasts, are shuttle vectors (vectors that can be constructed and amplified in bacteria—typically E. coli—and then "shuttled" to the organism of interest). They typically contain a plasmidic backbone (the bacterial moiety) and, for expression/secretion in Yarrowia, (1) a selection marker, (2) an expression/secretion cassette and (3) elements for transformation and maintenance into yeast cells. These different types of genetic elements are represented on a virtual vector in Fig. 2 and are detailed hereafter. The shuttle vector can be introduced in *E. coli* using standard molecular biology methods and transferred into the Yarrowia

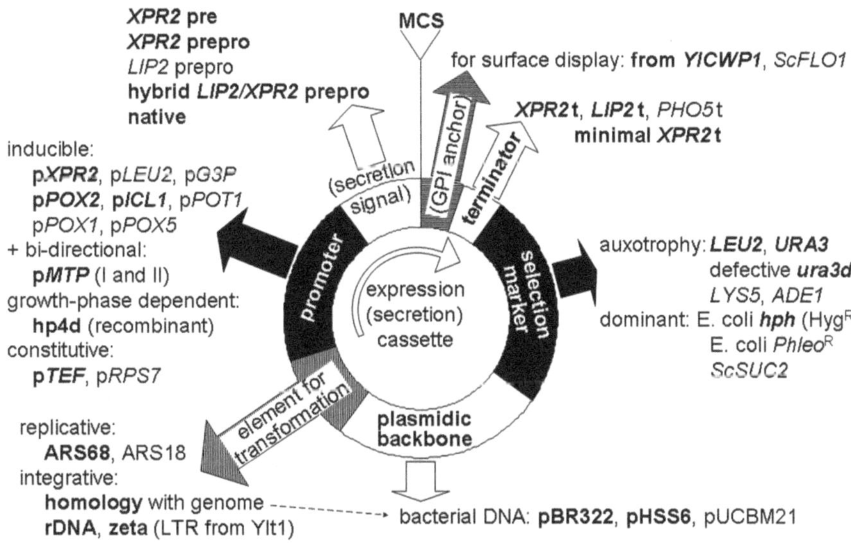

Fig. 2 General scheme of a vector for expression (with or without secretion) of heterologous proteins in Yarrowia. The different types of genetic elements required are represented on a virtual vector. Optional elements (signal peptide and GPI anchor) are indicated within *brackets*. The *arrows point*, for each part, to a list of the main genetic elements described in the literature. The genetic elements which have been used the most are in *bold case* (see the text for details). The homology with the genome which is used for the targeted integration of integrative vectors can coincide with a genetic element present in the vector for another purpose, such as the terminator region (of the expression cassette, or of the selection marker gene), or with the plasmidic backbone (if the same sequence has been integrated beforehand in the genome of the recipient strain)

recipient strain of interest using either a physical or a chemical method (depending on its integrative or replicative nature, as explained hereafter). Some examples of vectors developed in our laboratory or by some of our collaborators are shown in Table 2.

2.1.1 Integrative Versus Replicative Vectors

Two types of shuttle vectors can be used, differing by their mode of maintenance into yeast cells: episomal (replicative) vectors and vectors designed for integration into the yeast genome. None of the known Yarrowia strain bears any natural episome, but replicative plasmids using chromosomal replication origins have been designed. Several autonomously replicating sequences (ARS) were isolated (Fournier et al. 1991; Matsuoka et al. 1993). They exhibit unusual properties, due to the colocalisation of centromeric and replicative functions (Fournier et al. 1993). In consequence, ARS elements are not very attractive for the design of expression/secretion vectors, since the copy number of ARS-based vectors is only 1–3 per cell (Vernis et al. 1997; Madzak et al. 2000), and their gene expression level is correspondingly limited (Nicaud et al. 1991; Madzak et al. 2000). In addition, their high loss frequency imposes the maintenance of a selective pressure during cultivation, which is often incompatible with efficient industrial management. However, replicative vectors are useful when only a transient expression is required (e.g. pRRQ1 in Table 2). This is the case when using, for example, a gene disruption method based on the Cre-lox recombination system (Fickers et al. 2003), which requires the transient expression of Cre recombinase for efficient marker rescue. Replicative vectors can be efficiently introduced in Yarrowia cells using electroporation (Fournier et al. 1993).

In contrast, integrative vectors are much more adapted to heterologous production in Yarrowia and are widely used for that purpose. Integration of exogenous DNA into Yarrowia genome occurs almost exclusively by homologous recombination, with the notable exception of the zeta-based non-homologous integration system developed in our laboratory (cf. Sect. 2.6.2). Homologous recombination by single crossover is greatly stimulated by linearising the vector within the homology region, which also allows to target the integration to a precise genomic region. When using a sufficiently large region of homology (more than 300 bp on each side), a single complete copy of the vector integrates at the chosen site in more than 80 % of cases (Barth and Gaillardin 1996), while remaining events include multiple tandem integrations, gene conversions and out of site integrations. Integrated vectors exhibit a very high stability, as demonstrated by Hamsa and Chattoo (1994): they were retained without any rearrangement after 100 generations under non-selective cultivation conditions. Different genetic elements can be used for targeting, as exemplified in Fig. 2 and in Sects. 2.5, 2.6.1 and 2.6.2. Targeted integration allows to obtain very high transformation frequencies (10^5–10^6 transformants per μg of DNA—Xuan et al. 1988) when using chemical transformation methods. The preferred method for integrative transformation is the lithium acetate method described by Xuan et al. (1988), which has been simplified (one step) by Chen et al. (1997) and adapted for auto-cloning multicopy (defective) vectors in our laboratory (Madzak et al. 2005a). Until recently, electroporation was not used for integrating linearised vectors, due to its very low transformation efficiency with this kind of material. However, Wang et al. (2011) have recently developed a new electroporation procedure yielding efficiencies in the range of 10^4 transformants per μg of DNA for integrative vectors, making it an interesting alternative to chemical methods.

Although roughly ten times less efficient, the replacement of a chromosomal sequence using integration of a DNA fragment, with flanking homologous regions, by double crossover, can also easily be obtained (Barth and Gaillardin 1996). This method is notably used for gene deletions (i.e. Fickers et al. 2003).

In addition to their advantages over replicative vectors, in terms of easier handling and increased stability, integrative vectors also offer the possibility of multiple integrations, with correlated increase in gene expression (cf. Sect. 2.6).

2.1.2 Auto-cloning Versus Classical Shuttle Vectors

Shuttle vectors used for expression/secretion include some bacterial DNA, constituting the plasmidic backbone (cf. Fig. 2). This bacterial moiety basically consists in a bacterial replication origin and a marker gene for selection in *E. coli* (conferring resistance to an antibiotic, such as ampicillin, tetracycline or kanamycin). Following transformation with a classical shuttle vector, the presence of this bacterial DNA (especially of the antibiotic resistance gene) into the producing yeast strain could be a drawback to acceptance by regulatory authorities for commercial/ industrial applications. In order to alleviate this problem, our laboratory developed several "auto-cloning" expression vectors, which name reflects the fact that only Yarrowia DNA (together with the heterologous gene) is introduced into the recipient Yarrowia strain. Namely, the bacterial moiety of these vectors can be separated from the "yeast cassette" using restriction digestion followed by agarose gel electrophoresis, and the purified "yeast cassette" alone is used for transformation (Nicaud et al. 2000; Pignède et al. 2000). The resulting producing strain is devoid of bacterial DNA, which, together with Yarrowia GRAS status, should greatly facilitate regulatory approval. The "yeast cassette" which constitutes the transforming part of the auto-cloning vectors is composed of (1) an auxotrophy marker, (2) the expression cassette and (3) elements for integration into the recipient strain genome. Zeta sequences are generally used for this later purpose (cf. Sect. 2.6.2). Such auto-cloning vectors have been used for the production of homologous (Pignède et al. 2000) or heterologous proteins (Juretzek et al. 2001; Nicaud et al. 2002; Laloi et al. 2002; Nthangeni et al. 2004; Bourel et al. 2004; Shiningavamwe et al. 2006; Labuschagné and Albertyn 2007; Maharajh et al. 2008a; Roth et al. 2009). Examples of these vectors are given in Table 2.

A common point to both auto-cloning and classical shuttle vectors is that the first step of insertion of the heterologous gene is performed in *E. coli*. However, in a few rare cases, some shuttle vectors carrying either homologous or heterologous genes have encountered stability problems in *E. coli* (rearrangements, mutations—Babour et al. 2004, N. Houba-Hérin, personal communication). In order to palliate these problems, due to "poisonous" DNA sequences or to toxicity of leakily expressed proteins, we developed a "direct cloning" strategy. Stability problems occurred in *E. coli* during the cloning of maize cytokinin oxidase III (CKO3p—N. Houba-Hérin, personal communication) and were avoided by inserting the amplified Zm*CKO3* cDNA into a Yarrowia vector which, instead of being used to transform *E. coli* (bacterial cloning step), served as matrix for a PCR step. The Yarrowia vector used was designed to allow the synthesis of a PCR product mimicking an integrative expression vector linearised into a region homologous to the genome, in order to maximise transformation efficiency in Yarrowia. The PCR fragments obtained were used to transform directly Yarrowia, with an efficiency similar to that of a genuine Yarrowia expression vector (C. Madzak, unpublished data). This strategy, avoiding any bacterial step and leading to producing strains devoid of bacterial DNA, has been successfully used for expressing maize CKO3p in

Yarrowia (Massonneau et al. 2004). However, no comparison of the efficiency and precision of targeting between PCR fragments and linearised vectors was performed in this study.

2.2 Recipient Strains

At least in some yeast species, the choice of the host strain could be a main parameter for heterologous protein production efficiency, as exemplified by De Baetselier et al. (1991) in *S. cerevisiae*: a 100-fold difference range was found between strains. However, this subject has not really been studied in Yarrowia, and the strains used for heterologous production have only been selected more or less empirically for their capacity to secrete high levels of homologous proteins. Although very few comparative data are available for heterologous production in Yarrowia strains, Po1g strain was found around threefold more efficient than JM23SB strain for production of bovine (*Bos taurus taurus*) prochymosin (Madzak et al. 2000). A comprehensive list of Yarrowia strains which have been used for heterologous expression can be found in Madzak et al. (2004). A selection of host strains is shown in Table 1. In addition to E129 and the Po1 series of strains, used for heterologous expression, this table also includes the E150 strain which has been used for the Yarrowia genome sequencing project (Dujon et al. 2004). In contrast to E129 and E150 strains, which carry Ylt1 retrotransposon in their genome, Po1d and its derivatives are devoid of this transposable element. This point is relevant for the use of zeta-based vectors (cf. Sects. 2.6.1 and 2.6.2).

Until recently, Po1d (Le Dall et al. 1994) was one of the most frequently used recipient strains, due to interesting features regarding heterologous protein production: (1) high level of secretion; (2) deletion of AEP, a potent threat for secreted heterologous proteins; (3) auxotrophies due to non-revertant mutations; and (4) production of recombinant invertase (*SUC2* gene from *S. cerevisiae*), allowing the utilisation of sucrose as a new carbon source (cf. Table 1—Nicaud et al. 1989). The design of invertase-positive Yarrowia strains is particularly relevant for industrial applications, since they can grow efficiently on molasses, a cheap and abundant substrate (Wojtatowicz et al. 1997). The interesting features of Po1d strain have been retained in a series of derivatives (cf. Table 1) that have been further adapted for heterologous protein production (Madzak et al. 2000, 2004). Po1f, Po1g and Po1h were further deleted for the acid extracellular protease (AXP), eliminating all known sources of secreted proteasic activity. Po1g was fitted with an integrated pBR322 docking platform, rendering easy the further integration of pBR322-based vectors (cf. Sect. 2.5). Po1g and Po1h retain only one auxotrophy, allowing their transformants to be prototrophs for easier handling. The Po1 series of strains is now clearly prominent in the domain of heterologous production: one or the other strain was used as a host in most of the examples listed in Sect. 3.

A few wild-type Yarrowia strains isolated from marine environment have been used for peculiar applications, following the construction of Ura^- mutants using

5-fluoro-orotic acid (5-FOA) in order to allow their genetic engineering (Wang et al. 2009; Liu et al. 2010; Zhao et al. 2010; Cui et al. 2011). These strains were selected from a library of 78 marine Yarrowia strains for their interesting properties:

- ACA DC 50109 strain, exhibiting a high lipid content, was further engineered for use as single-cell oil producer (Zhao et al. 2010—cf. Sect. 4.2.4).
- SWJ-1b strain, exhibiting a high protein content (Wang et al. 2009), was further engineered for use as single-cell protein producer (Cui et al. 2011—cf. Sect. 4.2.3) or as citric acid producer (Liu et al. 2010—cf. Sect. 4.2.6).

2.3 Selection Marker Genes

Like in other yeasts, the genes that can be used for selection in Yarrowia are either auxotrophy or dominant markers. Auxotrophy markers are able to correct the growth defect of a corresponding auxotrophic strain, when dominant markers (acquisition of antibiotic resistance or new metabolic property) can be used in any strain, including wild-type prototrophic strains.

Attempts to develop dominant markers in Yarrowia have been hampered by the fact that this yeast is naturally resistant to most commonly used antibiotics. Yarrowia is however sensitive to the bleomycin/phleomycin group of antibiotics and to hygromycin B. Expression of heterologous genes conferring resistance to these antibiotics has been successful in Yarrowia (Gaillardin and Ribet 1987; Cordero Otero and Gaillardin 1996), but their use for selection was impaired by a high frequency of spontaneous resistance (Barth and Gaillardin 1996). However, more recently, the use of the recombinant hp4d promoter for expression of *E. coli* hygromycin B gene overcame this problem, allowing the use of this antibiotic for the selection of Yarrowia transformants (Fickers et al. 2003).

Another strategy for the design of a dominant marker has been heterologous expression of *S. cerevisiae SUC2* (Nicaud et al. 1989), rendering Yarrowia able to use sucrose as sole carbon source. This approach encountered technical problems, Yarrowia strains exhibiting residual growth on sucrose plates impurities (Barth and Gaillardin 1996). This work has nevertheless been useful for the design of invertase-producing strains (cf. Sect. 2.2).

Actually, the use of auxotrophy marker genes remains generally the best choice for the selection of Yarrowia transformants. The most commonly used are *LEU2* and *URA3*, especially since non-leaky and non-reverting *leu2* and/or *ura3* recipient strains have been made available. Namely, mutations *leu2-270* and *ura3-302* correspond to large internal deletions in the coding sequences (cf. Po1 series of strains in Table 1).

In order to enable the selection of multiple vector integrations and/or the subsequent amplification of expression cassettes, defective versions of the *URA3* marker have been designed by sequentially deleting its promoter (Le Dall et al. 1994).

Vectors carrying the *ura3d4* allele, retaining only 6 bp upstream from the ATG sequence, were no more able to confer an Ura^+ phenotype when present in a single copy, but were able to restore growth on selective medium in a multiple integration experiment (Le Dall et al. 1994—cf. Sects. 2.6.1 and 2.6.2). This defective selection marker has been used to select for multiple integration events for the production of homologous (Pignède et al. 2000) or heterologous proteins (Juretzek et al. 2001; Nicaud et al. 2002; Laloi et al. 2002; Nthangeni et al. 2004; Shiningavamwe et al. 2006; Maharajh et al. 2008a; Roth et al. 2009).

2.4 Expression Cassettes

In Yarrowia, like in other yeasts, efficient expression of heterologous genes could only be obtained using "sandwich vectors", namely, when the gene is inserted between yeast promoter and terminator sequences (Franke et al. 1988). This trio constitutes the basic "expression cassette", which can also optionally include a secretion signal (from a Yarrowia gene or possibly from the heterologous gene itself) and a membrane targeting signal (GPI anchor for surface display), as shown in Fig. 2.

2.4.1 Promoters

As stated in Sect. 1.2.2, the *XPR2* promoter (p*XPR2*) has been historically important during the development of Yarrowia research, since its early cloning and sequencing (Davidow et al. 1987c). However, its complex regulation hindered its industrial use: this strong promoter is active only at pH above 6, on media lacking preferred carbon and nitrogen sources, and its full induction requires high levels of peptones in the culture medium (Ogrydziak et al. 1977). This situation urged the search for new promoters, in which characteristics could fit the constraints of industrial production.

The functional dissection of p*XPR2* performed in our laboratory (Blanchin-Roland et al. 1994; Madzak et al. 1999) revealed that one of its upstream activating sequences (UAS) was poorly affected by environmental conditions. We used this element (UAS1B) to design a recombinant promoter: four tandem UAS1B copies were inserted upstream from a TATA box (minimal *LEU2* promoter). This recombinant promoter, termed hp4d for "hybrid promoter 4 direct" (copies), is almost independent from environmental conditions (pH, carbon and nitrogen sources, peptones) and is able to drive a strong expression in virtually any culture medium (Madzak et al. 1996, 2000). This is however not a constitutive promoter, since it retains some unidentified elements that drive a growth-phase-dependent gene expression. Namely, hp4d-driven heterologous gene expression was found to increase at the beginning of stationary phase (Madzak et al. 2000; Nicaud et al. 2002; Jolivalt et al. 2005; Kopecny et al. 2005; Madzak et al. 2005b). This peculiar

Table 1 Examples of strains used for production of heterologous proteins in Yarrowia

Strain CLIB Num.[a]	Genotype / Phenotype/characteristics	Reference
E129 CLIB 121	*MatA, lys11-23, leu2-270, ura3-302*[b], *xpr2-322* Lys$^-$, Leu$^-$, Ura$^-$, ΔAEP, Suc$^+$/carries Ylt1 retrotransposon (zeta-containing strain)	Barth and Gaillardin (1996)
E150 CLIB 122	*MatB, his1, leu2-270, ura3-302, xpr2-322* His$^-$, Leu$^-$, Ura$^-$, ΔAEP, Suc$^+$/carries Ylt1 retrotransposon, sequenced strain[c]	Barth and Gaillardin (1996)
Po1d CLIB 139	*MatA, leu2-270, ura3-302, xpr2-322* Leu$^-$, Ura$^-$, ΔAEP, Suc$^+$/no Ylt1 (zeta-free strain)	Le Dall et al. (1994)
Po1f[d] CLIB 724	*MatA, leu2-270, ura3-302, xpr2-322, axp1-2* Leu$^-$, Ura$^-$, ΔAEP, ΔAXP, Suc$^+$/no Ylt1, no extracellular protease	Madzak et al. (2000)
Po1g[e] CLIB 725	*MatA, leu2-270, ura3-302::URA3, xpr2-322, axp1-2* Leu$^-$, ΔAEP, ΔAXP, Suc$^+$/no Ylt1, no extracellular protease, pBR322 docking platform	Madzak et al. (2000)
Po1h[f] CLIB 882	*MatA, ura3-302, xpr2-322, axp1-2* Ura$^-$, ΔAEP, ΔAXP, Suc$^+$/no Ylt1, no extracellular protease	Madzak et al. (2004)

[a]The reference (CLIB Number) of the strain at the INRA's CIRM (International Center for Microbial Resources) Yeasts Library (Grignon, France) is indicated. Strains can be ordered from CIRM-Levures (http://www.inra.fr/internet/Produits/cirmlevures/page.php?page=home&lang=en)

[b]The alleles *leu2-270* and *ura3-302* correspond to non-reverting mutations, in which the major part of the coding sequences has been deleted. The allele *ura3-302* corresponds to *ura3::pXPR2:SUC2*, namely, to the disruption of *URA3* gene by the *SUC2* gene from *S. cerevisiae* under the control of the *XPR2* promoter. It confers the ability to grow on sucrose or mollasses (Nicaud et al. 1989), hence the corresponding Suc$^+$ property of the strain

[c]The genome of E150 strain has been sequenced by the Genolevures Consortium (Dujon et al. 2004). The data are available on the Genolevures website (http://www.genolevures.org/)

[d]Po1f was derived from Po1d by deletion (pop in, pop out) of *AXP1* gene

[e]Po1g was derived from Po1f by integration of pINA300′ vector (*URA3* gene in pBR322). This provides an integrated pBR322 docking platform for ulterior homologous integration of pBR-based vectors

[f]Po1h was derived from Po1f by gene conversion of *leu2-270* to *LEU2*

characteristic is particularly interesting for heterologous production, since it allows a partial dissociation of growth and expression phases to occur naturally, a situation permitting to avoid possible toxicity problems.

Two strong constitutive promoters have been isolated, by Muller et al. (1998), from Yarrowia *TEF1* (translation elongation factor-1α) and *RPS7* (ribosomal protein S7) genes. Their intended use was the isolation of enzyme genes by expression cloning, and p*TEF1* remains the prevalent promoter for this kind of applications. In contrast, these fully constitutive promoters are not recommended for heterologous production per se, since early expression of heterologous genes could be detrimental to cell growth.

Table 2 Examples of vectors for expression (secretion) of heterologous proteins in Yarrowia

Plasmid type	Characteristics	Drawbacks	Name	Expression cassette prom./secretion signal/term.	Selection marker	Targeting sequence	Reference
Replicative vector	Low copy number (1–3 copies), high transformation efficiency	Selective pressure required	pRRQ1	hp4d/ – /*XPR2*t	*LEU2*	no (ARS68)	Richard et al. (2001)
Monocopy integrative vectors (hp4d promoter)	Very high transformation efficiency, precise targeting, growth-phase-related expression	–	pINA1269 = pYLEX1 in YLEX™ kit (Yeastern Biotech Co., Taiwan)	hp4d/ – /*XPR2*t	*LEU2*	pBR322	Madzak et al. (2000)
			pINA1296 = pYLSC1 in YLEX™ kit (Yeastern Biotech Co., Taiwan)	hp4d/*XPR2* pre/*XPR2*t	*LEU2*	pBR322	Madzak et al. (2000)
Monocopy integrative vectors (inducible promoter)	Very high transformation efficiency, control of induction	–	p65IP	p*ICL1*/ – /*XPR2*t	*ura3d1*	rDNA	Juretzek et al. (2001)
			p66IP	p*ICL1*/ – /*XPR2*t	*ura3d1*	Zeta	Juretzek et al. (2001)
Multicopy integrative vectors (inducible promoter)	High copy number, control of induction	Low transformation efficiency	p64IP	p*ICL1*/ – /*XPR2*t	*ura3d4*	rDNA	Juretzek et al. (2001)
			p67IP	p*ICL1*/ – /*XPR2*t	*ura3d4*	Zeta	Juretzek et al. (2001)
Monocopy auto-cloning vectors (hp4d promoter)	Strain devoid of bacterial DNA, growth-phase-related expression	–	pINA1312	hp4d/ – /*XPR2*t	*ura3d1*	Zeta	Nicaud et al. (2002)
			pINA1317	hp4d/*XPR2* pre/*XPR2*t	*ura3d1*	Zeta	Nicaud et al. (2002)
Monocopy auto-cloning vector for	Surface display on 100 % cells strain devoid of bacterial	–	pINA1317-YlCWP110	hp4d/*XPR2* pre/	*ura3d1*	Zeta	Yue et al. (2008)

surface display (GPI anchor)	DNA, growth-phase-related expression			*YlCWP1* GPI anchor/ *XPR2*t			
Monocopy auto-cloning vectors (inducible promoter)	Strain devoid of bacterial DNA, control of induction	–	JMP62	p*POX2*/ – /*LIP2*t	*ura3d1*	Zeta	Nicaud et al. (2002)
			JMP61	p*POX2*/*LIP2* prepro/ *LIP2*t	*ura3d1*	Zeta	Nicaud et al. (2002)
Monocopy auto-cloning vectors (constitutive promoter)	Strain devoid of bacterial DNA, constitutive expression	–	pKOV96	p*TEF*/ – /*LIP2*t	*ura3d1*	Zeta	Labuschagné and Albertyn (2007)
Multicopy auto-cloning vectors (hp4d promoter)	Strain devoid of bacterial DNA, high copy number, growth-phase-related expression	Very low transformation efficiency	pINA1292	hp4d/ – /*XPR2*t	*ura3d4*	Zeta	Nicaud et al. (2002)
			pINA1297	hp4d/*XPR2* pre/*XPR2*t	*ura3d4*	Zeta	Nicaud et al. (2002)
Multicopy auto-cloning vectors (inducible promoter)	Strain devoid of bacterial DNA, high copy number, control of induction	Very low transformation efficiency	JMP64	p*POX2*/ – /*LIP2*t	*ura3d4*	Zeta	Nicaud et al. (2002)
			JMP63	p*POX2*/*LIP2* prepro/ *LIP2*t	*ura3d4*	Zeta	Nicaud et al. (2002)

Table 3 Production of heterologous proteins in Yarrowia (for each phylogenetic category, the works are globally sorted by publication/communication date, but this order is modulated in order to allow works on the same protein/organism to be grouped)

Organism/protein (MW) Vector[a]	Promoter	Secretion signal [Surface display signal][b] {Peroxisomal targeting}[c]	Cultivation mode: Production[d]	Reference[e]
Viruses				
Hepatitis B virus/pre-HBs antigen (30 kDa)				
Mono-int.	p*XPR2*	*XPR2* prepro	B: 85 mg/l[f]	Hamsa and Chattoo (1994)
Bacteriophage P1/Cre recombinase (41 kDa)				
Replic.	hp4d	None	SF: (+)	Richard et al. (2001)
–	–	–	SF: (+)	Fickers et al. (2003)
Eubacteria: Proteobacteria (Gram-negative)				
Escherichia coli/Tn5 phleomycin resistance gene (15 kDa)				
Mono-int.	p*LEU2*	None	SF: (+)	Gaillardin and Ribet (1987)
E. coli/β-galactosidase (116 kDa)				
Mono-int.	p*LEU2*	None	SF: (+)	Gaillardin and Ribet (1987)
–	p*XPR2*	None	SF: 400 U/l	Blanchin-Roland et al. (1994)
–	hp4d	None	SF: 420 U/l	Madzak et al. (2000)
–	p*G3P*	None	SF: 200 U/l	Juretzek et al. (2000a)
–	–	None + *G3P* intron	SF: 1,000 U/l (×5)	Juretzek et al. (2000a)
–	p*POX1*	None	SF: 1,000 U/l	Juretzek et al. (2000a)
–	p*POX5*	None	SF: 1,000 U/l	Juretzek et al. (2000a)
–	p*ICL1*	None	SF: 1,000 U/l	Juretzek et al. (2000a)
–	p*POX2*	None	SF: 2,500 U/l	Juretzek et al. (2000a)
–	p*POT1*	None	SF: 4,000 U/l	Juretzek et al. (2000a)
Replic.	hp4d	None	SF: 420 U/l	Madzak et al. (2000)
–	p*ICL1*	None	SF: 300 U/l	Juretzek et al. (2001)
Multi-int.	p*ICL1*	None	SF: 3,400 U/l (×11)	Juretzek et al. (2001)
Replic.	p*ALK1*	None	SF: 155 U/mg protein	Cho et al. (2010)
–	p*RPS7*	None	SF: 265 U/mg protein	Cho et al. (2010)
–	p*ICL1*	None	SF: 320 U/mg protein	Cho et al. (2010)
–	p*TEF1*	None	SF: 575 U/mg protein	Cho et al. (2010)
Replic.	p*TEF1*	None	SF: 140 U/l	Blazeck et al. (2011)

(continued)

Table 3 (continued)

Organism/protein (MW)		Secretion signal		
Vector[a]	Promoter	[Surface display signal][b] {Peroxisomal targeting}[c]	Cultivation mode: Production[d]	Reference[e]
–	p*EXP1*	None	SF: 87 U/l	Blazeck et al. (2011)
–	hp4d[l]	None	SF: 26 U/l	Blazeck et al. (2011)
–	hp8d[l]	None	SF: 200 U/l (×8)	Blazeck et al. (2011)
–	hp12d[l]	None	SF: 530 U/l (×20)	Blazeck et al. (2011)
–	hp16d[l]	None	SF: 640 U/l (×25)	Blazeck et al. (2011)
–	hp20d[l]	None	SF: 840 U/l (×32)	Blazeck et al. (2011)
–	hp28d[l]	None	SF: 1,200 U/l (×46)	Blazeck et al. (2011)
–	hp32d[l]	None	SF: 1,150 U/l (×44)	Blazeck et al. (2011)
E. coli/β-glucuronidase (68 kDa)				
Mono-int.	p*LEU2*	None	SF: (+)	Bauer et al. (1993)
Replic.	p*TEF1*	None	SF: (+)	Picataggio and Zhu (2005) (P)
–	p*GPM1*	None	SF: (+)	Picataggio and Zhu (2005) (P)
–	p*GPD*	None	SF: (++)	Picataggio and Zhu (2005) (P)
–	p*FBA1*	None	SF: (+)	Pollak and Zhu (2005) (P)
–	–	None + *FBA1* intron	SF: (++)	Pollak and Zhu (2005) (P)
–	p*GPAT*	None	SF: (+)	Xue and Zhu (2006a) (P)
–	p*YAT1*	None	SF: (++)	Xue and Zhu (2006b) (P)
–	p*EXP1*	None	SF: (+)	Sharpe et al. (2008) (P)
Mono-int.	p*TEF1*	None	SF: 650 nmol/min/mg	Hong et al. (2012)
–	p*GPM1*	None	SF: 650 nmol/min/mg	Hong et al. (2012)
–	p*TDH1*	None	SF: 1,600 nmol/min/mg (×2.5)	Hong et al. (2012)
–	p*FBA1*	None	SF: 3,600 nmol/min/mg (×5.5)	Hong et al. (2012)
–	–	None + *FBA1* intron	SF: 18,000 nmol/min/mg (×27.5)	Hong et al. (2012)
E. coli/hygromycin B resistance gene (41 kDa)				
Mono-int.	p*XPR2*	None	SF: (+)	Cordero Otero and Gaillardin (1996)
Replic.	p*XPR2*	None	SF: (+)	Cordero Otero and Gaillardin (1996)
–	hp4d	None	SF: (+)	Fickers et al. (2003)
E. coli/XylE catechol dioxygenase (59 kDa)				
Mono-int.	p*XPR2*	None	SF: (+)	Cordero PC
E. coli/amylolytic enzyme (85 kDa)				

(continued)

Table 3 (continued)

Organism/protein (MW) Vector[a]	Promoter	Secretion signal [Surface display signal][b] {Peroxisomal targeting}[c]	Cultivation mode: Production[d]	Reference[e]
Mono-int.	hp4d	Modified *XPR2* prepro	SF: 1 g/l	Libessart PC
Vitreoscilla stercoraria (purple photosynthetic bacteria)/single-chain haemoglobin VHb (18 kDa)				
Mono-int.	p*XPR2*	None	B: (+)	Bhave and Chattoo (2003)
Multi-int.	p*XPR2*	None	B: (+)	Bhave and Chattoo (2003)
Mono-int.	p*ICL1*	None	B: (+)	Bhave and Chattoo (2003)
Multi-int.	p*ICL1*	None	B: (+)	Bhave and Chattoo (2003)
Agrobacterium radiobacter/soluble epoxide hydrolase (33 kDa)				
Mono-int.	p*TEF1*	None	SF: (+)	Labuschagné (2005) (T)
Vibrio harveyi/haemolysin (46 kDa)				
Mono-int.	hp4d	*XPR2* pre [Yl*CWP1*]	SF: (+) Surface display 100 % cells	Yue et al. (2008)
Vibrio sp. QY101/alginate lyase (38 kDa)				
Mono-int.	hp4d	*XPR2* pre [Yl*CWP1*]	SF: 208 U/g (dw) Surface display	Liu et al. (2009)
Erwinia chrysanthemi/L-asparaginase (39 kDa)				
Mono-int.	p*POX2*	*LIP2* prepro	SF: (+)	Nicaud et al. (2009) (P)
Multi-int.	p*POX2*	*LIP2* prepro	SF: (++)	Nicaud et al. (2009) (P)
Pseudomonas aeruginosa/polyhydroxyalkanoate synthase (62 kDa)				
Mono-int.	p*POX2*	None {Bn*ICL*}	SF: (+) Peroxisomal targeting	Haddouche et al. (2010)
Eubacteria: Firmicutes (Gram-positive)				
Bacillus subtilis/endo-β-1,4-mannanase (41 kDa)				
Mono-int.	hp4d	*XPR2* pre [Sc*FLO1*]	SF: 62 IU/g (dw) Surface display	Yang et al. (2009)
Eubacteria: Actinobacteria (Gram-positive)				
Thermobifida fusca/thermostable α-amylase (65 kDa)				
Mono-int.	hp4d	*XPR2* pre	SF: 730 U/l	Yang et al. (2010b)
T. fusca/AXE thermostable esterase (28 kDa)				
Mono-int.	hp4d	*XPR2* pre	SF: 71 U/ml	Huang et al. (2011)
Arthrobacter sp. S37/endo-inulinase (79 kDa)				
Mono-int.	hp4d	*XPR2* pre	SF: 17 U/ml	Li et al. (2012)
Fungi: Ascomycetes				
Saccharomyces cerevisiae invertase (85 kDa)				
Mono-int.	p*XPR2*	*XPR2* pre	SF: (+)	Nicaud et al. (1989)
Mono-int.	p*FBA1*	Native + *FBA1* intron	SF: 45 U/l	Hong et al. (2012)
–	–	*XPR2* prepro + 13 aa + *FBA1* intron	SF: 39 U/l	Hong et al. (2012)

(continued)

Table 3 (continued)

Organism/protein (MW)		Secretion signal		
Vector[a]	Promoter	[Surface display signal][b] {Peroxisomal targeting}[c]	Cultivation mode: Production[d]	Reference[e]
Humicola insolens cellulase II (57 kDa)				
Replic.	p*XPR2*	Native	SF: 8 mg/l	Muller et al. (1998)
–	p*TEF1*	Native	SF: 7 mg/l	Muller et al. (1998)
–	p*RPS7*	Native	SF: 5 mg/l	Muller et al. (1998)
H. insolens xylanase I (27 kDa)				
Replic.	p*XPR2*	Native	SF: 2 mg/l	Muller et al. (1998)
–	p*TEF1*	Native	SF: 1 mg/l	Muller et al. (1998)
–	p*RPS7*	Native	SF: 250 μg/l	Muller et al. (1998)
Thermomyces lanuginosus lipase I (35 kDa)				
Replic.	p*XPR2*	Native	SF: (+)	Muller et al. (1998)
Aspergillus aculeatus cellulase I (29 kDa)				
Replic.	p*XPR2*	Native	SF: (+)	Muller et al. (1998)
A. aculeatus galactanase I (44 kDa)				
Replic.	p*XPR2*	Native	SF: (+)	Muller et al. (1998)
Mono-int.	hp4d	*XPR2* pre	SF: 3 mg/l	Swennen PC
A. aculeatus/polygalacturonase I (45 kDa)				
Replic.	p*XPR2*	Native	SF: (+)	Muller et al. (1998)
A. aculeatus/endo-β-1,4-mannanase (42 kDa)				
Mono-int.	hp4d	*LIP2* prepro	SF: 123 nkat/ml	Roth et al. (2009)
–	–	–	B: 685 nkat/ml (×6)	Roth et al. (2009)
Multi-int. (2)	hp4d	*LIP2* prepro	SF: 496 nkat/ml (×4)	Roth et al. (2009)
Multi-int. (9)	hp4d	Native	SF: 13,073 nkat/ml	Roth et al. (2009)
–	–	–	B: 6,719 nkat/ml (×0.5)	Roth et al. (2009)
–	–	–	FB: 26,139 nkat/ml (×2)	Roth et al. (2009)
–	–	–	Improved FB: 40,835 nkat/ml (×3)	van Zyl (2010) (P)
Trichoderma reesei endoglucanase I (45 kDa)				
Mono-int.	p*XPR2*	*XPR2* prepro	SF: (+/−)	Park et al. (2000)
–	–	Native	SF: 5 mg/l	Park et al. (2000)
–	–	–	FB: 100 mg/l (×20)	Park et al. (2000)
Arxula adeninivorans/glucoamylase (90 kDa)				
Mono-int.	hp4d	*XPR2* pre	SF: (+)	Swennen et al. (2002)
–	–	*XPR2* prepro	SF: (+)	Swennen et al. (2002)
Aspergillus oryzae/leucine aminopeptidase II (90 kDa)				
Mono-int.	hp4d	Hybrid *LIP2*/*XPR2* prepro	B: 320 U/l	Nicaud et al. (2002)
Multi-int.	hp4d	Hybrid *LIP2*/*XPR2* prepro	FB: 2,500 U/l (×8)	Nicaud et al. (2002)
Multi-int.	hp4d	Hybrid *LIP2*/*XPR2* prepro	FB: 28,000 U/l (×88)	Nicaud et al. (2002)

(continued)

Table 3 (continued)

Organism/protein (MW) Vector[a]	Promoter	Secretion signal [Surface display signal][b] {Peroxisomal targeting}[c]	Cultivation mode: Production[d]	Reference[e]
A. oryzae/tyrosinase (67 kDa)				
Mono-int.	hp4d	None	SF: 11 U/mg	Rao et al. (2011)
Alternaria alternata/Alta1p allergen (14 kDa)				
Replic.	p*MTPII*	Native	SF: (+)	Morin and Dominguez AR[g]
Fusarium moniliforme (aka *Gibberella fujikuroi*) /Δ15-desaturase (46 kDa)				
Mono-int.	p*GPD*	None	SF: (+)	Picataggio and Zhu (2005) (P)
F. moniliforme (aka *G. fujikuroi*)/Δ12-desaturase (53 kDa) (cf. also synthetic constructs)				
Mono-int.	p*FBA1*	None	SF: (+)	Xue and Zhu (2006b) (P)
–	p*GPD*	None	SF: (+)	Sharpe et al. (2008) (P)
–	p*YAT1*	None	SF: (+)	Sharpe et al. (2008) (P)
Candida albicans/epoxide hydrolase (34 kDa)				
Mono-int.	p*TEF1*	None	SF: (+)	Labuschagné (2005) (T)
Aspergillus niger/soluble epoxide hydrolase (CAB59813—44 kDa)				
Mono-int.	p*TEF1*	None	SF: (+)	Labuschagné 2005 (T)
A. niger SQ-6/soluble epoxide hydrolase (AAX78198—44 kDa)				
Mono-int.	p*TEF1*	None	SF: (+)	Labuschagné (2005) (T)
Aureobasidium pullulans 10/alkaline protease I (43 kDa)				
Mono-int.	hp4d	*XPR2* pre	SF: 49 U/ml	Ni et al. (2008)
A. pullulans HN2-3/alkaline protease II (43 kDa)				
Mono-int.	hp4d	*XPR2* pre [Yl*CWP1*]	SF: 691 U/g (dw) Surface display	Ni et al. (2009)
Aspergillus fumigatus/β-1,6-glucanase (50 kDa)				
Mono-int.	hp4d	*XPR2* pre	SF: 5 mg/l	Boisramé and Gaillardin (2009)
Candida antarctica/lipase B (34 kDa)				
Mono-int.	p*POX2*	*LIP2* prepro	SF: 510 U/l	Emond et al. (2010)
–	–	–	B: 5,090 U/l (×10) and 190 mg/l	Emond et al. (2010)
Saccharomycopsis fibuligera A11/acid protease (50 kDa)				
Mono-int.	hp4d	*XPR2* pre [Yl*CWP1*]	SF: 1,140 U/ml Surface display	Yu et al. (2010)
Kluyveromyces marxianus CBS 6556/exo-inulinase (59 kDa)				
Mono-int.	hp4d	*XPR2* pre	SF: 42 U/ml	Zhao et al. (2010)
–	–	–	B: 43 U/ml	Cui et al. (2011)
–	–	idem + [Yl*CWP1*]	SF: 23 U/mg cell dry weight	Liu et al. (2010)

(continued)

Table 3 (continued)

Organism/protein (MW)		Secretion signal		
Vector[a]	Promoter	[Surface display signal][b] {Peroxisomal targeting}[c]	Cultivation mode: Production[d]	Reference[e]
Williopsis saturnus (aka *Cyberlindnera saturnus*)/exo-β-1,3-glucanase (46 kDa)				
Mono int.	hp4d	*XPR2* pre	SF: 16 U/ml	Peng et al. (2011)
Fungi: Basidiomycetes				
Pycnoporus cinnabarinus/laccase I (54 kDa)				
Mono-int.	hp4d	Native	SF: (+/−)	Madzak et al. (2005b)
–	–	*XPR2* pre	SF: (+)	Madzak et al. (2005b)
–	–	*XPR2* prepro	SF: 8.5 mg/l	Madzak et al. (2005b)
–	–	–	B: 20 mg/l (×2.4)	Madzak et al. (2005b)
Trametes versicolor/laccase IIIb (58 kDa)				
Mono-int.	hp4d	Native	SF: 2.5 mg/l	Jolivalt et al. (2005)
–	–	*XPR2* pre	SF: (+)	Jolivalt et al. (2005)
Rhodosporidium paludigenum/epoxide hydrolase (46 kDa)				
Mono-int.	p*TEF1*	None	SF: (+)	Labuschagné (2005) (T)
Rhodosporidium toruloides/epoxide hydrolase (45 kDa)				
Mono-int.	p*TEF1*	None	SF: (+)	Labuschagné (2005) (T)
Cryptococcus neoformans/epoxide hydrolase (45 kDa)				
Mono-int.	p*TEF1*	None	SF: (+)	Labuschagné (2005) (T)
Rhodotorula araucariae/epoxide hydrolase (46 kDa)				
Mono-int.	p*TEF1*	None	SF: (+)	Labuschagné (2005) (T)
Mono-int.	hp4d	None	SF: 2,400 U/l	Maharajh et al. (2008a)
Multi-int.	hp4d	None	SF: 5,400 U/l (×2)	Maharajh et al. (2008a)
–	–	–	B: 22,750 U/l (×9.5)	Maharajh et al. (2008a)
–	–	–	FB: 194,700 U/l (×81)	Maharajh et al. (2008a)
–	–	–	+ exp. feed. rate: 206,000 U/l (×86)	Maharajh et al. (2008b)
Rhodotorula minuta/cytochrome P450 53B1 (58 kDa)				
Multi-int.	p*POX2*	None	SF: (+)	Shiningavamwe et al. (2006)
idem + surexpression of Yl*CPR*[h] under p*ICL1*			SF: (++)	Shiningavamwe et al. (2006)
Rhodotorula mucilaginosa/epoxide hydrolase (44 kDa)				
Mono-int.	p*TEF1*	None	SF: (+)	Labuschagné and Albertyn (2007)

(continued)

Table 3 (continued)

Organism/protein (MW) Vector[a]	Promoter	Secretion signal [Surface display signal][b] {Peroxisomal targeting}[c]	Cultivation mode: Production[d]	Reference[e]
Fungi: Zygomycetes + Mucoromycetes				
Mortierella alpina/Δ5-desaturase (50 kDa) (cf. also synthetic constructs)				
Mono-int.	p*TEF1*	None	SF: (+)	Xue and Zhu (2006b) (P)
Mono-int.	p*FAB1*	None + *FBA1* intron	SF: (+)	Xue and Zhu (2006b) (P)
M. alpina / Δ6-desaturase (51 kDa) (cf. also synthetic constructs)				
Mono-int.	hp4d	None	SF: (+)	Chuang et al. (2009, 2010)
M. alpina / Δ12-desaturase (44 kDa)				
Mono-int.	hp4d	None	SF: (+)	Chuang et al. (2009, 2010)
Rhizopus stolonifer/lipase (43 kDa)				
Mono-int.	hp4d	*XPR2* pre [Sc*FLO1*]	B: (+) Surface display	Song et al. (2011)
Rhizopus oryzae/lipase (30 kDa)				
Mono-int.	p*XPR2*	Native	SF: 3.23 U/ml	Yuzbashev et al. (2012)
–	–	*XPR2* prepro	SF: 1.45 U/ml	Yuzbashev et al. (2012)
–	–	*XPR2* pre + native pro	SF: 7.61 U/ml	Yuzbashev et al. (2012)
Fungi/Heterokonts: Oomycetes				
Pythium aphanidermatum / Δ17-desaturase (40 kDa) (cf. also synthetic constructs)				
Mono-int.	p*FAB1*	None + *FBA1* intron	SF: (+)	Sharpe et al. (2008) (P)
Mono-int.	p*EXP1*	None	SF: (+)	Sharpe et al. (2008) (P)
Protists: Euglenozoa				
Euglena gracilis / Δ9-elongase (30 kDa) (cf. also synthetic constructs)				
Mono-int.	p*GPAT*	None	SF: (+)	Sharpe et al. (2008) (P)
E. gracilis / Δ5-desaturase (50 kDa) (cf. also synthetic constructs)				
Mono-int.	p*FAB1*	None + *FBA1* intron	SF: (+)	Sharpe et al. (2008) (P)
Plants (Angiosperms): Monocotyledons				
Oryza sativa/α-amylase (45 kDa)				
Mono-int.	p*XPR2*	Native	SF: (+)	Park et al. (1997)
–	–	*XPR2* pre	SF: (–)	Park et al. (1997)
–	–	*XPR2* prepro	SF: (+)	Park et al. (1997)
–	–	Native	FB: 31,200 U/l	Chang et al. (1998)
–	–	–	One-step feeding FB: 88,000 U/l	Kim et al. (2000)
Zea mays/cytokinin oxidase III (55 kDa)				
Mono-int.	hp4d	Native	SF: (+/–)	Massonneau et al. (2004)

(continued)

Table 3 (continued)

Organism/protein (MW)		Secretion signal		
Vector[a]	Promoter	[Surface display signal][b] {Peroxisomal targeting}[c]	Cultivation mode: Production[d]	Reference[e]
–	hp4d	*XPR2* prepro	SF: (+)	Massonneau et al. (2004)
Z. mays/cytokinin oxidase I (55 kDa)				
Mono-int.	hp4d	*XPR2* prepro	SF: 12 mg/l	Kopecny et al. (2005)
Plants (Angiosperms): Dicotyledons				
Theobroma cacao/aspartic proteinase II (62 kDa)				
Multi-int.	hp4d	Hybrid *LIP2*/*XPR2* prepro	SF: (+)	Laloi et al. (2002)
Capsicum annuum/fatty acid hydroperoxide lyase (cytochrome P450) (55 kDa)				
Mono-int.	p*POX2*	None	SF: 1,200 U/l	Bourel et al. (2004)
Arabidopsis thaliana/soluble epoxide hydrolase (36 kDa)				
Mono-int.	p*TEF1*	None	SF: (+)	Labuschagné (2005) (T)
Insects				
Trichoplusia ni (Lepidoptere)/gut epoxide hydrolase (51 kDa)				
Mono-int.	p*TEF1*	None	SF: (+)	Labuschagné (2005) (T)
T. ni/microsomal epoxide hydrolase (51 kDa)				
Mono-int.	p*TEF1*	None	SF: (+)	Labuschagné (2005) (T)
Mammals (non-human)				
Bos taurus taurus/prochymosin (40 kDa)				
Mono-int.	p*LEU2*	*XPR2* prepro	SF: (+)	Franke et al. (1988)
–	p*XPR2*	*XPR2* pre + dipeptides	SF: (+)	Franke et al. (1988)
–	–	*XPR2* prepro	SF: (+)	Franke et al. (1988)
Replic.	p*XPR2*	*XPR2* prepro	SF: (+)	Nicaud et al. (1991)
Mono-int.	hp4d	*XPR2* prepro	SF: 20 mg/l	Madzak et al. (2000)
–	–	–	B: 160 mg/l (×8)	Madzak et al. (2000)
B. taurus taurus/cytochrome P450 17α (56 kDa)				
Replic.	p*ICL1*	None	SF: (+)	Juretzek et al. (2000b) (P)
Multi-int.	p*ICL1*	None	SF: (++)	Juretzek et al. (2000b) (P)
–	–	–	SF: (++)	Mauersberger et al. AR[i]
Sus scrofa domestica/α1-interferon (21 kDa)				
Mono-int.	p*XPR2*	*XPR2* prepro	SF: 40 U/l	Nicaud et al. (1991)
Replic.	p*XPR2*	*XPR2* prepro	SF: 120 U/l	Nicaud et al. (1991)
Mus musculus domesticus/interleukin 6 (20 kDa)				
Mono-int.	hp4d	*XPR2* prepro	SF: 15 mg/l	Sang PC
Lama glama glama/anti-ACE VHH single-chain antibody (30 kDa)				
Mono-int.	hp4d	*XPR2* pre	SF: (+)	Chartier PC

(continued)

Table 3 (continued)

Organism/protein (MW) Vector[a]	Promoter	Secretion signal [Surface display signal][b] {Peroxisomal targeting}[c]	Cultivation mode: Production[d]	Reference[e]
Human				
Homo sapiens/anaphylatoxin C5a (74 kDa)				
Mono-int.	p*XPR2*	*XPR2* prepro	SF: (+)	Davidow et al. (1987b) (P)
H. sapiens/blood coagulation factor XIIIa (80 kDa)				
Mono-int.	p*XPR2*	*XPR2* pre	SF: (−)	Tharaud et al. (1992)
–	–	*XPR2* prepro	SF: (−)	Tharaud et al. (1992)
–	–	*XPR2* pre + dipeptides	SF: 1 mg/l	Tharaud et al. (1992)
H. sapiens/proinsulin (10 kDa)				
Multi-int.	p*XPR2*	*XPR2* prepro	SF: (+)	James and Strick (1995) (P)
H. sapiens/insulinotropin (4 kDa)				
Multi-int.	p*XPR2*	*XPR2* prepro	SF: (+)	James and Strick (1995) (P)
H. sapiens/epidermal growth factor (6 kDa)				
Mono-int.	p*XPR2*	*XPR2* prepro	SF: 2 mg/l	Hamsa et al. (1998)
H. sapiens/tissue plasminogen activator (59 kDa)				
Mono-int.	p*XPR2*	*XPR2* prepro	SF: (+)	Franke PC
H. sapiens/α-foetoprotein (74 kDa)				
Mono-int.	hp4d	*XPR2* prepro	SF: 250 μg/l	Uchida PC
H. sapiens/β2-microglobulin (12 kDa)				
Mono-int.	hp4d	*XPR2* prepro	SF: 5 μg/l	Uchida PC
H. sapiens/soluble CD14 variant (48 kDa)				
Multi-int.	hp4d	Hybrid *LIP2*/*XPR2* prepro	FB: 500 mg/l	Gysler PC and AR[j]
H. sapiens/anti-Ras scFv single-chain antibody (30 kDa)				
Mono-int.	hp4d	*XPR2* pre	SF: 20 mg/l	Swennen et al. (2002)
–	–	*XPR2* prepro	SF: 20 mg/l	Swennen et al. (2002)
H. sapiens/anti-estradiol scFv single-chain antibody (30 kDa)				
Mono-int.	hp4d	*XPR2* pre	SF: (+)	Chartier PC
H. sapiens/cytochrome P450 1A1 (16 kDa)				
Mono-int.	p*POX2*	None	SF: 32 U (pM/min/dw)	Nthangeni et al. (2004)
idem + surexpression of Yl*CPR*[h] under p*ICL1*			SF: 48 U (×1.5)	Nthangeni et al. (2004)
idem + surexpression of Yl*CPR*[h] under p*POX2*			SF: 65 U (×2)	Nthangeni et al. (2004)
Multi-int.	p*POX2*	None	SF: 129 U (×4)	Nthangeni et al. (2004)
idem + surexpression of Yl*CPR*[h] under p*ICL1*			SF: 1,587 U (×50)	Nthangeni et al. (2004)
idem + surexpression of Yl*CPR*[h] under p*POX2*			SF: 1,645 U (×51)	Nthangeni et al. (2004)

(continued)

Table 3 (continued)

Organism/protein (MW)		Secretion signal		
Vector[a]	Promoter	[Surface display signal][b] {Peroxisomal targeting}[c]	Cultivation mode: Production[d]	Reference[e]
H. sapiens/microsomal epoxide hydrolase (51 kDa)				
Mono-int.	p*TEF1*	None	SF: (+)	Labuschagné (2005) (T)
H. sapiens/oestrogen receptor α (67 kDa)				
Replic.	p*ALK1*	None	SF: (+/−)	Cho et al. (2010)
–	p*RPS7*	None	SF: (+)	Cho et al. (2010)
–	p*ICL1*	None	SF: (+)	Cho et al. (2010)
–	p*TEF1*	None	SF: (++)	Cho et al. (2010)
H. sapiens/α2b-interferon (19 kDa) (cf. also synthetic constructs)				
Mono-int.	p*POX2*	*LIP2* prepro	SF: 5 mg/l	Gasmi et al. (2011a)
Synthetic constructs and variants				
From the **Proteobacteria** *Pantoea stewartii*/GGPP synthase[k] adapted to *Y. lipolytica* codon bias (34 kDa)				
Mono-int.	p*FBA1*	None + *FBA1* intron	SF: (+)	Sharpe et al. (2008) (P)
From the Proteobacteria *P. stewartii*/phytoene synthase adapted to *Y. lipolytica* codon bias (33 kDa)				
Mono-int.	p*GPD*	None + *GPD* intron	SF: (+)	Sharpe et al. (2008) (P)
From the Proteobacteria *P. stewartii*/phytoene desaturase adapted to *Y. lipolytica* codon bias (55 kDa)				
Mono-int.	p*EXP1*	None	SF: (+)	Sharpe et al. (2008) (P)
From the Proteobacteria *P. stewartii*/lycopene cyclase adapted to *Y. lipolytica* codon bias (42 kDa)				
Mono-int.	p*EXP1*	None	SF: (+)	Sharpe et al. (2008) (P)
Mono-int.	p*FBA1*	None + *FBA1* intron	SF: (+)	Sharpe et al. (2008) (P)
From the Proteobacteria *Pantoea agglomerans*/lycopene cyclase adapted to *Y. lipolytica* codon bias (43 kDa)				
Mono-int.	p*EXP1*	None	SF: (+)	Sharpe et al. (2008) (P)
From the Proteobacteria *Enterobacteriaceae* DC260/lycopene cyclase adapted to *Y. lipolytica* codon bias (43 kDa)				
Mono-int.	p*GPD*	None + *GPD* intron	SF: (+)	Sharpe et al. (2008) (P)
From the Proteobacteria *Brevundimonas vesicularis*/carotenoid ketolase adapted to *Y. lipolytica* codon bias (27 kDa)				
Mono-int.	p*GPAT*	None	SF: (+)	Sharpe et al. (2008) (P)
From the Proteobacteria *B. vesicularis*/carotenoid hydroxylase adapted to *Y. lipolytica* codon bias (18 kDa)				
Mono-int.	p*YAT1*	None	SF: (+)	Sharpe et al. (2008) (P)
Mono-int.	p*EXP1*	None	SF: (+)	Sharpe et al. (2008) (P)
Mono-int.	p*GPD*	None + *GPD* intron	SF: (+)	Sharpe et al. (2008) (P)

(continued)

Table 3 (continued)

Organism/protein (MW) Vector[a]	Promoter	Secretion signal [Surface display signal][b] {Peroxisomal targeting}[c]	Cultivation mode: Production[d]	Reference[e]
From the Proteobacteria *Agrobacterium aurantiacum*/carotenoid ketolase adapted to *Y. lipolytica* codon bias (27 kDa)				
Mono-int.	p*FBA1*	None + *FBA1* intron	SF: (+)	Sharpe et al. (2008) (P)
From the Proteobacteria *A. aurantiacum*/carotenoid hydroxylase adapted to *Y. lipolytica* codon bias (18 kDa)				
Mono-int.	p*FBA1*	None + *FBA1* intron	SF: (+)	Sharpe et al. (2008) (P)
From the **Ascomycetous Fungus** *Fusarium moniliforme*/Δ12-desaturase adapted to *Y. lipolytica* codon bias (53 kDa)				
Mono-int.	p*GPM*	None + *FBA1* intron	SF: (+)	Sharpe et al. (2008) (P)
Mono-int.	p*EXP1*	None	SF: (+)	Sharpe et al. (2008) (P)
From the Ascomycetous Fungus *Candida antarctica*/saturation mutagenesis of lipase B (34 kDa)				
Mono-int.	p*POX2*	*LIP2* prepro	SF: mean value of 20 mg/l	Emond et al. (2010)
–	–	–	B: mean value of 200 mg/l (×10)	Emond et al. (2010)
From the **Basidiomycetous Fungus** *Trametes versicolor*/directed mutagenesis of laccase IIIb (58 kDa)				
Mono-int.	hp4d	Native	SF: (+)	Madzak et al. (2006)
–	–	–	SF: (+)	Galli et al. (2011)
From the **Zygomycetous Fungus** *Mortierella alpina* / Δ5-desaturase adapted to *Y. lipolytica* codon bias (50 kDa)				
Mono-int.	p*EXP1*	None	SF: (+)	Sharpe et al. (2008) (P)
Mono-int.	p*GPAT*	None	SF: (+)	Sharpe et al. (2008) (P)
From the Zygomycetous Fungus *M. alpina* / Δ6-desaturase adapted to *Y. lipolytica* codon bias (51 kDa)				
Mono-int.	p*TEF1*	None	SF: (+)	Sharpe et al. (2008) (P)
Mono-int.	p*FBA1*	None + *FBA1* intron	SF: (+)	Sharpe et al. (2008) (P)
From the Zygomycetous Fungus *M. alpina* / elongase 1 adapted to *Y. lipolytica* codon bias (31 kDa)				
Mono-int.	p*GPD*	None + *GPD* intron	SF: (+)	Sharpe et al. (2008) (P)
Mono-int.	p*FBA1*	None + *FBA1* intron	SF: (+)	Sharpe et al. (2008) (P)
From the Zygomycetous Fungus *M. alpina*/$C_{16/18}$ elongase adapted to *Y. lipolytica* codon bias (35 kDa)				
Mono-int.	p*YAT1*	None	SF: (+)	Sharpe et al. (2008) (P)
Mono-int.	p*EXP1*	None	SF: (+)	Sharpe et al. (2008) (P)

(continued)

Table 3 (continued)

Vector[a]	Promoter	Secretion signal [Surface display signal][b] {Peroxisomal targeting}[c]	Cultivation mode: Production[d]	Reference[e]
From the **Oomycetous Fungus** *Pythium aphanidermatum*/Δ17-desaturase adapted to *Y. lipolytica* codon bias (40 kDa)				
Mono-int.	p*YAT1*	None	SF: (+)	Sharpe et al. (2008) (P)
From the **Haptophyteous Microalga** *Isochrysis galbana*/Δ5-desaturase adapted to *Y. lipolytica* codon bias (48 kDa)				
Mono-int.	p*YAT1*	None	SF: (+)	Sharpe et al. (2008) (P)
From the **Dinoflagelate** *Peridinium* sp./Δ5-desaturase adapted to *Y. lipolytica* codon bias (51 kDa)				
Mono-int.	p*YAT1*	None	SF: (+)	Sharpe et al. (2008) (P)
From the **Protist** (Heterokontophyta) *Thraustochytrium aureum*/elongase adapted to *Y. lipolytica* codon bias (30 kDa)				
Mono-int.	p*TEF1*	None	SF: (+)	Xue and Zhu (2006b) (P)
From the Protist (Euglenozoa) *Euglena gracilis*/Δ9-elongase adapted to *Y. lipolytica* codon bias (30 kDa)				
Mono-int.	p*FBA1*	None	SF: (+)	Sharpe et al. (2008) (P)
Mono-int.	p*FBA1*	None + *FBA1* intron	SF: (+)	Sharpe et al. (2008) (P)
Mono-int.	p*EXP1*	None	SF: (+)	Sharpe et al. (2008) (P)
Mono-int.	p*YAT1*	None	SF: (+)	Sharpe et al. (2008) (P)
Mono-int.	p*GPD*	None	SF: (+)	Sharpe et al. (2008) (P)
From the Protist *E. gracilis*/synthetic mutant of Δ8-desaturase (47 kDa)				
Mono-int.	p*FBA1*	None + *FBA1* intron	SF: (+)	Sharpe et al. (2008) (P)
Mono-int.	p*EXP1*	None	SF: (+)	Sharpe et al. (2008) (P)
Mono-int.	p*GPD*	None + *GPD* intron	SF: (+)	Sharpe et al. (2008) (P)
Mono-int.	p*YAT1*	None	SF: (+)	Sharpe et al. (2008) (P)
From the Protist *E. gracilis*/Δ5-desaturase adapted to *Y. lipolytica* codon bias (50 kDa)				
Mono-int.	p*EXP1*	None	SF: (+)	Sharpe et al. (2008) (P)
Mono-int.	p*YAT1*	None	SF: (+)	Sharpe et al. (2008) (P)

(continued)

Table 3 (continued)

Organism/protein (MW) Vector[a]	Promoter	Secretion signal [Surface display signal][b] {Peroxisomal targeting}[c]	Cultivation mode: Production[d]	Reference[e]
From the **Monocotyledonous Plant** *Zea mays*/directed mutagenesis of cytokinin oxidase I (55 kDa)				
Mono-int.	hp4d	*XPR2* prepro	SF: (+)	Kopecny et al. (2008)
–	–	–	SF: (+)	Kopecny et al. (2010)
From the **Cnidaria** *Aequorea victoria*/enhanced green fluorescent protein (EGFP—27 kDa)				
Mono-int.	hp4d	*XPR2* pre [Yl*CWP1*]	SF: (+) Surface display 100 % cells	Yue et al. (2008)
From the Cnidaria *Renilla* sp./green fluorescent protein adapted to mammalian codon bias (hrGFP—26 kDa)				
Replic.	p*FBA*	None	SF: fluorescence idem to control	Blazeck et al. (2011)
–	p*XPR2*	None	SF: fluorescence idem to control	Blazeck et al. (2011)
–	p*YAT1*	None	SF: fluo. × 1.2 relative to control	Blazeck et al. (2011)
–	p*GPAT*	None	SF: fluo. × 3.4 relative to control	Blazeck et al. (2011)
–	p*GPD*	None	SF: fluo. × 6.2 relative to control	Blazeck et al. (2011)
–	p*EXP1*	None	SF: fluo. × 9 relative to control	Blazeck et al. (2011)
–	p*TEF1*	None	SF: fluo. × 7.8 relative to control	Blazeck et al. (2011)
–	Reduced p*TEF1*[l]	None	SF: (+) = basal fluorescence	Blazeck et al. (2011)
–	idem + 8 UAS1B[l]	None	SF: fluo. × 15 relative to basal	Blazeck et al. (2011)
–	idem + 16 UAS1B[l]	None	SF: fluo. × 27 relative to basal	Blazeck et al. (2011)
–	hp4d[l]	None	SF: control for hrGFP mRNA level	Blazeck et al. (2011)
–	hp8d[l]	None	SF: ×11 relative to control mRNA	Blazeck et al. (2011)
–	hp16d[l]	None	SF: ×71 relative to control mRNA	Blazeck et al. (2011)
–	hp24d[l]	None	SF: ×90 relative to control mRNA	Blazeck et al. (2011)
–	hp32d[l]	None	SF: ×83 relative to control mRNA	Blazeck et al. (2011)

(continued)

Table 3 (continued)

Organism/protein (MW)		Secretion signal		
Vector[a]	Promoter	[Surface display signal][b] {Peroxisomal targeting}[c]	Cultivation mode: Production[d]	Reference[e]
From ***Homo sapiens***/α2b-interferon adapted to *Y. lipolytica* codon bias (19 kDa) (cf. native sequence for comparison)				
Mono-int.	p*POX2*	*LIP2* prepro	SF: 56 mg/l (×11)	Gasmi et al. (2011a)
–	–	idem + CACA sequence	SF: 84 mg/l (×16.5)	Gasmi et al. (2011a)
–	–	*LIP2* pre + dipeptides	SF: 95 mg/l (×19)	Gasmi et al. (2011a)
–	–	–	FB: 425 mg/l (×85)	Gasmi et al. (2011c)

[a]Expression/secretion vectors were either replicative (Replic.) or integrative ones. Centromeric replicative vectors can maintain 1–3 copies per cell (Vernis et al. 1997; Madzak et al. 2000). Integrative vectors were either classical shuttle vectors (in general based on pBR322), or auto-cloning vectors, from which a "yeast cassette" devoid of bacterial DNA can be purified for use in yeast transformation. Both vector types can integrate as monocopy (or at a low copy number) when a non-defective selection marker is used (Mono-int.) or as multicopy (in general, in the range of 10 copies per cell; Le Dall et al. 1994; Juretzek et al. 2001; Nicaud et al. 2000) when a promoter-defective selection marker gene is used (Multi-int.). In a few studies involving multicopy integrative vectors, the copy number was precisely determined and is indicated between brackets

[b]When a GPI anchor sequence was used for the surface display of the heterologous protein, its origin is indicated between square brackets

[c]When a peroxisomal targeting sequence was used for addressing the heterologous protein to peroxisomes, its origin is indicated between curly brackets

[d]Abbreviations used: SF for shake-flask , B for batch, FB for fed-batch cultivation, and "+ exp. feed. rate" for "+ exponential feeding rate". When no secretion signal was used, intracellular expression is indicated. When a secretion signal was present, secretion of the heterologous protein was always obtained (except for the case of hepatitis B antigen, which assembled intracellularly in Dane particles). Indicated production was estimated by measuring heterologous activity and thus corresponds to active heterologous protein (except for the case of *E. coli* amylolytic enzyme, which was produced under an inactive form). The symbols (–), (+/–), (+) and (++) were used to indicate a quantitative gradation in the successful production of a given protein, using different methodologies. Similarly, an increase factor is sometimes indicated between brackets

[e]Abbreviations used: PC for personal communication (cf. Acknowledgements), AR for abstract report from a congress, (P) for a patent and (T) for a PhD thesis

[f]Despite the presence of a secretion signal, hepatitis B antigen assembled intracellularly in the form of Dane particles

[g]Morin and Dominguez, abstract report from TYLIM (3rd *Yarrowia lipolytica* International Meeting, Dresden, July 2002)

[h]Cytochrome P450 (CYP450p) monooxygenases require association with cytochrome P450 reductase (CPRp) to achieve optimal activities. Native Yl*CPR* was overexpressed in order to increase recombinant CYP450p activity

[i]Mauersberger et al., abstract report from TYLIM (3rd *Yarrowia lipolytica* International Meeting, Dresden, July 2002)

[j]Gysler et al., abstract report from TYLIM (3rd *Yarrowia lipolytica* International Meeting, Dresden, July 2002)

[k]GGPP synthase stands for geranylgeranyl pyrophosphate synthase

[l]Blazeck et al. (2011) constructed a large array of recombinant promoters, carrying various copy numbers (from 1 to 32) of the UAS1B enhancer from p*XPR2*, inserted upstream of a minimal p*LEU2*, or of different variant versions of p*TEF1*. They named the former series "pUAS1Bn-Leum", where *n* is the number of UAS1B copies. As pUAS1B4-Leum is similar to hp4d promoter, we have substituted this shorter notation (hpnd, where n is the number of UAS1B copies) to the one used by Blazeck and co-workers. For the purpose of clarity, we have also selected, in this table, only a subset of the results they described. Similarly, we have simplified the results they obtained, using the recombinant promoters based on different variant versions of p*TEF1*, by indicating only the best results, without detailing the sizes of the p*TEF1* used

For this reason, the controlled expression of heterologous genes during cultivation was desirable, and several laboratories searched for inducible promoters which, in contrast to p*XPR2*, could respond to an easily controllable factor. Dominguez et al. (1998) reported the use of the metallothionein promoter, inducible by metallic salts, which also presents the unique property of being bidirectional. However, the peculiar nature of the inducer impairs its industrial use. Our laboratory essentially focused on Yarrowia capacity to grow on hydrophobic substrates, and several promoters were isolated from key genes from fatty acid pathways (Juretzek et al. 2000a): isocitrate lyase (*ICL1*), 3-oxo-acyl-CoA thiolase (*POT1*) and acyl-CoA oxidases (*POX1*, *POX2* and *POX5*). These promoters have been tested for β-galactosidase expression, during growth on various carbon sources, in comparison with the glycerol-3-phosphate dehydrogenase (*G3P*) promoter, with p*XPR2* and with hp4d (Juretzek et al. 2000a). This study revealed that p*ICL1*, p*POT1* and p*POX2* were the strongest inducible promoters available in Yarrowia (cf. Sect. 3.1 and Table 3). The comparison with hp4d was in strong disfavour of this latter, but the short cultivation time used in this study did not allow this growth-phase-dependent promoter to develop its full potential. A better estimation of hp4d performance, for β-galactosidase expression, showed it to be similar to that of induced p*XPR2* (Madzak et al. 2000—cf. Sect. 3.1 and Table 3). More recently, the compared use of p*POX2* and hp4d for production of Yarrowia native lipase *LIP2* led to similar yields (Nicaud et al. 2002), obtained during growth phase on oleic acid for the former and during stationary phase (on glucose) for the latter. The two *POT1* and *POX2* promoters are highly inducible by fatty acids and alkanes and repressed by glucose and glycerol, when p*ICL1* is strongly inducible by fatty acids, alkanes, ethanol and acetate, but not tightly repressed by glucose and glycerol. Despite their high efficiency, these new promoters could present some drawbacks for industrial use: the hydrophobic nature of p*POT1* and p*POX2* inducers may be incompatible with efficient production and/or purification of some heterologous proteins. In contrast, they are well adapted for use in engineering Yarrowia hydrophobic substrate pathways (Bourel et al. 2004; Nthangeni et al. 2004; Shiningavamwe et al. 2006; Haddouche et al. 2010). Genome data and transcriptome analyses are still expected to provide the "perfect" inducible Yarrowia promoter, combining user-friendly inducer with possibility of complete repression.

In a very recent work, Hong et al. (2012) compared intracellular expression of *E. coli* β-glucuronidase gene directed by p*TEF1* and by several promoters from glycolytic genes (newly described or previously described only in the cited patents), in high-glucose medium: p*FBA1* (from fructose 1,6-bisphosphate aldolase gene—Pollak and Zhu 2005), p*TDH1* (from glyceraldehyde-3-phosphate dehydrogenase gene) and p*GPM1* (from phosphoglycerate mutase gene—Picataggio and Zhu 2005). Activity of p*FBA1* was respectively 2.2- and 5.5-fold higher than those from p*TDH1* and p*GPM1*, the activity of this latter being similar to that of p*TEF1*. Interestingly, the use of p*FBA1* with the native *FBA1* intron allowed a further increase of expression by a factor of five (cf. Sect. 3.1).

At last, a very important improvement in the design of strong and tunable Yarrowia promoters has been carried out recently by Blazeck et al. (2011), based on the concept of the hybrid hp4d promoter. These authors have constructed a large array of recombinant promoters, carrying various copy numbers (from 1 to 32) of the UAS1B enhancer from p*XPR2*, inserted upstream of a minimal p*LEU2*, or of different versions of p*TEF1*. They named the former series "pUAS1Bn-Leum", where n is the number of UAS1B copies, but, as pUAS1B4-Leum is similar to hp4d promoter, we preferred to use here, in the text and in Table 3, the shorter notation of "hpnd", where n is the number of UAS1B copies. One of the key points of this study is that the newly designed promoters, together with some endogenous Yarrowia promoters, have been characterised at the single-cell level using a flow cytometry fluorescence-based assay, in addition to their analysis at transcriptional and whole-cell levels. Unless specified, intracellular expression of the reporter genes was achieved using centromeric replicative vectors. Under these conditions, only one (encoding hrGFP) of the four fluorescence genes tested (the others being encoding yECitrine, EGFP and mStrawberry) drove a detectable fluorescence (being optimised for expression in mammalian cells, hrGFP gene was the one exhibiting the highest codon adaptation index for Yarrowia). The hrGFP reporter gene was used to evaluate the promoter strengths of seven previously identified endogenous Yarrowia promoters, including four ones cited previously only in patents: p*GPAT* (from glycerol-3-phosphate o-acyl transferase gene—Xue and Zhu 2006a), p*YAT1* (from ammonium transporter gene—Xue and Zhu 2006b), p*EXP1* (from export protein gene—Sharpe et al. 2008) and p*GPD* (from glyceraldehyde-3-phosphate dehydrogenase gene—Picataggio and Zhu 2005). The relative fluorescence ordering was p*EXP1* > p*TEF1* > p*GPD* > p*GPAT* > p*YAT1* > p*XPR2* > p*FBA1* (cf. Table 3). Concerning the recombinant "hpnd" promoters, an exponential increase in fluorescence was observed when n increased from 1 to 8. Then, the increase became linear through $n = 19$, and fluorescence seemed to saturate through $n = 32$ (Blazeck et al. 2011). The complete "hpnd" promoter library exhibited a range of more than 400-fold in terms of mRNA levels when n varied from 1 to 32, and the strongest promoters in this set exhibited fluorescence levels eightfold higher than those of usual endogenous promoters. Twelve promoters selected from the "hpnd" promoter library, together with endogenous p*EXP1* and p*TEF1*, were also tested with *E. coli lacZ* as the reporter gene: a maximum value of 1,200 Miller units was obtained for "hp28d" in the β-galactosidase assay, and the overall data showed a strong positive statistical correlation with those obtained using hrGFP gene (Blazeck et al. 2011—cf. Table 3). However, a discrepancy could be observed between β-galactosidase activity observed for "hp4d" in this study (26 Miller units) and that previously obtained with the original hp4d on a replicative vector (420 Miller units—Madzak et al. 2000). Similarly, Madzak et al. (2000) obtained a similar *lacZ* expression when using integrative or replicative vectors, when Blazeck et al. (2011) described a twofold higher hrGFP fluorescence level when using integrative vectors instead of replicative ones. Juretzek et al. (2000a, 2001) have also previously reported a threefold higher *lacZ* expression directed by p*ICL1* when using integrative instead of replicative vectors (cf. Table 3). Could all these discrepancies simply be explained by an unexpectedly strong influence of the vector design?

As one could legitimately worry about the genetic stability of hybrid promoters carrying such a high number of tandem repeats, this point was tested on the basis of sequence fidelity after nonselective serial subculturing. Two strains carrying selected promoter constructs, "hp12d" and "hp16d", were subcultured for 36 generations: 17 out of 20 progeny isolates from the former and 20 out of 20 from the latter were positively confirmed by sequence and restriction enzyme analyses (Blazeck et al. 2011). Thus, the "hpnd" promoters seem suitably stable in Yarrowia for academical applications, but their long-term stability level may be insufficient for industrial ones. Namely, GMP (Good Manufacturing Practice) guidelines (website: http://www.gmp-compliance.org/eca_link_navigator.html) generally impose very strict standards and require producing strains to be stable over hundreds of generations.

At last, Blazeck et al. (2011) tested recombinant promoters carrying either 8 or 16 tandem UAS1B sequences upstream from a series of different longer/reduced versions of p*TEF1*, for hrGFP gene expression. When the enhancer elements were added to the *TEF1* promoter elements, a substantial fluorescence increase was observed. The enhancement obtained with 8 UAS1B was roughly half that observed with 16 of them. Moreover, this enhancing effect occurred for all p*TEF1* variants used, from the more minimal to the full-length p*TEF1* (including naturally occurring UAS elements) and even for longer variants of *TEF1* promoter.

In addition to providing a series of tunable hybrid promoters, including some of unprecedented strength, this work also suggests that endogenous promoters in Yarrowia are enhancer-limited and that this limitation can be partially or fully alleviated through addition of tandem UAS copies. Importantly, Yarrowia strains containing strong hybrid promoters did not exhibit any growth defect, demonstrating that transcription factor availability was not a limiting factor and that even the strongest promoters did not deplete measurably the transcription elements. Thus, Blazeck et al. (2011) opened the way to the design of fine-tuned promoters, in which tandem UAS copies could serve as synthetic transcriptional amplifiers, for increasing generically the expression levels of selected endogenous promoters.

2.4.2 Secretion Signals

The secretion of a heterologous protein into the culture medium is advantageous in both academical and industrial contexts, since it greatly simplifies activity testing and production/purification processes. For that purpose, a secretion signal should be transcriptionally fused upstream from the mature sequence of the protein of interest, in order to target the nascent polypeptide to the secretion pathway. Either homologous (from Yarrowia genes) or heterologous (native) secretion signals have already been used successfully (cf. examples in Sect. 3.3). Until recently, the *XPR2* prepro region was by far the most widely used secretion signal. This sequence was shown to target the early steps of protein secretion to the co-translational pathway of translocation (He et al. 1992; Yaver et al. 1992). It includes a pre region (the secretion signal per se), a dipeptide stretch and a pro region which is required for AEP transit, acting as an internal chaperone allowing the mature part of AEP to adopt a conformation compatible with secretion (Fabre et al. 1991, 1992). The removal of *XPR2* pre region occurs in RE and that of the pro region probably in late Golgi apparatus (reviewed in Beckerich et al. 1998). Retaining the pro region for heterologous expression allowed to maintain the environment of the initiator ATG codon and was intended to address the recombinant protein to co-translational secretion pathway. However, it was shown that the *XPR2* pre region alone (with or without the following dipeptide stretch) was sufficient to drive efficient heterologous secretion (Franke et al. 1988; Nicaud et al. 1989; Tharaud et al. 1992; Swennen et al. 2002). Moreover, the maturation of the heterologous protein was occasionally found to be more complete when only the pre region was used (Swennen et al. 2002). In consequence, the pre region alone is, in general, preferably used now for directing secretion (Jolivalt et al. 2005; Yue et al. 2008; Ni et al. 2008, 2009; Yang et al. 2009, 2010b; Liu et al. 2009; Boisramé and Gaillardin 2009; Yu et al. 2010; Zhao et al. 2010; Huang et al. 2011; Song et al. 2011; Li et al. 2012). Additionally, its smaller size is more advantageous for vector design. Alternative secretion signals have also been used with success: the prepro region from Yarrowia *LIP2* gene (Pignède et al. 2000; Roth et al. 2009; Emond et al. 2010) and a hybrid between *XPR2* and *LIP2* prepro regions (Nicaud et al. 2002; Laloi et al. 2002). More recently, Gasmi et al. (2011a) tested different secretion signals issued from *LIP2* gene and found that the *LIP2* pre region alone (followed by the X-Ala/X-Pro stretch) performed better than the complete *LIP2* prepro region. This suggests that, similarly to the case of *XPR2*-derived signals, the *LIP2* pro region is not necessary for secretion of heterologous proteins in Yarrowia.

2.4.3 Targeting to Microbodies

The possibility to target heterologous proteins to cellular microbodies, such as peroxisomes, is particularly interesting for the purpose of engineering Yarrowia metabolism. In a recent work, Haddouche et al. (2010) described the targeting to Yarrowia peroxisomes of a fusion protein linking the polyhydroxyalkanoate synthase from *Pseudomonas aeruginosa* to the C-terminus of the glyoxysomal isocitrate lyase from *Brassica napus*. This latter sequence has been previously shown to direct efficient peroxisomal targeting of the same heterologous protein in *S. cerevisiae* (Poirier et al. 2001). However, its C-terminal tripeptide SKM was modified by Haddouche et al. (2010) into AKI, in order to better fit the supposed Yarrowia terminal tripeptide consensus. Namely, such a C-terminal AKI has been described in the peroxisomal trifunctional enzyme hydratase/dehydrogenase/epimerase from *Candida tropicalis*, a yeast more closely related to Yarrowia than *S. cerevisiae* (Aitchison et al. 1991).

2.4.4 Surface Display Systems

The techniques for displaying heterologous proteins on the surface of yeast cells have received increasing attention in the past decade (reviewed by Pepper et al. 2008 and Shibasaki et al. 2009). Surface display has many potential applications in biotechnological, industrial and medical domains, such as immobilised biocatalysis, bioconversion, bioremediation or biosensor and live vaccine development. It also constitutes an interesting tool for ultra-high-throughput screening of new enzymatic activities. Among many advantages, using so-called arming yeasts for bioconversion allows an easy separation of product from catalyst.

Surface display systems generally operate via transcriptional fusion of the heterologous gene to a DNA sequence acting as a signal for addition of a glycosylphosphatidylinositol (GPI) anchor. GPI anchors are post-translational modifications that fasten the modified protein in the outer leaflet of the cell membrane of eukaryotic cells. Positioned at the C-terminus of the anchored protein, they are complex structures including a phosphoethanolamine linker, a glycan core and a phospholipid tail (reviewed by Paulick and Bertozzi 2008).

The first report of a surface display expression system in Yarrowia was made by Yue et al. (2008), using the GPI anchor domain of Yl*CWP1* (a Yarrowia cell wall protein, previously isolated and characterised by Jaafar and Zueco 2004). Yue et al. (2008) described the successful display of active EGFP (enhanced green fluorescent protein) and of active *Vibrio harveyi* haemolysin on 100 % of transformed Yarrowia cells. This latter example can be favourably compared with the previous display of *V. harveyi* haemolysin on *S. cerevisiae* cells using the pYD1 Yeast Display Vector Kit (Invitrogen, USA), for which the heterologous protein was detected on only a third of transformed cells (Zhu et al. 2006). This difference could possibly be due to a higher expression (driven by hp4d promoter) in the Yarrowia system (Yue et al. 2008). The Yl*CWP1*-based Yarrowia surface display system has been used since then for displaying several other proteins for various applications (Ni et al. 2009; Liu et al. 2009, 2010; Yu et al. 2010—cf. Sect. 4.2.6). The corresponding pINA1317-YlCWP110 vector (cf. Table 2) makes use of the 110 C-terminal amino acids from Yl*CWP1* as GPI anchor signal. A variant surface display vector, using a larger (131 amino acids) C-terminal fragment from Yl*CWP1*, has also been tested for some heterologous proteins displayed, with similar results (ZM. Chi, personal communication, C. Madzak, unpublished data). Thus, the 110 C-terminal fragment fully encompasses the Yl*CWP1* GPI anchor signal, and using a larger fragment does not seem justified in regard to available data. However, the observation, by Liu et al. (2009), that adding a C-terminal 6×His tag to a displayed alginate lyase was able to increase 2.4-fold its specific activity, strongly suggests that adding a spacer element could possibly improve surface display of some heterologous proteins.

More recently, the successful use of the GPI anchor domain from *S. cerevisiae FLO1* gene (fused in N-terminus of target protein), for surface display of a *Bacillus subtilis* mannanase on Yarrowia cells, has been described by Yang et al. 2009 (cf. Sect. 4.2.6). This alternative surface display system has been used recently to arm a Yarrowia strain with *Rhizopus stolonifer* lipase, for use in wastewater bioremediation (Song et al. 2011).

2.5 Development of a Commercial Yarrowia Expression Kit

A number of vectors for expression (secretion) of heterologous proteins in Yarrowia are available (cf. some examples in Table 2). When we projected to include some expression/secretion vectors in a commercial kit, we took into consideration transformation efficiency and ease of handling. We choose two pBR322-based monocopy integrative vectors with hp4d promoter: pINA1269 and pINA1296 (cf. Table 2). The integration of these shuttle vectors by single crossover can be efficiently directed, by linearisation within the pBR322 backbone, into the integrated pBR322 docking platform of Po1g strain. This process ensures a very high transformation efficiency, in the range of 10^5 colonies per μg of transforming DNA. The targeting of a unique copy at a precisely known genomic site makes this system particularly adapted to genetic engineering of proteins: consequences of mutations on enzyme activity can be assessed directly on the supernatant of transformant cultures.

The use of the strong recombinant hp4d promoter allows a growth-phase-related expression: without the need of any inducer, expression mainly develops during the stationary phase, on a high variety of media. This peculiar characteristic allows a partial dissociation of growth and expression phases, particularly adapted to efficient heterologous protein production. The protein of interest could be expressed intracellularly, or secreted using its native secretion signal, when inserted into pINA1269 (pYLEX1 in the kit). Alternatively, it could be secreted using Yarrowia *XPR2* pre secretion signal, when inserted into pINA1296 (pYLSC1 in the kit). The corresponding Po1g recipient strain is engineered for growth on sucrose or mollasses and deleted for extracellular proteases (cf. Sect. 2.2 and Table 1). YLEX™ Expression Kit can be purchased from Yeastern Biotech Co. (Taiwan—website: http://www.yeastern.com/) or from Gentaur (Belgium—website: http://www.gentaur.com/).

2.6 Amplification of Expression Cassette Copy Number

Considering the low amplification of expression attainable using replicative vectors in Yarrowia, the preferred strategy for increasing heterologous gene copy number consists in multiple integrations of a vector (or expression cassette) into the genome.

2.6.1 Homologous Multiple Integrations

First attempts for increasing copy number in Yarrowia used targeted integration into ribosomal DNA (rDNA) cluster, with a defective marker for selection of multiple integrants (James and Strick 1995; Le Dall et al. 1994). In contrast to other yeasts, rDNA in Yarrowia is organised in several dispersed clusters, with a total number of rDNA units estimated to be higher than 200 (Casaregola et al. 1997). The most effective of defective selection markers, *ura3d4* allele (cf. Sect. 2.3), was chosen among a series of promoter-deleted alleles designed by Le Dall et al. (1994) and constitutes the basis of all multicopy vectors now available (cf. Table 2). It allowed toobtain strains carrying up to 60 integrated copies of a vector carrying a fragment of rDNA G unit for targeting and *XPR2* as a reporter gene (Le Dall et al. 1994). However, these high copy number integrants were stable only under cultivation conditions non-inducing for *XPR2* expression. Under inducing conditions, deamplification occurred during cultivation, the number of copies eventually stabilising around ten. The amount of AEP secreted in culture medium increased linearly with gene dosage, up to around ten copies, but overproduction above this level seemed to have deleterious effects (Le Dall et al. 1994). It is probable that stability of high copy number integrants could depend on the nature of the protein produced.

Interestingly, multiple integrations can also be selected for, using a defective selection marker, when using a single integration site, as demonstrated by the Juretzek et al. (2001): vectors carrying *ura3d4* allele were integrated either in rDNA or zeta sequences (see Sect. 2.6.2) or at *XPR2* locus. Multicopy transformants were obtained in all cases, regardless of the target used for integration, thus indicating the determining role of the defective selection marker in the process. As also observed previously by Le Dall et al. (1994), multiple integrations occurred almost always in tandemly repeated copies, at only one or two sites. Similarly, transformant copy number was also directed towards stabilisation around 10–13 copies during cultivation, by an amplification/deamplification process, probably reflecting the optimum auxotrophy complementation driven by *ura3d4* allele (Juretzek et al. 2001).

Very recently, a new method involving homologous multiple integrations for the obtention of stable multicopy transformants has been described in a patent from our laboratory (Nicaud et al. 2009). The general idea is to provide a genetically engineered strain, deleted for major homologous secreted proteins (for alleviating metabolic burden), in which several loci have been selected for targeting integrative vectors. Correct integrations could be checked easily using reporter systems, since they confer a new phenotype to the transformed strain (inactivation of a function, detectable on plate). This strain could be sequentially transformed by a series of integrative vectors with different auxotrophy or dominant markers, with the possible use of marker rescue (Nicaud et al. 2009). This system allows to obtain a final number of stable copies of the gene of interest ranging from three to ten (Nicaud et al. 2009).

2.6.2 Zeta Sequences, Non-homologous Integration and amplification

Among possible targets for multiple integration, retrotransposons constitute an interesting possibility. The first one to be described in Yarrowia was Ylt1 (Schmid-Berger et al. 1994), from Ty3-*gypsy* group, which is characterised by its very large (714 bp) long terminal repeats (LTRs) termed zeta sequences. Surprisingly, Ylt1 is not present in all Yarrowia wild-type strains. Among laboratory strains, Ylt1 was found in wild American isolate YB423 and in its derivative strains from different inbreeding programmes (strains' origins reviewed in Barth and Gaillardin 1996). In contrast, it was not detected in wild German isolate H222 nor in wild French isolate W29 and its derivative Po1d (Casaregola et al. 2000; Juretzek et al. 2001).

In Ylt1-carrying strains, the retrotransposon is present in at least 35 copies dispersed in the whole genome. Additionally, zeta LTRs are also present as solo elements, in at least 30, and up to 60 copies for some strains (Schmid-Berger et al. 1994; Juretzek et al. 2001). Thus, zeta sequences could provide at least 100 targeting sites per genome of Ylt1-carrying strain, constituting a powerful tool for multiple homologous integration. Targeted zeta-based auto-cloning vectors have been used successfully to overexpress Yarrowia extracellular lipase (Pignède et al. 2000) and to express heterologous genes (Juretzek et al. 2001). Lipase-overexpressing transformants with around ten integrated copies of zeta-based vector were found to maintain this copy number after 2 weeks of exponential growth (120 generations), in both non-inducing and lipase-inducing cultivation conditions (Pignède et al. 2000).

Surprisingly, zeta sequences were shown to exhibit a completely unexpected property: they are able to enhance non-homologous integration into Yarrowia strains devoid of Ylt1 retrotransposon (Nicaud et al. 2000), providing a process for non-homologous Yarrowia transformation. In addition to the variability in heterologous production, linked to copy number, which is found in multicopy integrants, the use of non-homologous integration is another major source of heterogeneity among transformants. Namely, some integration sites can impair cell growth or heterologous protein expression and/or secretion. Influence of integration site on gene expression in yeast has been investigated in *S. cerevisiae* by Thompson and Gasson (2001), and more recently by Flagfeldt et al. (2009), who both found differences of the level of an order of magnitude between sites leading to lowest and highest expression of *E. coli* β-galactosidase. Similar effects were observed in Yarrowia strains carrying randomly integrated zeta-based expression cassettes: transformants with similar copy numbers could exhibit very different activity levels, and expression of both the defective selection marker (*ura3d4*) and the heterologous gene from a single integrated cassette could be increased, probably due to regulatory elements close to integration site (Roth et al. 2009). This problem highlights the need to screen all transformants obtained for heterologous activity, since results from a genetic screening could not be used directly to predict production efficiency.

Thus, the use of a non-homologous transformation system requires testing a high number of independent transformants, in order to be able to select a good producer strain. However, the obtention of numerous multicopy transformants using a non-homologous integration system could be technically difficult: transformation efficiency of Ylt1-free strains by zeta-based vectors is lower than with classical integrative vectors, and the use of a defective selection marker decreases the transformation frequency by several orders of magnitude. Yarrowia can be very efficiently transformed by classical integrative vector (10^5–10^6 transformants per μg of DNA). The use of *ura3d4* defective allele for selection lowers dramatically transformation efficiency, even in the case of vectors integrating by homology with the genome: multicopy zeta-based vectors used in Ylt1-carrying (zeta-containing) strains generate only around 10^2 transformants per μg (Juretzek et al. 2001). Transformation efficiency of the same vectors in Ylt1-free strains (non-homologous integration) is reduced still further: around 10 transformants per μg (Juretzek et al. 2001).

Despite its limitations, the use of zeta-based vectors for non-homologous transformation of Ylt1-free strains constitutes an interesting method for the obtention of multicopy producer strains, since non-homologously integrated copies are more dispersed in the genome than those targeted by homology, leading to better stability of high copy number integrants (Nicaud et al. 2000). This method has been successfully used to produce several heterologous proteins (Juretzek et al. 2001; Nicaud et al. 2002; Laloi et al. 2002; Nthangeni et al. 2004; Shiningavamwe et al. 2006; Maharajh et al. 2008a; Roth et al. 2009; van Zyl 2010).

However, integration of multiple copies of an expression cassette into the host genome may increase metabolic burden. Namely, some authors reported important decreases in biomass and growth rate of Yarrowia producing strains, as a result of multicopy integration of zeta-based cassettes (Roth et al. 2009, C. Madzak, unpublished data). Some results obtained during cultivation scale-up of an *Aspergillus aculeatus* mannanase-producing strain, carrying nine integrated zeta-based cassettes, seemed to show that not all biomass contributed to enzyme production (cf. Sect. 3.5), leading Roth et al. (2009) to hypothesise a shift in metabolic flux towards growth, at the detriment of enzyme expression, without ruling out a possible instability of the strain leading to a loss of productivity. Moreover, very recently, new data on non-homologously transformed multicopy strains have casted doubt on their long-term stability, especially for high copy numbers (Nicaud et al. 2009). Regulations about GMO from GMP guidelines (website: http://www.gmp-compliance.org/eca_link_navigator.html) impose very strict standards: genomic modifications of the producing strain should be described precisely, and be stable over hundreds of generations. Some producing strains obtained using multicopy zeta-based vectors have failed to obtain marketing authorisations, since their stability level has not been judged sufficient for industrial applications. These limitations have prompted the development of alternative amplification strategies (Nicaud et al. 2009—cf. end of Sect. 2.6.1).

2.6.3 Multiple Cassette Vectors

Another strategy for increasing heterologous production is to include two, or even three, expression cassettes in a single vector. This method is also adapted for coordinated co-expression of different heterologous genes, especially if gene dosage is important. Possible problems could be the vector size limit (probably around 12 Kb) and recombinations between repeated sequences. The double use of a same promoter has already been described several years ago for some vectors, in which hp4d promoter was used for expressing both a dominant marker gene and Cre recombinase (Fickers et al. 2003). No particular stability problem has been reported with these double-hp4d vectors, which are currently used in our laboratory for marker rescue; these replicative vectors are however used only for transient expression. More interestingly, in the series of "hpnd" promoters developed by Blazeck et al. (2011), "hp12d" and "hp16d" have been shown to be fairly stable, at least for academical applications (cf. Sect. 2.4.1), despite the high number of UAS1B tandem repeats they contain.

More recently, several patents on genetic engineering of Yarrowia metabolic pathways have described triple and even quadruple expression cassettes in a single vector, implying different promoters and heterologous genes (Xue and Zhu 2006b; Sharpe et al. 2008). At last, very recently, two expression cassettes encoding two different heterologous desaturases, but both using hp4d as promoter, have been inserted, in the same orientation, in an integrative vector used for engineering Yarrowia fatty acid pathway (Chuang et al. 2009, 2010—cf. Sect. 4.2.4). However, for the moment, no report has been made of using this technique to increase the copy number of a same expression cassette: it remains to be determined if this situation (longer internal homology encompassing promoter, gene and terminator sequences) could impair vector stability.

2.7 *High-Throughput Screening Processes*

Developing high-throughput screening procedures is important for both screening of new biocatalysts through expression cloning and improving selected enzymes through directed enzyme evolution. Bordes et al. (2007) have developed such a process for Yarrowia. Although the chosen enzyme was the homologous extracellular lipase YlLip2p, this work opened the way for high-throughput screening of heterologous proteins in Yarrowia. The procedure was optimised for protein expression in 96-well microplates. Transformation efficiency and targeting of the zeta-based vector used were increased by constructing a Yarrowia strain carrying an integrated zeta docking platform. Following optimisation, both transformation efficiency (8,000 transformants per μg of DNA) and variability were shown to be compatible with high-throughput screening requirements.

As lipases play an important role in asymmetric biocatalysis, improving their performances or tailoring them for novel substrates is a challenge in protein engineering. Due to their peculiar characteristics (high stability in immobilised form at elevated temperature, unique substrate specificities), the two lipases (A and B) from *Candida antarctica* constitute promising targets for high-throughput screening procedures. Recently, the above Yarrowia high-throughput screening process has been developed into an efficient recombinant expression system used to engineer a His-tagged *C. antarctica* lipase B (Emond et al. 2010). The production of the recombinant lipase could be scaled up in batch reactor to reach high yields (up to 200 mg and more than 5,000 units/l), despite the use of a single-copy integration strategy. A small library of lipase variants has been created using saturation mutagenesis of residue A281 and has been screened for lipase activity towards *para*-nitrophenyl butyrate. This process yielded several mutants displaying improved catalytic efficiencies compared to the wild-type enzyme (Emond et al. 2010). The authors compared the performances of their high-throughput screening system, for the engineering of *C. antarctica* lipase B, to its counterpart in *S. cerevisiae* (Zhang et al. 2003): the time necessary for enzyme production in microplates was much shorter (2 days versus 7 days), and the scaled-up yields were much higher (200 mg versus 80 mg/l). This demonstrated the high potential of this Yarrowia high-throughput process for the reliable and rapid screening of libraries as well as for the efficient production of interesting mutants.

3 Review of Heterologous Protein Production

Table 3 presents a comprehensive survey of the literature concerning production of heterologous proteins in Yarrowia. In an attempt of exhaustivity, we chose to include some unpublished results, described only in patents or congress reports, when their contribution seemed pertinent. The examples amount to 104 heterologous proteins, from a large range of phylogenetic origins (60 species, from viruses and eubacteria to human). The vector type and molecular tools used for protein production are briefly indicated, and the yield is given whenever available. In all cases (with the sole exception of *E. coli* amylolytic enzyme, which was produced in an inactive glycosylated form—N. Libessart, personal communication), production was quantified by measuring heterologous activity and thus corresponded to active protein yield. The size of the proteins produced is ranging from 4 kDa (human insulinotropin—James and Strick 1995) to 116 kDa for intracellularly expressed *E. coli* β-galactosidase (first report by Gaillardin and Ribet 1987). The largest heterologous proteins secreted, with a size of 90 kDa, were *Arxula adeninivorans* glucoamylase (Swennen et al. 2002) and *Aspergillus oryzae* leucine aminopeptidase (Nicaud et al. 2002).

As can be seen from Table 3, production yield could vary greatly from one example to another. As was also observed in other production systems, each protein remains a particular case, and predicting the results remains difficult. However, globally, most proteins have been produced efficiently, even using only monocopy expression and non-optimised cultivation conditions. Namely, yields of heterologous proteins secreted in shaked flasks from monocopy transformants were often of the mg to tens of mg per litre level, especially for fungal proteins (Muller et al. 1998; Park et al. 2000; Madzak et al. 2005b; Jolivalt et al. 2005) but also for plant (Kopecny et al. 2005), mammalian (Madzak et al. 2000, BC. Sang, personal communication) and human proteins (Tharaud et al. 1992; Swennen et al. 2002; Gasmi et al. 2011a). It is to note that shake-flask production of maize CKO1p from a hp4d-based monocopy vector (Kopecny et al. 2005) was sufficient to allow crystallisation of the recombinant enzyme for structural analysis (Kopecny et al. 2004).

As shown from several examples, scale-up to batch or fed-batch cultivation increased production by a factor of 10- to 20-fold (cf. Sect. 3.5). In a few examples, very competitive secretion yields were obtained from monocopy transformants cultivated in batch, such as for *C. antarctica* lipase expressed using p*POX2* and secreted using *LIP2* prepro (200 mg/l—Emond et al. 2010).

Additionally, two- to tenfold increases were observed when using multicopy expression vectors (Juretzek et al. 2001; Nicaud et al. 2002; Maharajh et al. 2008a; Roth et al. 2009). A few examples demonstrated that the increase obtained using the combined use of multicopy vectors and of cultivation optimisation could be as high as two orders of magnitude (Nicaud et al. 2002; Roth et al. 2009). Such combined strategies were reported to allow the production of 500 mg/l of human CD14 (C. Gysler, personal communication and abstract report from 2002) and even of gram per litre yields of a fungal mannanase (van Zyl 2010 and personal communication), a level observed previously with homologous proteins only (cf. Sect. 3.5).

For the sake of readability, informations on Yarrowia strains used were omitted in Table 3. The Po1 series of strains (Po1d and its derivatives) are clearly the recipient strains which have been the most employed. The overall amount of data globally demonstrates the efficiency and versatility of Yarrowia as host for heterologous production and offers a few elements for comparison with other yeast host systems (cf. Sect. 3.6).

3.1 Promoters and Transcription Efficiency

The 104 examples of heterologous proteins listed in Table 3 have been expressed using various promoters, but only a few of them have been widely used (promoters described only in patents being excluded):

- The recombinant hp4d promoter, for 40 heterologous proteins
- The peptone-inducible *XPR2* promoter, for 24 heterologous proteins
- The constitutive *TEF1* promoter, for 22 heterologous proteins
- The fatty acid-inducible *POX2* or *ICL1* promoters, for 13 heterologous proteins

XPR2 promoter has been for a long time the only strong promoter available in Yarrowia, hence the high number of early examples. However, its use has been discontinued since more competitive promoters have been described: hp4d (Madzak et al. 1996, 2000), p*TEF1* (Muller et al. 1998), p*POX2* and p*ICL1* (Juretzek et al. 2000a). The use of hp4d has particularly developed recently, due to some interesting characteristics: its versatility allows various culture conditions, and its growth-phase-dependent expression allows partial dissociation of growth and production phases. The satisfying results obtained with various heterologous proteins show that this only partial dissociation is however sufficient to ensure high production levels (with the possible exception of highly toxic proteins). The recently developed series of strong and tunable "hpnd" promoters (Blazeck et al. 2011—cf. Sect. 2.4.1) and the subsequent possibility to enlarge this concept to fine-tuning other Yarrowia endogenous promoters will probably play a major role in the future of heterologous production.

The example of intracellular expression of *E. coli* β-galactosidase offers a basis for comparing promoter strength. A group of efficient promoters, driving expressions of hundreds of standard Miller units, is composed of p*G3P*, p*XPR2* and hp4d (in order of increasing efficiency—Juretzek et al. 2000a; Blanchin-Roland et al. 1994; Madzak et al. 2000). Another group is constituted by inducible promoters from fatty pathways, which allow to attain, under inducing cultivation conditions, levels of thousands of standard Miller units: p*POX1*, p*POX5*, p*ICL1*, p*POX2* and p*POT1* (in order of increasing efficiency—Juretzek et al. 2000a). Surprisingly, compared to their outstanding performance with hrGFP gene expression (cf. Sect. 2.4.1), the series of strong "hpnd" promoters developed by Blazeck et al. (2011) did not seem to perform as efficiently with β-galactosidase activity (yields around thousand standard Miller units), compared to the best endogenous promoters. This discrepancy remains to be explained, but comparing more carefully the various methodologies used, especially for cultivation and induction conditions may possibly be a clue.

Interestingly, the use of p*G3P* with the native *G3P* upstream intron fused to the heterologous protein allowed to increase β-galactosidase expression by a factor of five (Juretzek et al. 2000a). The same strategy, consisting in retaining an upstream intron with the corresponding Yarrowia promoter, has also been used by several authors in order to increase expression of heterologous proteins by p*FBA1* (Pollak and Zhu 2005; Xue and Zhu 2006b; Sharpe et al. 2008; Hong et al. 2012) and p*GPD* (Sharpe et al. 2008). In the very recent work of Hong et al. (2012), the use of p*FBA1* with the native *FBA1* intron allowed to increase *E. coli* β-glucuronidase expression by a factor of five, a result similar to that previously observed for another promoter/protein couple by Juretzek et al. (2000a). This possible intron effect was analysed further by constructing a p*GPM1*::*FBA1* chimeric promoter (p*GPM1* plus 5′ region of *FBA1* gene—Hong et al. 2012). This chimeric promoter showed a significant increase in promoter strength compared to p*GPM1*, confirming the role of the intron-containing *FBA1* gene 5′ region in transcriptional enhancement. Such an effect of positive correlation between intron and mRNA stability is known as intron-mediated enhancement (IME) and has been reported in various fungi and "higher" eukaryotes (invertebrates, plants and mammals—briefly reviewed in Hong et al. 2012). IME is not effective with all introns, and its mechanism remains elusive. It has been suggested that the splicing of an upstream intron could make genes more receptive to RNA polymerase II, affecting the frequency of gene transcription and reducing transcription-associated genome instability (Niu and Yang 2011). As Yarrowia contains more intronic genes (15 %—Mekouar et al. 2010) than all other yeast species sequenced up to now, IME could possibly play an important role in the regulation of gene expression in this yeast.

In conclusion, the examples from Table 3 could give some indications for choosing the optimal promoter for a given application:

- p*TEF1*, if constitutive expression is needed, such as for expression cloning
- p*ICL1*, p*POX2* or p*POT1*, for highly inducible expression, as long as the nature of the inducer (fatty acids or alkanes for all of them, plus ethanol or acetate for the former) is not incompatible with production/purification process
- hp4d, or one of the stronger "hpnd" promoters for strong growth-phase-dependent expression in various cultivation conditions, with no induction required

In addition, the recent observation, by Blazeck et al. (2011), that endogenous Yarrowia promoters are enhancer-limited and could be improved by adding tandem UAS copies (cf. Sect. 2.4.1) will probably trigger the engineering of endogenous promoters selected for their intrinsic properties.

3.2 Codon Usage Bias and Translation Efficiency

A survey of the examples from Table 3 shows that the lowest production levels were observed for some human proteins: factor XIIIa (Tharaud et al. 1992), EGF (Hamsa et al. 1998) and especially α-foetoprotein and β2-microglobulin (K. Uchida, personal communication). Such low yields could possibly be due to important differences in codon usage between *Homo sapiens* and Yarrowia, leading to poor translation efficiency of some human genes in this host. The main difference between the two codon biases is found for arginine: among the six possible Arg codons, the four preferred in *H. sapiens* are the less employed in Yarrowia. Codon usage of this latter is peculiar among yeasts, being rather different from those of *S. cerevisiae* and most yeasts, and more similar to those from *Aspergillus* genus (cf. tables of codon usage at http://www.kazusa.or.jp/codon/index.html). The rather recent cost-effective availability of synthetic genes will probably boost the use of genes adapted to Yarrowia codon bias. Until recently, only a few patents had reported the expression of codon-optimised heterologous genes in Yarrowia, but no comparison with unmodified sequences has been shown (Xue and Zhu 2006b; Sharpe et al. 2008). To our knowledge, the only study allowing such a comparison is the recent work from Gasmi et al. (2011a), which showed that adaptation to Yarrowia codon bias could increase expression of human α2b-interferon by a factor of 11. Thus, if the interest of codon optimisation remains to be further evaluated, it already seems to constitute a promising approach.

In this same study, Gasmi et al. (2011a) tested the effect of inserting a CACA sequence immediately upstream from the initiation codon and found that this led to a 1.5-fold increase in recombinant interferon activity. Thus, it seems that the AUG context could play a significant role in translation efficiency in Yarrowia, as previously described in many eukaryotes (reviewed in Kozak 1999). Such observations have led to the concept of Kozak consensus sequence, a preferred choice in AUG upstream and downstream context, playing a major role in the initiation of translation. Nakagawa et al. (2008) analysed the nucleotide sequences around the initiation codon among 47 eukaryote species, including animals, fungi, plants and protists. They observed that preferred AUG contexts differed between species, roughly reflecting their evolutionary relationships. There was a strong bias for A/G at position -3 and A/C at position -2, a consensus consistent with the high frequency of CACA sequence immediately upstream from AUG observed, in Yarrowia, in highly expressed genes (Gasmi et al. 2011a). However, more experimental data would be required to evaluate the role that AUG context could play in translation efficiency in Yarrowia and to determine the Kozak consensus of this yeast.

3.3 Secretion Signals, Maturation and Secretion Efficiency

Among the examples from Table 3, 54 heterologous proteins have been secreted using various secretion signals:

- 40 heterologous proteins with a signal derived from *XPR2* gene (*XPR2* pre, with or without dipeptides, *XPR2* prepro)
- 25 heterologous proteins with their own native signal
- 5 heterologous proteins with a signal derived from *LIP2* gene (*LIP2* pre with dipeptides, *LIP2* prepro)
- 3 heterologous proteins with a hybrid signal *LIP2*/*XPR2* prepro

When secretion signals were used, the secretion of the heterologous protein was always obtained at least with one of them, with the exception of the peculiar case of hepatitis B antigens, which were retained intracellularly due to their assembly into Dane particles (Hamsa and Chattoo 1994). For *S. cerevisiae* invertase, the first attempts of secretion using *XPR2* pre region were not fully successful, since most of the enzyme (90 %) was retained in the periplasm, like in the native organism (Nicaud et al. 1989). However, very recent attempts using either the native secretion signal, or the *XPR2* prepro region with 13 additional amino acids from mature *XPR2*, have led to complete invertase secretion (Hong et al. 2012). In the few cases for which secretion efficiency has been estimated by comparing heterologous protein activity inside and outside the cells, intracellular activity remained undetectable when high levels were observed in culture supernatant (Franke et al. 1988; Nicaud et al. 1991; Madzak et al. 2005b; Kopecny et al. 2005).

Native secretion signals, mainly from fungal origin, have often been tested for directing the export of cognate heterologous proteins in Yarrowia. In most cases, they were able to drive successful secretion. Muller et al. (1998) produced six fungal proteins (from *A. aculeatus*, *Humicola insolens* and *Thermomyces lanuginosus*) using their native signal peptide. Several other examples were described more recently, not only for fungi, which are relatively close to Yarrowia in terms of phylogeny (*T. reesei*, *Alternaria alternata*, *Trametes versicolor*, *Pycnoporus cinnabarinus*, *A. aculeatus* and *Rhizopus oryzae*—Park et al. 2000, Morin and Dominguez abstract report from 2002, Jolivalt et al. 2005; Madzak et al. 2005b; Roth et al. 2009; Yuzbashev et al. 2012), but also for plants (rice α-amylase—Park et al. 1997, maize CKO1p—Massonneau et al. 2004). Secretion driven by native signal peptides can even be more efficient than with Yarrowia signals. It was the case for five fungal proteins: *T. reesei* endoglucanase I (Park et al. 2000), *T. versicolor* laccase IIIb (Jolivalt et al. 2005), *A. aculeatus* endo-β-1,4-mannanase (Roth et al. 2009), *S. cerevisiae* invertase (Hong et al. 2012) and *R. oryzae* lipase (Yuzbashev et al. 2012). For the *A. aculeatus* mannanase, the higher efficiency of the native signal can only be inferred from the comparison of results obtained with multicopy transformants, a 2-copies one using *LIP2* prepro and a 9-copies one using the native signal, which exhibit a 26-fold higher activity (a difference much higher than expected from respective copy numbers). For the *R. oryzae* lipase, a hybrid signal linking *XPR2* pre region to the native pro region of the lipase was also tested and was the most efficient for secretion of the recombinant enzyme (Yuzbashev et al. 2012). However, despite these positive examples, the use of Yarrowia secretion signals appears a more reliable choice, especially for proteins from "higher" eukaryotes, like those from mammals.

A few studies compared the efficiency of different secretion signals derived from *XPR2* gene for a same promoter/heterologous protein combination. Interestingly, some of the concerned proteins were difficult to produce under an active form in heterologous expression systems:

- For rice α-amylase, *XPR2* prepro was able to drive secretion, but not *XPR2* pre (Park et al. 1997).
- For human factor XIIIa, which is naturally secreted by an unknown signal-peptide-independent mechanism (Muesch et al. 1990), Tharaud et al. (1992) observed that both *XPR2* pre and prepro were unable to direct secretion. In contrast, factor XIIIa could be re-routed through Yarrowia secretory pathway when using *XPR2* pre together with the adjacent XA/XP dipeptide region. However, previously, *XPR2* pre + dipeptides and *XPR2* prepro were found similarly efficient for bovine prochymosin secretion (Franke et al. 1988).
- For human anti-Ras single-chain antibody (scFv, containing disulfide bridges), and *A. adeninivorans* glucoamylase, secretion efficiencies of *XPR2* pre and

prepro regions were found similar (Swennen et al. 2002). However, the use of *XPR2* prepro led to an incomplete maturation of recombinant scFv.
- For *P. cinnabarinus* laccase I (a glycoprotein with disulfide bridges and four associated coppers), secretion efficiencies of *XPR2* pre and prepro regions were also found similar (Madzak et al. 2005b).

In order to evaluate the precision of recombinant protein maturation (processing of pre or pro region), N-terminal amino acid sequences of some heterologous proteins were determined: bovine prochymosin secreted using *XPR2* prepro, maize CKO1p secreted using *XPR2* prepro and *P. cinnabarinus* laccase I secreted using either *XPR2* pre or *XPR2* prepro were found to be correctly processed (Franke et al. 1988; Kopecny et al. 2004, 2005; Madzak et al. 2005b). For human anti-Ras scFv secreted using either *XPR2* pre or *XPR2* prepro, amino-terminal sequences showed a precise processing, but maturation was incomplete with *XPR2* prepro, with some pro-scFv remaining unprocessed (Swennen et al. 2002). In a few cases, abnormal processing was observed when *XPR2* prepro region was used to direct secretion: recombinant rice α-amylase and human EGF exhibited amino-terminal sequences differing from those of native proteins (Park et al. 1997; Hamsa et al. 1998). On the basis of secondary structure prediction, Park et al. (1997) showed that Xpr6p protease, required for pro-AEP processing, could not access correctly to dibasic cleavage site in pro-amylase fusion, promoting an aberrant upstream cleavage.

In conclusion, as *XPR2* pro sequence do not appear to be necessary for secretion of heterologous proteins, and has been associated to maturation defects, the use of *XPR2* pre sequence is now generally preferred (this was the secretion signal chosen for use in Yarrowia expression kit). However, *XPR2* pro region was associated to the highest heterologous production yield observed in Yarrowia: a bacterial amylolytic enzyme was secreted, using a monocopy vector based on hp4d promoter, to a level estimated to 1 g/l in shaked flask (N. Libessart, personal communication). This result was obtained with a mutated *XPR2* pro sequence, but, as the protein was inactive, the possible reasons for such high yield have not been investigated further. This however raised the interesting possibility of increasing substantially secretion efficiency by modifying signal sequences.

In this context, Gasmi et al. (2011a) compared recently the efficiencies of the *LIP2* prepro region and of the *LIP2* pre region alone (followed by the X-Ala/X-Pro stretch) for the production of a codon-optimised human α2b-interferon. The use of this latter secretion signal allowed a 1.7-fold increase in interferon activity, leading to the conclusion that, as seen previously for *XPR2*, the pro sequence from *LIP2* gene does not seem necessary for secretion of heterologous proteins in Yarrowia.

3.4 Glycosylation and Overglycosylation

For some of the example of Table 3, the overall glycosylation pattern of the recombinant protein has been checked using Western blotting combined to endoglycosidase treatment. None of heterologous proteins secreted in Yarrowia exhibited heavy hyperglycosylation, but moderate overglycosylation (in the range of 10–20 additional kDa) was often observed. This was the case for *S. cerevisiae* invertase, for human factor XIIIa, for *T. reesei* endoglucanase I, for maize CKO1p, for *P. cinnabarinus* laccase I, for *T. versicolor* laccase IIIb and for *A. aculeatus* endo-β-1,4-mannanase (Nicaud et al. 1989; Tharaud et al. 1992; Park et al. 2000; Kopecny et al. 2005; Madzak et al. 2005b; Jolivalt et al. 2005; Roth et al. 2009), for all of which enzymatic activities were however not impaired. In the case of *P. cinnabarinus* laccase I, the observed 20 kDa overglycosylation was lesser than that previously observed in another yeast, *P. pastoris*, but higher than that obtained in a filamentous fungi, *A. niger* (Madzak et al. 2005b—cf. Sect. 3.6).

As stated in Sect. 1.2.3, only a few studies analysed precisely glycosylation in Yarrowia. Among homologous proteins, YlLip2p carried majorly oligosaccharide chains of 8 or 9 mannoses (notwithstanding minor heavier isoforms—Jolivet et al. 2007). Concerning heterologous production, human tissue plasminogen activator secreted in Yarrowia was found to carry only short chains of 8–10 residues (A. Franke, personal communication). Song et al. (2007) provided recently more precise informations: oligosaccharidic chains assembled on either homologous YlLip2p or recombinant *T. reesei* endoglucanase I were found to be similar, exhibiting heterogeneous sizes ranging from 8 to 12 mannoses. This same study also concluded the probable absence of immunogenic terminal α-1,3-mannose linkage in Yarrowia, based on the observations that Man8GlcNAc2 was converted into Man5GlcNAc2 after α-1,2-mannosidase treatment and that no open reading frame homologous to *S. cerevisiae MNN1* (α-1,3-mannosyltransferase) appeared in the genome (Song et al. 2007). Data about glycosylation in a Yl*OCH1*-deleted Yarrowia strain, from the same study, are described in Sect. 4.5.

The example of maize cytokinin oxidase I secretion in Yarrowia has also provided some insight on glycosylation: combined results from MALDI-TOF and 3D structural analysis have shown the presence of about 84 hexose units dispatched between five N-glycosylated sites, among the eight potential N-glycosylation sites predicted from the sequence (Kopecny et al. 2005). The average length of sugar chain per site was thus of 17 hexose units. Interestingly, these results obtained in Yarrowia could be compared to those observed for the same protein secreted in *P. pastoris*: the total extent of N-glycosylation was similar in this other yeast (Bilyeu et al. 2001), but for only four sites, which did not overlap exactly those observed in Yarrowia (Malito et al. 2004). Thus, all potential N-glycosylation sites were not automatically glycosylated, and their choice could vary between different host yeasts. Similarly, it was observed that the two potential sites of bovine prochymosin, not glycosylated in the native organism, were also not glycosylated in Yarrowia (Madzak et al. 2000).

3.5 Cultivation Conditions and Scale-Up

Although, in most cases, heterologous proteins could be secreted efficiently under standard Yarrowia cultivation conditions (shake-flask at 28 °C, YPD rich medium), some of them could show unusual sensitivity to medium composition and/or growth temperature and require some peculiar conditions in order to be produced in substantial amounts. This was the case for *P. cinnabarinus* laccase I, a complex metalloenzyme (glycoprotein with two disulphide bridges and four associated coppers), for which production in Yarrowia was found highly variable depending on culture medium composition (Madzak et al. 2005b). Lac1p production was shown to be strongly dependent on both growth rate and cell morphology: it was significant only in media where growth rate was low and when yeast cell form was predominant (absence of dimorphic transition to hyphae). Among eight media tested, the best results were obtained using PPB 20 Cit (Madzak et al. 2005b), a sucrose-based medium with 20 mM citrate buffer, which combined slow growth rate with complete lack of mycelium formation. Interestingly, raising citrate buffer concentration to 50 and then 200 mM, causing both an increase in growth rate (citrate being metabolised) and a transition from yeast cell form to mycelium (citrate being a positive effector of mycelium formation—Ruiz-Herrera and Sentandreu 2002), provoked a drastic decrease of laccase production (Madzak et al. 2005b). This case highlighted the interest of using hp4d, since this versatile promoter allowed to test different media of various compositions. The observed link between growth rate and efficient secretion of a complex protein was not surprising, since it has been observed in yeast that high growth temperatures and synthesis rates were detrimental to correct protein folding (reviewed in Romanos et al. 1992). However, such an effect was not observed in a similar case, the production of another fungal laccase (LacIIIbp from *T. versicolor*), for which secretion in rich YPD medium, where growth rate was high, gave satisfactory results (Jolivalt et al. 2005).

The case of Lac1p points to a link between dimorphism and protein secretion in Yarrowia: the yield of recombinant enzyme was significant only when the yeast form was predominant, and mycelium formation was detrimental to production. A similar observation has been made more recently, during expression of a codon-optimised human α2b-interferon gene in Yarrowia (Gasmi et al. 2011b). This is not surprising that protein secretion could be linked to dimorphism since several genes described in the secretion pathway of Yarrowia are also implicated in morphological transition (Swennen and Beckerich 2007). However, not all secreted heterologous proteins are sensitive to this dimorphism effect: in addition to LacIIIbp cited above, another glycoprotein, maize CKO1p, was produced as efficiently in PPB medium with 50 mM citrate buffer (occurrence of yeast to hyphae dimorphic transition) than in YPD (predominant yeast cell form—Kopecny et al. 2004, 2005). Comparing these examples shows that each heterologous protein remains a particular case and that predicting results could be difficult.

Although previous use of Yarrowia for citric acid and single-cell protein production has generated data on its cultivation in large bioreactors, conditions are generally not adaptable to heterologous protein production. More recently, optimisation of medium and cultivation conditions, for batch and fed-batch cultivation, have been performed for production of homologous (YlLip2p—Nicaud et al. 2002) or heterologous proteins (Park et al. 2000; Madzak et al. 2000;

Nicaud et al. 2002; Maharajh et al. 2008a, b; Roth et al. 2009; van Zyl 2010; Emond et al. 2010; Gasmi et al. 2011c).

As shown in Table 3, batch cultivation drove an eightfold increase, over shake-flask cultivation, for expression of *E. coli* β-galactosidase (up to 160 mg/l—Madzak et al. 2000), a sixfold one for secretion of *A. aculeatus* endo-β-1,4-mannanase (Roth et al. 2009) and a tenfold one for secretion of *C. antarctica* lipase (up to 200 mg/l—Emond et al. 2010), all from monocopy transformants. The increase obtained using batch cultivation could however be lesser, as for *P. cinnabarinus* laccase I (2.4-fold, up to 20 mg/l—Madzak et al. 2005b). Such growth conditions can even be detrimental, as was observed with a multicopy transformant carrying nine integrated copies of *A. aculeatus* mannanase expression cassette: a twofold decrease in secreted activity was observed, compared to shake-flask cultivation (Roth et al. 2009). The authors hypothesised that integration of multiple copies may have resulted in increased metabolic load on the producing strain, as indicated by the lower biomass concentration attained (decrease of 50 %) compared to a monocopy transformant (Roth et al. 2009). Fermenters being more suited for growth than shake-flasks, it was possible that, under potentially stressing conditions (high copy number), metabolic flux could be redirected towards growth at the cost of enzyme production. Possible occurrence of other factors, like a loss of productivity due to genetic instability of the multicopy producing strain in non-selective medium, was however not ruled out (Roth et al. 2009). Nevertheless, the use of fed-batch cultivation allowed to obtain a twofold increase in secreted activity, over shake-flask cultivation, for the same multicopy transformant (Roth et al. 2009), and even a threefold increase, when using more recently further improved fed-batch conditions (van Zyl 2010). Namely, this latter author described in a patent that a limited growth rate (preferably in the range of 0.035–0.039 per hour) was able to increase both growth rate and heterologous enzyme production, thereby optimising the productivity of the fermentation. Finally, despite the lesser than expected effect of scale-up, this research group has been able to obtain a very high activity from a multicopy transformant for which the native mannanase secretion signal was used: at first more than 26,000 nkat/ml (Roth et al. 2009), hence a more than 200-fold increase over a monocopy transformant using *LIP2* prepro, cultivated in shake-flask, and then almost 41,000 nkat/ml (van Zyl 2010), hence a more than 300-fold increase over the same monocopy transformant. This represents a production of heterologous enzyme in the range of the gram per litre, approaching the levels obtained previously only for homologous proteins, such as AEP.

Fed-batch cultivation was shown to drive a 20-fold increase, over shake-flask cultivation, during production of *T. reesei* endoglucanase I (up to 100 mg/l—Park et al. 2000), but only a 4.5-fold increase for production of active human interferon (Gasmi et al. 2011a, c). Nicaud et al. (2002) described the optimisation of *A. oryzae* leucine amino peptidase production, by the combined use of multicopy expression (eightfold increase) and of batch or fed-batch cultivation in defined media. Enzyme secretion was 11-fold higher when fed-batch (constant rate of 47 g of glucose per litre after the end of batch phase) was used. The global production increase obtained was of 88-fold between a monocopy integrant grown in batch and a multicopy integrant grown in fed-batch (Nicaud et al. 2002). Similar results (global 90-fold increase) were observed in the same conditions for homologous YlLip2p (Nicaud et al. 2002). Combined use of multicopy expression and fed-batch cultivation was also reported to allow the production of 500 mg/l of human CD14 (C. Gysler, personal communication and abstract report from 2002).

Chang et al. (1998) described the secretion of 31,200 units/l of rice α-amylase activity, driven by *XPR2* promoter and rice native secretion signal, using fed-batch cultivation of a Yarrowia monocopy integrant in defined media. They increased further the volumetric productivity for this enzyme, by a factor of three, by using a cyclic fed-batch process: cultivation was started initially as fed-batch, and, as cell concentration reached 60 g/l, culture broth was harvested from bioreactor, and fresh production medium was provided, the remaining cells serving as inoculum for the next fed-batch cycle. This "withdrawal and refill" procedure could be repeated for several (eight in the described example) cycles (Chang et al. 1998). More recently, some of the same authors were able to increase further rice α-amylase production, from the same Yarrowia producing strain, while simplifying cultivation process: fed-batch cultivation in a 10 % glycerol medium, followed by one-step feeding with concentrated medium (60 % glycerol) enabled to obtain a yeast cell density higher than 100 g/l, together with secretion of 88,000 units/l of rice α-amylase activity (Kim et al. 2000).

Maharajh et al. (2008a) expressed *Rhodotorula araucariae* epoxide hydrolase in Yarrowia, in order to use producing strains as whole-cell biocatalysts for preparation of enantiopure bioactive drug intermediates (cf. Sect. 4.2.2). Using a combination of multicopy expression and fed-batch cultivation in defined media, they obtained epoxide hydrolase activity of 194,700 units/l, corresponding to 1,760 units/g of dry weight, a level two orders of magnitude higher than in wild-type *R. araucariae* (Maharajh et al. 2008a). Although only a twofold increase in expression was obtained with the multicopy integrant, scale-up to fed-batch cultivation allowed a 36-fold increase of produced activity (81-fold increase compared to a monocopy integrant grown in shake-flask). A maximal yeast cell density of 100 g/l during fed-batch cultivation was described in this study (Maharajh et al. 2008a). Some of the same authors were able to increase further recombinant epoxide hydrolase production, by applying exponential glucose feed rates (optimal rate was 0.06 per hour) to fed-batch cultivation: 206,000 units/l were obtained, corresponding to an 86-fold increase compared to a monocopy integrant grown in shake-flask (Maharajh et al. 2008b).

In conclusion, fine-tuning cultivation conditions can be of major importance for efficient production of some complex heterologous proteins. In most cases, scale-up could be expected to increase productivity by a factor of one order of magnitude. When used in combination with multicopy expression, optimised scale-up could allow a production increase of two orders of magnitude.

3.6 Comparison with Competitor Expression Systems

As already stated in Sect. 1.1, from the comparative study by Muller et al. (1998), Yarrowia compares favourably to other yeasts as an expression/secretion system. In addition, some of the more recent examples from Table 3, for which a precise comparison of Yarrowia with other expression systems has been performed, reinforce this conclusion. Concerning intracellular expression in shake-flask, fungal epoxide hydrolases (EH) provide some elements of comparison between different expression systems, although concerned species are sometime different: multicopy expression of *R. araucariae* EH in Yarrowia resulted in heterologous activity fourfold higher than that observed for *Rhodotorula glutinis* EH expressed in *E. coli*, and more than one order of magnitude higher than for the same *R. glutinis* EH in *P. pastoris* (Maharajh et al. 2008a). Activity obtained for *R. glutinis* EH expressed in *P. pastoris* was tenfold higher than in the wild-type native organism, when activity obtained for *R. araucariae* EH expressed in Yarrowia was more than two orders of magnitude higher than in the corresponding wild-type native organism.

Concerning comparison of secretion yields between Yarrowia and other expression systems, most examples concerned, interestingly, human proteins or complex glycoproteins:

- Secretion of human α-foetoprotein was twofold higher in Yarrowia than in *P. pastoris* (K. Uchida, personal communication).
- Secretion of human anti-Ras scFv was twofold higher in Yarrowia than in *K. lactis* (Swennen et al. 2002).
- Secretion of maize cytokinin oxidase I was performed in two yeast expression systems (Yarrowia and *P. pastoris*). Although precise comparison of secreted activity levels is difficult due to disparities in data presentation, production was in the same range in both yeasts (Bilyeu et al. 2001; Kopecny et al. 2005). Overall glycosylation level was also similar in the two yeasts, but the average length of mannose chains was slightly shorter in Yarrowia (on five of eight potential N-glycosylation sites, instead of four in P. pastoris—cf. Sect. 3.4).
- Secretion of *P. cinnabarinus* laccase I could be compared in two yeast (Yarrowia and *P. pastoris*) and one fungal (*A. niger*) expression systems. The *P. pastoris* system was not competitive in this case, since transformation efficiency was very low (25 transformants per μg, versus 10^4 in Yarrowia), recombinant laccase was hyperglycosylated (addition of 40 kDa, versus 20 in Yarrowia) and laccase activity was more than threefold lower than in Yarrowia (Otterbein et al. 2000; Madzak et al. 2005b). In contrast, the *A. niger* system produced high levels of active recombinant laccase (70 mg/l in shake-flask, versus 8.5 for Yarrowia), with a molecular mass similar to that of the native protein (Record et al. 2002). However, transformation is problematic in the *A. niger* system (cointegration of two vectors by ectopic recombination), efficiency is low (100 transformants per μg) and random recombination integration system requires analysis of a high number of transformants (Record et al. 2002). These characteristics constitute a

major drawback for performing genetic engineering on heterologous genes in *A. niger*, when Yarrowia is in contrast very competitive in this domain (cf. Sect. 4.1).

- Secretion of *A. aculeatus* endo-β-1,4-mannanase could also be compared in two yeast (Yarrowia and *S. cerevisiae*) and one fungal (*A. niger*) expression systems. Surprisingly, at regard of previous observations, this enzyme was more heavily overglycosylated in Yarrowia than in both *S. cerevisiae* and *A. niger* (around 20 additional kDa—Setati et al. 2001; Roth et al. 2009; van Zyl et al. 2009). However, maximal activity produced in Yarrowia cultivated in shake-flask (13,100 nkat/ml—Roth et al. 2009) largely outpassed that produced in *S. cerevisiae* (around 500 nkat/ml—Setati et al. 2001). Activity produced in Yarrowia and *A. niger* was of a similar level for cultivation in shake-flask (13,100 versus 16,600 nkat/ml—Roth et al. 2009; van Zyl et al. 2009). No scale-up having been performed for *A. niger* cultivation, it was not possible to check if the maximal activity produced in Yarrowia grown in fed-batch (more than 40,000 nkat/ml—van Zyl 2010) could be challenged.
- Secretion of *C. antarctica* lipase in Yarrowia, using a single-copy integration strategy for high-throughput screening, yielded up to 200 mg/l (Emond et al. 2010), when only 80 mg/l was obtained in *S. cerevisiae* (Zhang et al. 2003).
- Secretion of *Thermobifida fusca* thermostable α-amylase in Yarrowia (730 units/l—Yang et al. 2010b) was higher than that observed in *P. pastoris* (510 units/l—Yang et al. 2010a) and than that produced in *E. coli* (350 units/l—Yang and Liu 2007).
- More spectacularly, secretion of *T. fusca* thermostable esterase in Yarrowia was 140 times higher than that observed in *P. pastoris* (Huang et al. 2011). This difference appears even more spectacular when one recalls that this example compares monocopy integration in Yarrowia (using the YLEX™ kit) to probably multiple copies in *P. pastoris*.

One could also recall that Yarrowia surface display system, using Yl*CWP1* GPI anchor, was shown to be much more efficient that the one commercialised for *S. cerevisiae*: 100 % of transformed Yarrowia cells were displaying *V. harveyi* haemolysin, when only a third of transformed *S. cerevisiae* cells were doing so (Yue et al. 2008; Zhu et al. 2006—cf. Sect. 2.4.4).

In conclusion, one can resume as follows the advantages of Yarrowia over *S. cerevisiae* as an expression system:

- Higher secretion capacity (up to gram per litre levels for both homologous and heterologous proteins—cf. Sect. 3.5 and Table 3)
- More efficient secretion of large heterologous proteins (up to 90 kDa—cf. introduction of Sect. 3)
- Protein secretion pathway following mainly co-translational pathway, like in "higher" eukaryotes (cf. Sect. 1.2.2)
- Probable absence of immunogenic terminal α-1,3-mannose linkage (cf. Sects. 1.2.3 and 3.4)

- No excessive hyperglycosylation, and overglycosylation generally more moderate (cf. Sects. 1.2.3 and 3.4)
- More efficient surface display system (cf. Sect. 2.4.4)

The advantages listed above are also found in at least some "non-conventional" yeast competitor systems (*P. pastoris*, *H. polymorpha*, *K. lactis*, *S. occidentalis* and *A. adeninivorans*). However, Yarrowia system exhibits some other characteristics that also confer some advantages over most of these yeast competitors and especially over the most used, *P. pastoris*.

- GRAS status (cf. Sect. 1.2.1)
- Better regularity of performance (cf. Sect. 1.1)
- Higher yields in most cases, especially for complex glycoproteins (e.g. laccases)
- Choice of different promoters (constitutive, growth-phase-dependent, inducible, tunable) adapted to various applications (cf. Sect. 2.4.1)
- Possibility of very high transformation efficiency (10^5–10^6 transformants per μg of DNA—cf. Sect. 2.1.1)
- Possibility of monocopy targeted integration, useful for genetic engineering (cf. Sects. 2.1.1, 2.5 and 4.1)
- Possibility of constructing producing strains devoid of bacterial DNA (cf. Sect. 2.1.2)
- Possibility of surface display, interesting for use of whole cells for bioconversion (cf. Sects. 2.4.4 and 4.2)
- Possibility of scale-up, growth to high cell density (higher than 100 g/l) and high-throughput screening (cf. Sects. 2.7 and 3.5)

Some recent examples described above show that the *A. niger* fungal expression system could be remarkably efficient in terms of production and glycosylation levels. However, technical complexity and relative inefficiency of transformation process, together with absence of targeted recombination, limit the use of this fungus and render it unadapted to genetic engineering. Thus, although *A. niger* could outpass Yarrowia in the process of selecting a good producer strain, the overall potentialities of the two expression systems cannot be compared.

4 Conclusions and Prospects

We tried to exemplify in this review the high potential of Yarrowia expression system. Available tools include easy-to-handle expression/secretion vectors and recipient strains and surface display systems, together with amplification systems and potential for scale-up and high-throughput screening. The peculiar characteristics of Yarrowia secretion pathway, closer by some aspects to that of "higher" eukaryotes than to that of most yeasts, make it a particularly interesting host for production of complex (i.e. therapeutic) proteins. In this conclusion, we will focus on a few applications rendered possible by some interesting characteristics of the Yarrowia system and on its potential for future developments.

4.1 Use of Yarrowia for Genetic Engineering

Genetic engineering of proteins requires their efficient production in a heterologous host and, preferably, a targeted integration system allowing the comparison of transformants, together with the possibility of high-throughput screening. Some elements of Yarrowia expression system appear to be particularly adapted to this purpose: pBR322-based expression/secretion vectors, such as those included in the YLEX™ kit (cf. Sect. 2.5 and Table 2), could provide a very high transformation efficiency and a precise targeting of monocopy integration events. When using such vectors, the consequences of directed or random mutations on the recombinant protein activity can be assessed easily, directly from culture surpernatant. These pBR322-based vectors have been used for genetic engineering of two heterologous proteins, a fungal laccase and a maize cytokinin oxidase, using a structure-based directed mutagenesis strategy.

Fungal laccases are used industrially for paper bleaching and have many other potential applications, either environmental (biodegradation of xenobiotics) or industrial (i.e. development of oxygen cathodes in biocells, biosensors, organic synthesis—Mayer and Staples 2002). As reviewed in Jolivalt et al. (2005), their genetic engineering has been at first impaired by difficulties encountered for their production in heterologous hosts: low production levels and hyperglycosylation problems (i.e. in *Aspergillus oryzae*, *S. cerevisiae* and *P. pastoris*) or low transformation efficiency and inadapted ectopic recombination system (i.e. in *A. niger*). Yarrowia expression system allowed to overcome these obstacles by providing high transformation efficiency, targeted monocopy integration and efficient secretion of active laccases (Madzak et al. 2005b; Jolivalt et al. 2005). It is currently used to engineer *T. versicolor* LacIIIbp, a very stable enzyme with one of highest known redox potential (0.780 V/ENH), into an optimised recombinant enzyme with wider substrate specificity, and more neutral optimal pH, in order to develop bioremediation processes of waters or soils polluted by xenobiotics (Mougin et al. 2003). Some promising results have been obtained: a D206N mutant laccase was found to exhibit a shift of 1.4 units of optimal pH towards neutrality (Madzak et al. 2006), and mutations of several Phe residues flanking the substrate entrance door were able to modify laccase specificity (C. Jolivalt, personal communication). In particular, it was shown very recently that replacement of some large Phe residues with smaller Ala ones remodelled the substrate entrance door, making the mutants more efficient for oxidation of hindered substrates, such as xenobiotics (Galli et al. 2011). These preliminary results constitute a first step in laccase optimisation for environmental applications.

In the case of maize CKO1p, structure-based directed mutagenesis was intended to allow a fine analysis of the enzyme reactional mechanism (Kopecny et al. 2008). A series of mutations have been performed on a dozen of residues, possibly involved in functional conformation, quinone binding, substrate binding or interaction with FAD cofactor (Kopecny et al. 2008). The amounts of recombinant CKO1 secreted in Yarrowia in shake-flask, using a pBR322-based vector with hp4d promoter, were sufficient to allow the crystallisation and structural analysis of both wild-type (Kopecny et al. 2004, 2005) and mutated enzymes (Kopecny et al. 2008). Compared crystallographic analysis and biochemical study of these mutants is currently providing new insight into the reactional mechanism of CKO1 (Kopecny et al. 2010), which is a main actor in plant development (Werner and Schmulling 2009).

Besides heterologous proteins, Yarrowia has been used for optimisation of homologous extracellular lipase YlLip2p, which has prompted the recent development of a high-throughput screening process (Bordes et al. 2007—cf. Sect. 2.7). This approach has now developed into a new secretion system for the lipase B from *C. antarctica* (Emond et al. 2010), as a prerequisite for future heterologous protein engineering experiments. A saturation mutagenesis library focused on residue A281 was constructed, and a reliable screening protocol was defined and validated. Several mutants with catalytic efficiencies higher than that of wild-type enzyme were isolated by high-throughput screening and rapidly purified and characterised (Emond et al. 2010).

4.2 Use of Yarrowia as Whole-Cell Biocatalyst

Expression of heterologous proteins, either intracellularly or using a surface display system, allows the use of Yarrowia as a whole-cell biocatalyst, either for bioconversion or for applications involving molecular recognition (e.g. live vaccine, biosensor). Among the examples of Yarrowia strain engineering by intracellular expression of new heterologous genes, most involved modification of lipid metabolism pathways. These experiments took advantage of the very effective uptake of lipids, characteristic of this yeast, to enable the efficient transport of hydrophobic substrates to catalytic sites within the cells (Fickers et al. 2005). This efficiency of hydrophobic compound uptake in Yarrowia has been attributed to the presence of protrusions on the cell surface, forming channels connected to the cell interior (Fickers et al. 2005). Some major examples of use of recombinant Yarrowia strains for bioconversion will be examined further in this section.

4.2.1 Cytochrome-P450-Expressing Yarrowia Strains

Several heterologous cytochrome P450 (CYP450) genes from various phylogenetic origins (fungus *Rhodotorula minuta*, plant *Capsicum annuum*, mammals *Bos taurus taurus* and *H. sapiens*) have been expressed intracellularly in Yarrowia, for use of producing strains as whole-cell biocatalysts (Shiningavamwe et al. 2006; Bourel et al. 2004; Juretzek et al. 2000b; Nthangeni et al. 2004). Eukaryotic CYP450s constitute a superfamily of membrane-bound, haem-containing monooxygenases linked to metabolism of drugs, steroid hormones and fatty acids (Werck-Reichhart and Feyereisen 2000). Efficient heterologous production of CYP450s is of interest for both biotechnological and therapeutic applications. It represents an essential step towards investigating their role in xenobiotics and therapeutic drug degradation pathways, and in their use in bioconversion, for synthesis of hydroxylated biomolecules of industrial interest (Werck-Reichhart and Feyereisen 2000). Fatty acid-inducible promoters (p*POX2*, p*ICL1*) have been used for CYP450 expression in Yarrowia (Shiningavamwe et al. 2006; Bourel et al. 2004; Juretzek et al. 2000b; Nthangeni et al. 2004). Fungal and mammalian cytochrome P450 monooxygenases requiring a cofactor, their association with cytochrome P450 reductase (CPR) is necessary to achieve optimal activities (Werck-Reichhart and Feyereisen 2000). Concomitant overexpression of homologous Yl*CPR* in both *R. minuta* and *H. sapiens* CYP450-expressing Yarrowia strains was shown to increase recombinant CYP450 activity (Shiningavamwe et al. 2006; Nthangeni et al. 2004—cf. Table 3). These studies demonstrated the efficiency of Yarrowia for expression of fungal or human CYP450s and its in vivo efficiency for shuttling electrons from homologous YlCprp to heterologous CYP450s. In contrast, plant CYP450 (from *C. annuum*) is a fatty acid hydroperoxide lyase, which needs no cofactor (Bourel et al. 2004). Yarrowia strains expressing *C. annuum* CYP450 were shown to have an interesting potential, as whole-cell biocatalysts, for biosynthesis of volatile aldehyde flavouring molecules of high interest in the aroma industry (Bourel et al. 2004).

4.2.2 Epoxide-Hydrolase-Expressing Yarrowia Strains

Another category of heterologous enzymes for which Yarrowia offers an interesting intracellular expression system, allowing whole-cell bioconversion, is constituted by epoxide hydrolases. These enzymes catalyse the formation of enantiopure epoxides and diols, which serve as high-value intermediates in the fine chemicals and pharmaceutical industries (Fretland and Omiecinski 2000). Several heterologous EH genes from various phylogenetic origins (bacteria *Agrobacterium radiobacter*, fungi *A. niger*, *Rhodosporidium paludigenum*, *Rhodosporidium toruloides*, *Cryptococcus neoformans*, *R. araucariae*, *Rhodotorula mucilaginosa*, plant *Arabidopsis thaliana*, insect *Trichoplusia ni*, mammal *H. sapiens*) have been expressed intracellularly in Yarrowia (Labuschagné et al. 2004; Maharajh et al. 2008a, b; Labuschagné and Albertyn 2007). These enzymes are either soluble (among the EH cited here, it was the case for those of *A. Radiobacter*, *A. niger* and *A. thaliana*) or membrane-bound (the seven others—Labuschagné 2005). Epoxides being hydrophobic substrates, it was hypothesised that Yarrowia could act as an effective whole-cell biocatalyst for their bioconversion, protrusion channels forming an efficient uptake system for them. This prediction was confirmed by the interesting results obtained, especially with fungal EHs (Maharajh et al. 2008a, b; Labuschagné and Albertyn 2007). Intracellular expression of *R. mucilaginosa* EH in Yarrowia, using a monocopy vector based on *TEF1* promoter, showed that the engineered strain could exhibit unsurpassed performances: whole-cell biotransformation of a racemic substrate revealed an activity and an enantiomeric selectivity far superior to any other reported in literature using wild-type organisms (Labuschagné and Albertyn 2007). Namely, enantiomeric excess of 97 % and yield of 55 % were obtained from Yarrowia strains expressing *R. mucilaginosa* EH versus enantiomeric excesses of 62–65 % and yields of 33–36 % from wild-type *Rhodotorula* sp. (Labuschagné and Albertyn 2007). The level of heterologous activity obtained by intracellular expression of *R. araucariae* EH in Yarrowia, using a multicopy vector based on hp4d promoter, was also much higher than those previously reported for fungal EH in other expression systems (Maharajh et al. 2008a—cf. Sect. 3.6). Optimisation of cultivation conditions for recombinant Yarrowia strains expressing *R. araucariae* EH was performed by Maharajh et al. (2008a, b) and demonstrated their high potential for whole-cell enantiomeric bioconversion of racemic epoxides (cf. Sect. 3.5).

4.2.3 Inulinase-Expressing Yarrowia Strains as Single-Cell Protein Producers

Inulin is a natural compound present as a reserve carbohydrate in the roots and tubers of many plants, such as Jerusalem artichoke, chicory, dahlia and yacon (reviewed in Chi et al. 2009). This carbohydrate, corresponding to over 70 % of the dried material from the tubers, is a mixture of oligo- and polysaccharides chains of various lengths, composed of fructose units (linked by β-(2-1)-D-fructosyl-fructose bonds) and terminated with a glucose unit at reducing end (Chi et al. 2009). Inulinases are able to hydrolyse it into fructose and glucose. Inulin and its hydrolysates have many biotechnological applications: they have been used to produce single-cell protein, single-cell oil, citric acid, bioethanol, ultra-high fructose syrup, 2,3-butanediol, lactic acid, mannitol and sorbitol (reviewed in Chi et al. 2009). The fact that no oleaginous yeast, including Yarrowia, produces any inulinase activity has impaired the use of cheap inulin sources as a renewable raw material in a number of processes. However, recently, the use of recombinant inulinase-expressing Yarrowia strains has been described for various applications: single-cell oil production (cf. Sect. 4.2.4), citric acid production (using an arming Yarrowia strain—cf. Sect. 4.2.6) and single-cell protein production.

SWJ-1b strain, isolated from fish gut, was selected from a library of 78 marine Yarrowia strains for its high protein content (48 % of cell dry weight—Wang et al. 2009). A uracil mutant of this strain was used to express an exo-inulinase from *Kluyveromyces marxianus* and cultivated in batch on Jerusalem artichoke (*Helianthus tuberosus*) tuber extract, allowing crude protein to reach 54 % of cell dry weight (Cui et al. 2011). This example demonstrated the potential of engineered SWJ-1b strain to actively convert inulin into single-cell protein, enhancing further the interest of using Yarrowia as single-cell protein producer.

4.2.4 Yarrowia Strains as Single-Cell Oil Producers

Essential fatty acids are unsaturated fatty acids which are not synthesised by mammals (i.e. *H. sapiens*), but are essential for health and should be supplied by diet. Among them, γ-linolenic acid (GLA) is an omega-6 fatty acid which exerts various beneficial effects, including therapeutic impact on inflammatory processes and breast cancer prevention (Das 2008; Menendez et al. 2005). Since natural sources of GLA are expensive and unreliable, there is a strong economical interest in the production of single-cell oil enriched in this essential fatty acid. Recently, Chuang et al. (2009, 2010) reported the first use of an oleaginous yeast, Yarrowia, as host for the synthesis of GLA-rich single-cell oil. They constructed a Yarrowia strain co-expressing two genes involved in GLA biosynthesis in the fungus *Mortierella alpina*, encoding Δ12- and Δ6-desaturases. Engineered Yarrowia cells producing simultaneously both enzymes (using a multiple cassette vector—cf. Sect. 2.6.3) were able to produce and accumulate up to 20 % of GLA in their total cellular lipids (44 % in triacylglycerol fraction—Chuang et al. 2009, 2010). In this study, a higher amount of lipids was obtained when cells were grown in nitrogen-reduced media, in agreement with previous data showing that a high C/N molar ratio during growth increased lipid accumulation by oleaginous fungi (Papanikolaou et al. 2004).

Recently, ACA-DC 50109 strain was selected from a library of marine Yarrowia strains for its high lipid content. A fungal exo-inulinase (from *K. marxianus*) was expressed in a uracil mutant of this strain, and the recombinant Yarrowia strain was shown to be able to perform the direct conversion of inulin-containing materials (Jerusalem artichoke tuber extract—cf. Sect. 4.2.3) into single-cell oil (Zhao et al. 2010), with the accumulation of 51 % (w/w) oil into yeast cells. Over 91 % of the fatty acids produced were $C_{16:0}$, $C_{18:2}$ and mainly $C_{18:1}$ (more than 58 %), a lipid composition consistent with requirements for biodiesel production (Zhao et al. 2010).

Growth conditions optimisation and molecular/genetic engineering are expected to allow further improvement of yeast-oil yield and composition. For example, such studies could benefit from the use of recently described mutant Yarrowia strains able to accumulate larger quantities of lipids (up to 60 %—Nicaud et al. 2010).

Genetic engineering of fatty acid metabolic pathways in Yarrowia, using intracellular expression of heterologous desaturases and elongases of various phylogenetic origins (mainly from fungi, but also from a microalga, a dinoflagellate and protists—cf. Table 3), has been the subject of a high number of patents (Picataggio and Zhu 2005; Xue and Zhu 2006b; Xue et al. 2007; Sharpe et al. 2008; Hong et al. 2009). Sharpe et al. (2008) revendicated the use of engineered Yarrowia strains (able to produce carotenoids, together with unsaturated fatty acids and antioxidants) in animal and human food industries.

Anecdotally, Wang et al. (2009) reported that the transformation of a uracil mutant of marine SWJ-1b Yarrowia strain (isolated from fish gut) with a surface display vector expressing EGFP was able to modify the fatty acid content of the resulting EGFP-displaying strain, thus increasing its potential interest as food component for fish-farming industry (single-cell protein and oil provider).

4.2.5 Use of Yarrowia Strains for Polyhydroxyalkanoate Synthesis

Polyhydroxyalkanoates (PHAs) are linear polyesters which can be produced by bacterial fermentation of sugar or lipids. They are biodegradable and can be used in the production of bioplastics, with a wide range of potential applications, including use in medical and pharmaceutical industries (Chen and Wu 2005). Recently, some recombinant Yarrowia strains expressing the PHA synthase gene from *P. aeruginosa* in their peroxisomes (cf. Sect. 2.4.3) were developed in our laboratory and found able to produce PHA in various yields, depending on their POX genotype (Haddouche et al. 2010). It was also further shown that the redirection of the fatty acid flux towards β-oxidation, by deletion of the neutral lipid synthesis pathway, was able to lead to a significant increase in PHA levels (more than 7 % of the cell dry weight—Haddouche et al. 2011).

4.2.6 Arming Yarrowia Strains

The recent development of surface display systems in Yarrowia (Yue et al. 2008; Yang et al. 2009—cf. Sect. 2.4.4) offers new possibilities for its use as whole-cell biocatalyst for either bioconversion or applications involving molecular recognition (e.g. live vaccine, biosensor). Using Yl*CWP1* GPI anchor signal, alkaline protease II from marine yeast *Aureobasidium pullulans* has been displayed on Yarrowia cells, which were applied to production of bioactive peptides from different cheap sources of proteins (Ni et al. 2009). Peptides produced by proteolysis of single-cell proteins from a marine yeast were found to exhibit the highest angiotensin-converting-enzyme inhibitory activity, while peptides produced from Spirulina (*Arthrospira platensis*) powder exhibited the highest antioxidant activity. More recently, a fungal acid protease was also displayed, for use of the recombinant arming Yarrowia strain as a rennet substitute in milk clotting (Yu et al. 2010). Similarly, a fungal exo-inulinase (from *K. marxianus*) was displayed on a marine Yarrowia strain (uracil mutant from SWJ-1b strain), for use in production of citric acid from inulin (Liu et al. 2010), with important applications in food industry. Alginate lyase from marine bacterium *Vibrio* sp. has also been displayed on Yarrowia cells, which were applied to hydrolysis of sodium alginate and its constituents, poly-β-D-mannuronate and poly-α-L-guluronate (Liu et al. 2009). Low-molecular-weight oligosaccharides such as those produced by displayed alginate lyase have been shown to exhibit bioactive properties, such as antioxidation, cytokine activation and prevention of cardiovascular/cerebrovascular diseases (Liu et al. 2009). Interestingly, displayed alginate lyase was found to exhibit increased (2.4-fold) specific activity when a C-terminal 6×His tag was used, suggesting that this added element could act as a spacer for better surface display (Liu et al. 2009).

Surface display of haemolysin from marine pathogenic bacterium *V. harveyi* on Yarrowia cells offered an interesting example of an application involving molecular recognition: the development of live vaccines. Haemolysin-armed Yarrowia cells were found to exhibit haemolytic activity towards erythrocytes from flounder (*Paralichthys olivaceus*), a marine fish (Yue et al. 2008). As demonstrated in a previous study, immunisation of turbot (*Scophthalmus maximus*) with *S. cerevisiae* cells displaying haemolysin was able to provide fishes with a significant protection against *V. harveyi* infection (Zhu et al. 2006). Thus, Yarrowia cells displaying haemolysin constitute good candidates for the development of live vaccines for fish-farming industry, for which *Vibrio* sp. constitutes a major threat (Li and Woo 2003). The use of Yarrowia for such applications would benefit from its GRAS status (retained with the use of auto-cloning surface display vectors) and from its natural adaptation to marine environment (van Uden and Fell 1968; Butinar et al. 2005).

As an alternative to Yl*CWP1* GPI anchor signal, Yang et al. (2009) described the surface display of *B. subtilis* endo-β-1,4-mannanase on Yarrowia cells, using the *S. cerevisiae FLO1* GPI anchor signal (cf. Sect. 2.4.4). This work could constitute a preliminary step in the use of mannanase-armed Yarrowia strains as whole-cell biocatalyst for pulp bleaching or in food industry (e.g. extraction of vegetable oil from leguminous seeds, reduction of viscosity in coffee extracts). In order to combine the intrinsic lipolytical properties from Yarrowia with the high activity of some fungal lipases, the same *S. cerevisiae* GPI anchor signal has been recently used to construct Yarrowia whole-cell biocatalysts displaying *R. stolonifer* lipase for effective treatment of oily wastewaters (Song et al. 2011). Within 72 h of whole-cell treatment, in open activated sludge bioreactor, 97 % of oil was removed when using lipase-armed yeast, while it was only 87 % when using control Polg strain, demonstrating how genetic engineering could benefit to Yarrowia use for bioremediation.

These examples demonstrate the potential of whole-cell Yarrowia displaying new heterologous activities for use in biotechnology, bioremediation and food and pharmaceutical industries.

4.3 Use of Yarrowia as Biosensor

Biosensors are analytical devices for detecting compounds of interest, which combine a biological component with a physicochemical detector. They have many potential applications in various domains, among which are the therapeutic (e.g. glucose monitoring in diabetes patients) and environmental (e.g. measuring levels of toxic substances before and after bioremediation) ones. Interestingly, a wild-type Yarrowia strain, isolated from a diesel oil contaminated soil, has been used in a microbial biosensor for detection of middle-chain alkanes (Alkasrawi et al. 1999), hydrocarbons constituting both a health hazard and a serious threat to aquatic and terrestrial ecosystems. More recently, Cho et al. (2010) described a biosensor for the detection of environmental estrogens, based on a genetically engineered Yarrowia strain: an expression vector was used to produce intracellularly human oestrogen receptor α, in a strain carrying a chromosome-integrated *lacZ* reporter gene under the control of oestrogen response elements (EREs). This biosensor strain constitutes a highly sensitive detection system for endocrine disruptors, such as natural or synthetic hormones, pesticides and commercial chemicals. This work demonstrated the sensitivity and reproducibility of the Yarrowia biosensor for characterising environmental estrogens. Development of biosensors based on genetically engineered Yarrowia strains expressing heterologous proteins will probably be boosted in near future by the availability of surface display systems (cf. Sect. 2.4.4).

4.4 Projects for "Universal" vectors

Despite the overall good regularity of performance of Yarrowia expression/secretion system, some heterologous genes could however occasionally encounter production problems (low expression or secretion level, incomplete or abnormal maturation, excessive or inadequate glycosylation), raising the question to know if another yeast system would have allowed to obtain better results. Prediction of results being difficult, it would be useful to handle a whole range of host organisms in order to find the more adapted to a particular case. However, testing independently several expression systems constitutes an important investment of time and money. A unique expression/secretion vector that could be targeted to various candidate hosts would greatly facilitate such preliminary comparison.

The CoMed™ vector system (Steinborn et al. 2006) has been designed for such purpose: this versatile system consists in vectors built up in a modular way, comprising ARS sequences or rDNA targeting sequences, dominant or auxotrophy selection markers and an expression cassette (e.g. *A. adeninivorans TEF1* promoter, MCS, *S. cerevisiae PHO5* terminator). The combination of a rDNA targeting sequence (e.g. derived from *A. adeninivorans*) and of a suitable selection marker (e.g. *A. adeninivorans TEF1* promoter directing expression of *E. coli hph* gene, conferring resistance to hygromycin B) enabled transformation of host strains from a wide range of yeast species (Steinborn et al. 2006; Boer et al. 2007). After selection of the more appropriate host, the CoMed™ vector can be further adapted for optimised production by exchanging some "universal" modules with species-specific ones. A CoMed™ vector expressing GFP (green fluorescent protein) gene was used, as proof of concept, to transform successfully different yeast species: *S. cerevisiae*, *A. adeninivorans* (either budding or filamentous cells), *H. polymorpha*, *P. pastoris*, *Debaryomyces polymorphus* and *Debaryomyces hansenii* (Steinborn et al. 2006). Another CoMed™ vector, using *A. adeninivorans LEU2* gene for selection, was used for expression/secretion of human interleukin-6 in a range of auxotrophic hosts: *A. adeninivorans*, *H. polymorpha* and *S. cerevisiae*. Even though the heterologous protein was secreted in all cases, important differences were observed in precursor processing: only *A. adeninivorans* was able to perform faithful maturation, when the two other yeasts produced N-terminally truncated proteins (Boer et al. 2007). Although no report of use of CoMed™ vector in Yarrowia is described yet in literature, such a vector using *A. adeninivorans* rDNA, for targeting, and *TEF1* promoter directing expression of *E. coli hph* gene, for selection, was shown to enable Yarrowia transformation (G. Kunze, personal communication).

More recently, a new "universal" vector system, derived from the CoMed™ one, has been developed: Xplor 2 consists in a modulable auto-cloning vector system, from which bacterial DNA can be removed prior to transformation. Integration of a smaller "yeast cassette" increased transformation efficiency and stability of integrants in *A. adeninivorans* (Boer et al. 2009).

The recent development of these "universal" vector systems will doubtlessly provide rapidly new elements for the comparison of different yeast host systems.

4.5 Projects for Engineered and "Humanised" Strains

Further improvements of Yarrowia expression/secretion system could be realised through strain engineering. Identification of bottlenecks in secretion pathway, together with better knowledge of genes involved in quality control and protein degradation (Swennen et al. 2010), should benefit to the development of improved recipient strains.

As processing of *XPR2* prepro region was sometimes found to be incomplete, Swennen et al. (2002) hypothesised that increasing YlXpr6p endoprotease levels could improve maturation efficiency. Indeed, YlXpr6p overexpression was found to enhance the amount of mature human anti-Ras scFv secreted using *XPR2* prepro region (D. Swennen, personal communication). Sohn et al. (1998) analysed the effects of YlPmr1p, a Golgi-specific Ca^{2+}-ATPase, on secretion of both homologous and heterologous proteins. An inhibitory effect on rice α-amylase secretion was observed (like for Sc*PMR1*-disrupted S. cerevisiae—Sohn et al. 1998). In contrast, secretion of *T. reesei* endoglucanase was not affected, except for a reduction of its overglycosylation (Sohn et al. 1998). Pleiotropic effects on dimorphism and pH regulation during cultivation were also observed, which can explain the opposite impacts on AEP and AXP activities, for which pH is a major regulator. Maturation of AEP was also affected, probably due to impairment of Ca^{2+}-dependent YlXpr6p endoproteasic activity. Overexpression of chaperones and/or foldases has led to important secretion improvements in *S. cerevisiae* (Shusta et al. 1998). However, in Yarrowia, their overexpression has been impaired by plasmidic instability problems (J-M. Beckerich, unpublished data). In the near future, complete genome data, together with transcriptome and proteome analyses, will doubtlessly foster studies on global response of Yarrowia strains challenged with the stress of overproducing heterologous proteins.

Besides quantitative improvements of secretion, it would be highly desirable to develop engineered strains with qualitative improvement of glycosylation, namely, with glycosylation patterns closer to those of mammals (i.e. *H. Sapiens*). Such "humanised" yeast strains are already available for a few other species, i.e. *P. pastoris* and *H. polymorpha* (Hamilton et al. 2006; Hamilton and Gerngross 2007; Oh et al. 2008).

Genetic organisation of N-glycosylation of proteins in Yarrowia begins to be understood (Barnay-Verdier et al. 2008). A few of the implied genes have been identified and their product characterised; three α-1,6-mannosyltransferases (Yl*MNN9*, Yl*ANL1* and Yl*OCH1*) have been cloned, and correspondingly disrupted Yarrowia strains have been constructed (Jaafar et al. 2003; Barnay-Verdier et al. 2004). More recently, Song et al. (2007) have investigated the effects of Yl*OCH1* deletion on N-glycosylation of two secreted glycoproteins: a homologous lipase, YlLip2p, and a heterologous protein, *T. reesei* endoglucanase I. For both proteins, glycosylation was reduced in the Yl*OCH1*-deleted strain, which synthesised only the core oligosaccharide Man8GlcNAc2, when the wild-type strain synthesised oligosaccharides with heterogeneous sizes, from Man8GlcNAc2 to Man12GlcNAc2 (Song et al. 2007). These results demonstrate that Yl*OCH1*-deleted strain exhibits a defect in the addition of the first α-1,6-mannose residue onto the core oligosaccharide, Man8GlcNAc2. Thus, YlOch1p appears to be a key enzyme responsible for adding the first mannose onto the core oligosaccharide in Yarrowia (Barnay-Verdier et al. 2004; Song et al. 2007). Therefore, Yl*OCH1*-deleted strain could be used as host for producing glycoproteins lacking outer-chain mannoses (Song et al. 2007). It also constitutes a first step in further development of engineered strains for the production of therapeutic glycoproteins carrying human-compatible N-linked oligosaccharides.

In the same study, Song et al. (2007) observed, using MALDI-TOF mass spectrometry analysis, that N-linked glycans assembled on secreted glycoproteins contained phosphates and that some minor peaks in HPLC profiles of oligosaccharides from wild-type Yarrowia strain could be mannosylphosphorylated forms of Man7GlcNAc2 to Man9GlcNAc2. Identification of a gene involved in the mannosylphosphorylation of N-linked oligosaccharides in Yarrowia has firstly been described in a patent deposited by the same research team (Park et al. 2008): Yl*MPO1* was shown to be necessary to mannosylphosphorylation of secreted proteins, and a Yl*MPO1*-deleted strain, secreting glycoproteins devoid of mannosylphosphate, was revendicated for use in therapeutic protein production. More recently, the corresponding experimental work has been fully described by Park et al. (2011): the function of five Yarrowia genes showing a significant sequence homology with either *MNN4* or *MNN6* from *S. cerevisiae* has been analysed. These two genes are required for mannosylphosphorylation of N- and O-glycans in *S. cerevisiae*, a process adding negative charges on cell surface (Jigami and Odani 1999). The disruption of Yl*MPO1* (homologous to Sc*MNN4*) resulted in the complete disappearance of the acidic sugar moiety in both N- and O-linked glycan profiles, when the quadruple disruption of all Sc*MNN6* homologues had no effect (Park et al. 2011). Thus, YlMpo1p performs a major role in mannosylphosphorylation of oligosaccharides in Yarrowia. Park et al. (2011) consequently suggested that a Yl*och1*Δ Yl*mpo1*Δ double mutant would be a useful host for producing glycoproteins lacking yeast-specific hypermannosylation and mannosylphosphorylation.

Further strain engineering, based on above results and on adaptation of those from *P. pastoris* or *H. Polymorpha*, should provide "humanised" Yarrowia strains, able to produce glycoproteins with human-compatible N-linked oligosaccharides for therapeutic applications.

Acknowledgements We wish to thank all colleagues and collaborators who accepted to communicate unpublished data: Dr. Martine Chartier (Laboratoire d'Ingénierie des Systèmes Macromoléculaires, CNRS, Marseille, France), Prof. Zhen-Ming Chi (Unesco Chinese Center of Marine Biotechnology and Institute of Marine Biodiversity and Evolution, Ocean University of China, Qingdao, China), Dr. Ricardo Cordero (former PhD student from our laboratory), Dr. Art Franke (Pfizer, USA), Dr. Christof Gysler (Nestec, Switzerland), Dr. Nicole Houba-Hérin (Laboratoire de Biologie Cellulaire, INRA, Versailles, France), Dr. Claude Jolivalt (ENSCP Chimie ParisTech, Laboratoire Charles Friedel, Paris, France), Dr. Gotthard Kunze (Leibniz-Institut für Pflanzengenetik und Kulturpflanzenforschung, Gatersleben, Germany), Dr. Nathalie Libessart (Roquette Frères, France), Dr. Bi-Ching Sang (Pharmingen, USA), Dr. Dominique Swennen (our laboratory), Dr. Kohji Uchida (Oriental Yeast, Japan) and Dr. Petrus J. van Zyl (Bioprocess Development, CSIR Biosciences, Modderfontein, South Africa).

References

Aitchison JD, Murray WW, Rachubinski RA (1991) The carboxyl-terminal tripeptide Ala-Lys-Ile is essential for targeting *Candida tropicalis* trifunctional enzyme to yeast peroxisomes. J Biol Chem 266(34):23197–23203

Alkasrawi M, Nandakumar R, Margesin R, Schinner F, Mattiasson B (1999) A microbial biosensor based on *Yarrowia lipolytica* for the off-line determination of middle-chain alkanes. Biosens Bioelectron 14(8–9):723–727

Babour A, Beckerich JM, Gaillardin C (2004) Identification of an UDP-Glc:glycoprotein glucosyltransferase in the yeast *Yarrowia lipolytica*. Yeast 21:11–24

Bankar AV, Kumar AR, Zinjarde SS (2009) Environmental and industrial applications of *Yarrowia lipolytica*. Appl Microbiol Biotechnol 84(5):847–865

Barnay-Verdier S, Boisramé A, Beckerich JM (2004) Identification and characterization of two alpha-1,6-mannosyltransferases, Anl1p and Och1p, in the yeast *Yarrowia lipolytica*. Microbiology 150:2185–2195

Barnay-Verdier S, Beckerich JM, Boisramé A (2008) New components of *Yarrowia lipolytica* Golgi multi-protein complexes containing the alpha-1,6-mannosyltransferases YlMnn9p and YlAnl1p. Curr Genet 54(6):313–323

Barth G, Gaillardin C (1996) Yarrowia lipolytica. In: Wolf K (ed) Nonconventional yeasts in biotechnology: a handbook. Springer, Heidelberg, pp 313–388

Barth G, Gaillardin C (1997) Physiology and genetics of the dimorphic fungus *Yarrowia lipolytica*. FEMS Microbiol Rev 19:219–237

Bauer R, Paltauf F, Kohlwein SD (1993) Functional expression of bacterial β-glucuronidase and its use as a reporter system in the yeast *Yarrowia lipolytica*. Yeast 9:71–75

Beckerich JM, Boisramé-Baudevin A, Gaillardin C (1998) *Yarrowia lipolytica*: a model organism for protein secretion studies. Int Microbiol 1:123–130

Bhave SL, Chattoo BB (2003) Expression of Vitreoscilla hemoglobin improves growth and levels of extracellular enzyme in *Yarrowia lipolytica*. Biotechnol Bioeng 84(6):658–666

Bilyeu KD, Cole JL, Laskey JG, Riekhof WR, Esparza TJ, Kramer MD, Morris RO (2001) Molecular and biochemical characterization of a cytokinin oxidase from maize. Plant Physiol 125:378–386

Blanchin-Roland S, Cordero Otero R, Gaillardin C (1994) Two upstream activation sequences control the expression of the *XPR2* gene in the yeast *Yarrowia lipolytica*. Mol Cell Biol 14:327–338

Blazeck J, Liu L, Redden H, Alper H (2011) Tuning gene expression in *Yarrowia lipolytica* by a hybrid promoter approach. Appl Environ Microbiol 77(22):7905–7914

Boer E, Steinborn G, Matros A, Mock HP, Gellissen G, Kunze G (2007) Production of interleukin-6 in *Arxula adeninivorans*, *Hansenula polymorpha* and *Saccharomyces cerevisiae* by applying the wide-range yeast vector (CoMed) system to simultaneous comparative assessment. FEMS Yeast Res 7(7):1181–1187

Boer E, Piontek M, Kunze G (2009) Xplor 2 – an optimized transformation/expression system for recombinant protein production in the yeast *Arxula adeninivorans*. Appl Microbiol Biotechnol 84(3):583–594

Boisramé A, Gaillardin C (2009) Heterologous expression and characterization of a beta-1, 6-glucanase from *Aspergillus fumigatus*. Appl Microbiol Biotechnol 82(4):663–669

Boisramé A, Kabani M, Beckerich JM, Hartmann E, Gaillardin C (1998) Interaction of Kar2p and Sls1p is required for efficient co-translational translocation of secreted proteins in the yeast *Yarrowia lipolytica*. J Biol Chem 273(47):30903–30908

Bordes F, Fudalej F, Dossat V, Nicaud JM, Marty A (2007) A new recombinant protein expression system for high-throughput screening in the yeast *Yarrowia lipolytica*. J Microbiol Methods 70(3):493–502

Bourel G, Nicaud JM, Nthangeni B, Santiago-Gomez P, Belin JM, Husson F (2004) Fatty acid hydroperoxide lyase of the green bell pepper : cloning in *Yarrowia lipolytica* and biogenesis of volatile aldehydes. Enzyme Microb Technol 35:293–299

Brown JD, Hann BC, Medzihradszky KF, Niwa M, Burlingame AL, Walter P (1994) Subunits of the *Saccharomyces cerevisiae* signal recognition particle required for its functional expression. EMBO J 13(18):4390–4400

Buckholz RG, Gleeson MAG (1991) Yeast systems for the commercial production of heterologous proteins. Biotechnology 9:1067–1072

Butinar L, Santos S, Spencer-Martins I, Oren A, Gunde-Cimerman N (2005) Yeast diversity in hypersaline habitats. FEMS Microbiol Lett 244:229–234

Casaregola S, Feynerol C, Diez M, Fournier P, Gaillardin C (1997) Genomic organization of the yeast *Yarrowia lipolytica*. Chromosoma 106:380–390

Casaregola S, Neuvéglise C, Lépingle A, Bon E, Feynerol C, Artiguenave F, Winckler P, Gaillardin C (2000) Genomic exploration of the hemiascomycetous yeasts: 17. *Yarrowia lipolytica*. FEBS Lett 487:95–100

Cereghino JL, Cregg JM (2000) Heterologous protein expression in the methylotrophic yeast *Pichia pastoris*. FEMS Microbiol Rev 24:45–66

Cereghino GPL, Cereghino JL, Ilgen C, Cregg JM (2002) Production of recombinant proteins in fermenter cultures of the yeast *Pichia pastoris*. Curr Opin Biotechnol 13:329–332

Chang CC, Ryu DD, Park CS, Kim JY (1998) Improvement of heterologous protein productivity using recombinant *Yarrowia lipolytica* and cyclic fed-batch process strategy. Biotechnol Bioeng 59(3):379–385

Chen GQ, Wu Q (2005) The application of polyhydroxyalkanoates as tissue engineering materials. Biomaterials 26:6565–6578

Chen DC, Beckerich JM, Gaillardin C (1997) One-step transformation of the dimorphic yeast *Yarrowia lipolytica*. Appl Microbiol Biotechnol 48(2):232–235

Chi ZM, Chi Z, Zhang T, Liu GL, Yue LX (2009) Inulinase expressing microorganisms and applications of inulinases. Appl Microbiol Biotechnol 82:211–220

Cho EM, Lee HS, Eom CY, Ohta A (2010) Construction of high sensitive detection system for endocrine disruptors with yeast n-alkane-assimilating Yarrowia lipolytica. J Microbiol Biotechnol 20(11):1563–1570

Chuang LT, Chen DZ, Chen YH, Nicaud JM, Madzak C, Huang YS (2009) Production of functional alpha-linolenic acid (GLA) by expression of fungal delta12- and delta6-desaturases genes in the oleaginous yeast *Yarrowia lipolytica*. In: Hou CT, Shaw JF (eds) Biocatalysis and agricultural biotechnology. CRC, Boca Raton, pp 163–180

Chuang LT, Chen DC, Nicaud JM, Madzak C, Chen YH, Huang YS (2010) Co-expression of heterologous desaturase genes in *Yarrowia lipolytica*. N Biotechnol 27(4):277–282

Cordero Otero R, Gaillardin C (1996) Efficient selection of hygromycin-B-resistant *Yarrowia lipolytica* transformants. Appl Microbiol Biotechnol 46:143–148

Cregg JM, Cereghino JL, Shi J, Higgins DR (2000) Recombinant protein expression in *Pichia pastoris*. Mol Biotechnol 16:23–52

Cui W, Wang Q, Zhang F, Zhang SC, Chi ZM, Madzak C (2011) Direct conversion of inulin into single cell protein by the engineered *Yarrowia lipolytica* carrying inulinase gene. Process Biochem 46(7):1442–1448

Das UN (2008) Essential fatty acids and their metabolites could function as endogenous HMG-CoA reductase and ACE enzyme inhibitors, anti-arrhythmic, anti-hypertensive, anti-atherosclerotic, anti-inflammatory, cytoprotective, and cardioprotective molecules. Lipids Health Dis 7:37–54

Davidow LS, Dezeeuw JR (1985) Process for transformation of *Yarrowia lipolytica*. European Patent Application EP0138508 (A1)

Davidow LS, Franke AE, De Zeeuw JR (1987a) Expression and secretion of heterologous proteins by *Yarrowia lipolytica* transformants. European Patent Application EP0220864 (A1)

Davidow LS, Franke AE, De Zeeuw JR (1987b) New *Yarrowia lipolytica* transformants used for expression and secretion of heterologous proteins, especially prorennin and human anaphylatoxin C5a. European Patent Application EP86307839 (A1)

Davidow LS, O'Donnell MM, Kaczmarek FS, Pereira DA, DeZeeuw JR, Franke AE (1987c) Cloning and sequencing of the alkaline extracellular protease gene of *Yarrowia lipolytica*. J Bacteriol 169(10):4621–4629

De Baetselier A, Vasavada A, Dohet P, Ha-Thi V, De Beukelaer M, Erpicum T, De Clerck L, Hanotier J, Rosemberg S (1991) Fermentation of a yeast producing *A. niger* glucose oxidase: scale-up, purification and characterization of the recombinant enzyme. Biotechnology 9:559–561

Dean N (1999) Asparagine-linked glycosylation in the yeast Golgi. Biochim Biophys Acta 1426:309–322

Dominguez A, Ferminan E, Sanchez M, Gonzalez FJ, Perez-Campo FM, Garcia S, Herrero AB, San Vicente A, Cabello J, Prado M, Iglesias FJ, Choupina A, Burguillo FJ, Fernandez-Lago L, Lopez MC (1998) Non-conventional yeasts as hosts for heterologous protein production. Int Microbiol 1:131–142

Dominguez A, Ferminan E, Gaillardin C (2000) *Yarrowia lipolytica*: an organism amenable to genetic manipulation as a model for analyzing dimorphism in fungi. Contrib Microbiol 5:151–172

Dujon B, Sherman D, Fischer G, Durrens P, Casaregola S, Lafontaine I, De Montigny J, Marck C, Neuveglise C, Talla E, Goffard N, Frangeul L, Aigle M, Anthouard V, Babour A, Barbe V, Barnay S, Blanchin S, Beckerich JM, Beyne E, Bleykasten C, Boisrame A, Boyer J, Cattolico L, Confanioleri F, De Daruvar A, Despons L, Fabre E, Fairhead C, Ferry-Dumazet H, Groppi A, Hantraye F, Hennequin C, Jauniaux N, Joyet P, Kachouri R, Kerrest A, Koszul R, Lemaire M, Lesur I, Ma L, Muller H, Nicaud JM, Nikolski M, Oztas S, Ozier-Kalogeropoulos O, Pellenz S, Potier S, Richard GF, Straub ML, Suleau A, Swennen D, Tekaia F, Wesolowski-Louvel M, Westhof E, Wirth B, Zeniou-Meyer M, Zivanovic I, Bolotin-Fukuhara M, Thierry A, Bouchier C, Caudron B, Scarpelli C, Gaillardin C, Weissenbach J, Wincker P, Souciet JL (2004) Genome evolution in yeasts. Nature 430:35–44

Emond S, Montanier C, Nicaud JM, Marty A, Monsan P, André I, Remaud-Siméon M (2010) New efficient recombinant expression system to engineer *Candida antarctica* lipase B. Appl Environ Microbiol 76(8):2684–2687

Fabre E, Nicaud JM, Lopez MC, Gaillardin C (1991) Role of the proregion in the production and secretion of the *Yarrowia lipolytica* alkaline extracellular protease. J Biol Chem 266:3782–3790

Fabre E, Tharaud C, Gaillardin C (1992) Intracellular transit of a yeast protease is rescued by trans-complementation with its prodomain. J Biol Chem 267:15049–15055

Fickers P, Le Dall MT, Gaillardin C, Thonart P, Nicaud JM (2003) New disruption cassettes for rapid gene disruption and marker rescue in the yeast *Yarrowia lipolytica*. J Microbiol Methods 55(3):727–737

Fickers P, Benetti PH, Wache Y, Marty A, Mauersberger S, Smit MS, Nicaud JM (2005) Hydrophobic substrate utilisation by the yeast *Yarrowia lipolytica*, and its potential applications. FEMS Yeast Res 5:527–543

Flagfeldt DB, Siewers V, Huang L, Nielsen J (2009) Characterization of chromosomal integration sites for heterologous gene expression in *Saccharomyces cerevisiae*. Yeast 26(10):545–551

Fournier P, Guyaneux L, Chasles M, Gaillardin C (1991) Scarcity of ARS sequences isolated in a morphogenesis mutant of the yeast *Yarrowia lipolytica*. Yeast 7:25–36

Fournier P, Abbas A, Chasles M, Kudla B, Ogrydziak DM, Yaver D, Xuan JW, Peito A, Ribet AM, Feynerol C, He F, Gaillardin C (1993) Colocalization of centromeric and replicative functions on autonomously replicating sequences isolated from the yeast *Yarrowia lipolytica*. Proc Natl Acad Sci USA 90:4912–4916

Franke AE, Kaczmarek FS, Eisenhard ME, Geoghegan KF, Danley DE, De Zeeuw JR, O'Donnell MM, Gollaher MG, Davidow LS (1988) Expression and secretion of bovine prochymosin in *Yarrowia lipolytica*. In: Pierce G (ed) Developments in industrial microbiology, vol 29. Elsevier, Amsterdam, pp 43–57

Fretland AJ, Omiecinski CJ (2000) Epoxide hydrolases: biochemistry and molecular biology. Chem Biol Interact 129(1–2):41–59

Gaillardin C, Ribet AM (1987) *LEU2* directed expression of β-galactosidase activity and phleomycin resistance in *Yarrowia lipolytica*. Curr Genet 11:369–375

Gaillardin C, Heslot H, Ribet AM (1985) Vecteur de transformation de la levure *Yarrowia lipolytica*, procédé de transformation et levure transformée. French Patent Application FR2566424 (A1)

Galli C, Gentili P, Jolivalt C, Madzak C, Vadalà R (2011) How is the reactivity of laccase affected by single-point mutations? Engineering laccase for improved activity towards sterically demanding substrates. Appl Microbiol Biotechnol 91(1):123–131

Gasmi N, Fudalej F, Kallel H, Nicaud JM (2011a) A molecular approach to optimize hIFN α2b expression and secretion in *Yarrowia lipolytica*. Appl Microbiol Biotechnol 89(1):109–119

Gasmi N, Ayed A, Nicaud JM, Kallel H (2011b) Design of an efficient medium for heterologous protein production in *Yarrowia lipolytica*: case of human interferon alpha 2b. Microb Cell Fact 10:38–50

Gasmi N, Ayed A, Ammar BB, Zrigui R, Nicaud JM, Kallel H (2011c) Development of a cultivation process for the enhancement of human interferon alpha 2b production in the oleaginous yeast, *Yarrowia lipolytica*. Microb Cell Fact 10:90–100

Gellisen G, Hollenberg CP (1997) Applications of yeast in gene expression studies: a comparison of *Saccharomyces cerevisiae*, *Hansenula polymorpha* and *Kluyveromyces lactis* – a review. Gene 190:87–97

Gellissen G, Kunze G, Gaillardin C, Cregg JM, Berardi E, Veenhuis M, van der Klei I (2005) New yeast expression platforms based on methylotrophic *Hansenula polymorpha* and *Pichia pastoris* and on dimorphic *Arxula adeninivorans* and *Yarrowia lipolytica* – a comparison. FEMS Yeast Res 5(11):1079–1096

Goochee CF, Gramer MJ, Andersen DC, Bahr JB, Rasmussen JR (1991) The oligosaccharides of glycoproteins: bioprocess factors affecting oligosaccharide structure and their effect on glycoprotein properties. Biotechnology 9:1347–1355

Grinna LS, Tschopp JF (1989) Size distribution and general structural features of N-linked oligosaccharides from the methylotrophic yeast *Pichia pastoris*. Yeast 5:107–115

Haddouche R, Delessert S, Sabirova J, Neuvéglise C, Poirier Y, Nicaud JM (2010) Roles of multiple acyl-CoA oxidases in the routing of carbon flow towards β-oxidation and polyhydroxyalkanoate biosynthesis in *Yarrowia lipolytica*. FEMS Yeast Res 10(7):917–927

Haddouche R, Poirier Y, Delessert S, Sabirova J, Pagot Y, Neuvéglise C, Nicaud JM (2011) Engineering polyhydroxyalkanoate content and monomer composition in the oleaginous yeast *Yarrowia lipolytica* by modifying the ß-oxidation multifunctional protein. Appl Microbiol Biotechnol 91(5):1327–1340

Hamilton SR, Gerngross TU (2007) Glycosylation engineering in yeast: the advent of fully humanized yeast. Curr Opin Biotechnol 18(5):387–392

Hamilton SR, Davidson RC, Sethuraman N, Nett JH, Jiang Y, Rios S, Bobrowicz P, Stadheim TA, Li H, Choi BK, Hopkins D, Wischnewski H, Roser J, Mitchell T, Strawbridge RR, Hoopes J, Wildt S, Gerngross TU (2006) Humanization of yeast to produce complex terminally sialylated glycoproteins. Science 313:1441–1443

Hamsa PV, Chattoo BB (1994) Cloning and growth-regulated expression of the gene encoding the hepatitis B virus middle surface antigen in *Yarrowia lipolytica*. Gene 143:165–170

Hamsa PV, Kachroo P, Chattoo BB (1998) Production and secretion of biologically active human epidermal growth factor in *Yarrowia lipolytica*. Curr Genet 33:231–237

He F, Beckerich JM, Gaillardin C (1992) A mutant of 7SL RNA in *Yarrowia lipolytica* affecting the synthesis of a secreted protein. J Biol Chem 267(3):1932–1937

Hong SP, Sharpe P, Xue Z, Yadav N, Zhu Q (2009) High eicosapentaenoic acid oils from improved strains of *Yarrowia lipolytica*. World Patent Application WO2010147907 (A1)

Hong SP, Seip J, Walters-Pollak D, Rupert R, Jackson R, Xue Z, Zhu Q (2012) Engineering *Yarrowia lipolytica* to express secretory invertase with strong FBA1(IN) promoter. Yeast 29:59–72

Huang YC, Chen YF, Chen CY, Chen WL, Ciou YP, Liu WH, Yang CH (2011) Production of ferulic acid from lignocellulolytic agricultural biomass by *Thermobifida fusca* thermostable esterase produced in *Yarrowia lipolytica* transformant. Bioresour Technol 102(17):8117–8122

Jaafar L, Zueco J (2004) Characterization of a glycosylphosphatidylinositol-bound cell-wall protein (GPI-CWP) in *Yarrowia lipolytica*. Microbiology 150(Pt 1):53–60

Jaafar L, Leon M, Zueco J (2003) Identification of the *MNN9* gene of *Yarrowia lipolytica* (Yl*MNN9*) and phenotype analysis of a mutant ylmnn9Δ strain. Yeast 20:633–644

James L, Strick C (1995) Multiple integrative vectors and *Yarrowia lipolytica* transformants. World Patent Application WO9506739 (A1)

Jigami Y, Odani T (1999) Mannosylphosphate transfer to yeast mannan. Biochim Biophys Acta 1426(2):335–345

Jolivalt C, Madzak C, Brault A, Caminade E, Malosse C, Mougin C (2005) Expression of laccase IIIb from the white-rot fungus *Trametes versicolor* in the yeast *Yarrowia lipolytica* for environmental applications. Appl Microbiol Biotechnol 66(4):450–456

Jolivet P, Bordes F, Fudalej F, Cancino M, Vignaud C, Dossat V, Burghoffer C, Marty A, Chardot T, Nicaud JM (2007) Analysis of *Yarrowia lipolytica* extracellular lipase Lip2p glycosylation. FEMS Yeast Res 7(8):1317–1327

Juretzek T, Wang HJ, Nicaud JM, Mauersberger S, Barth G (2000a) Comparison of promoters suitable for regulated overexpression of β-galactosidase in the alkane-utilizing yeast *Yarrowia lipolytica*. Biotechnol Bioprocess Eng 5:320–326

Juretzek T, Mauersberger S, Barth G (2000b) Recombinant haploid or diploid Yarrowia lipolytica cells for the functional heterologous expression of cytochrome P450 systems. World Patent Application WO0003008

Juretzek T, Le Dall M, Mauersberger S, Gaillardin C, Barth G, Nicaud JM (2001) Vectors for gene expression and amplification in the yeast *Yarrowia lipolytica*. Yeast 18:97–113

Kim JW, Park TJ, Ryu DD, Kim JY (2000) High cell density culture of *Yarrowia lipolytica* using a one-step feeding process. Biotechnol Prog 16(4):657–660

Kim MW, Rhee SK, Kim JY, Shimma Y, Chiba Y, Jigami Y, Kang HA (2004) Characterization of N-linked oligosaccharides assembled on secretory recombinant glucose oxidase and cell wall mannoproteins from the methylotrophic yeast *Hansenula polymorpha*. Glycobiology 14 (3):243–251

Kopecny D, Briozzo P, Joly N, Houba-Hérin N, Madzak C, Pethe C, Laloue M (2004) Purification, crystallization and preliminary X-ray diffraction study of a recombinant cytokinin oxidase from *Zea mays*. Acta Crystallogr D Biol Crystallogr 60(8):1500–1501

Kopecny D, Pethe C, Sebela M, Houba-Hérin N, Madzak C, Majira A, Laloue M (2005) High-level expression and characterization of *Zea mays* cytokinin oxidase/dehydrogenase in *Yarrowia lipolytica*. Biochimie 87(11):1011–1022

Kopecny D, Popelkova H, Madzak C, Briozzo P, Sebela M, Frébort I, Majira A, Laloue M, Houba-Hérin N (2008) Site-directed mutagenesis of maize cytokinin oxidase/dehydrogenase. Physiol Plant 133(S3):189–190

Kopecny D, Briozzo P, Popelkova H, Sebela M, Koncitikova R, Spichal L, Nisler J, Madzak C, Frébort I, Laloue M, Houba-Hérin N (2010) Phenyl- and benzylurea cytokinins as competitive inhibitors of cytokinin oxidase/dehydrogenase: a structural study. Biochimie 92(8):1052–1062

Kozak M (1999) Initiation of translation in prokaryotes and eukaryotes. Gene 234(2):187–208

Labuschagné M (2005) Cloning and expression of epoxide hydrolases: *Yarrowia lipolytica* as an attractive heterologous expression host. PhD Thesis, University of Free State, Bloemfontein, South Africa

Labuschagné M, Albertyn J (2007) Cloning of an epoxide hydrolase encoding gene from *Rhodotorula mucilaginosa* and functional expression in *Yarrowia lipolytica*. Yeast 24:69–78

Labuschagné M, Botes AL, Albertyn J (2004) Cloning and sequencing of an epoxide hydrolase gene from *Rhodosporidium paludigenum*. DNA Seq 15:202–205

Laloi M, Mac Carthy J, Morandi O, Gysler C, Bucheli P (2002) Molecular and biochemical characterisation of two aspartic proteinases TcAP1 and TcAP2 from *Theobroma cacao* seeds. Planta 215:754–762

Le Dall MT, Nicaud JM, Gaillardin C (1994) Multiple-copy integration in the yeast *Yarrowia lipolytica*. Curr Genet 26:38–44

Li J, Woo NY (2003) Pathogenicity of Vibrios in fish: an overview. J Ocean Univ Qingdao 2:117–128

Li Y, Liu GL, Wang K, Chi ZM, Madzak C (2012) Overexpression of the endo-inulinase gene from *Arthrobacter* sp. S37 in *Yarrowia lipolytica* and characterization of the recombinant endo-inulinase. J Mol Catal B Enzym 74(1–2):109–115

Liu G, Yue L, Chi Z, Yu W, Chi Z, Madzak C (2009) The surface display of the alginate lyase on the cells of *Yarrowia lipolytica* for hydrolysis of alginate. Mar Biotechnol 11:619–626

Liu XY, Chi Z, Liu GL, Wang F, Madzak C, Chi ZM (2010) Inulin hydrolysis and citric acid production from inulin using the surface-engineered *Yarrowia lipolytica* displaying inulinase. Metab Eng 12(5):469–476

Madzak C, Blanchin-Roland S, Gaillardin C (1996) Upstream activating sequences and recombinant promoter sequences functional in Yarrowia and vectors containing them. European Patent Application EP0747484 (A1)

Madzak C, Blanchin-Roland S, Cordero Otero R, Gaillardin C (1999) Functional analysis of upstream regulating regions from the *Yarrowia lipolytica XPR2* promoter. Microbiology 145:75–87

Madzak C, Tréton B, Blanchin-Roland S (2000) Strong hybrid promoters and integrative expression/secretion vectors for quasi-constitutive expression of heterologous proteins in the yeast *Yarrowia lipolytica*. J Mol Microbiol Biotechnol 2(2):207–216

Madzak C, Gaillardin C, Beckerich JM (2004) Heterologous protein expression and secretion in the non-conventional yeast *Yarrowia lipolytica*: a review. J Biotechnol 109(1–2):63–81

Madzak C, Nicaud JM, Gaillardin C (2005a) *Yarrowia lipolytica*. In: Gellissen G (ed) Production of recombinant proteins: microbial and eucaryotic expression systems. Wiley-VCH, Weinheim, pp 163–189

Madzak C, Otterbein L, Chamkha M, Moukha S, Asther M, Gaillardin C, Beckerich JM (2005b) Heterologous production of a laccase from the basidiomycete *Pycnoporus cinnabarinus* in the dimorphic yeast *Yarrowia lipolytica*. FEMS Yeast Res 5(6–7):635–646

Madzak C, Mimmi MC, Caminade E, Brault A, Baumberger S, Briozzo P, Mougin C, Jolivalt C (2006) Shifting the optimal pH of activity for a laccase from the fungus *Trametes versicolor* by structure-based mutagenesis. Protein Eng Des Sel 19(2):77–84

Maharajh D, Roth R, Lalloo R, Simpson C, Mitra R, Gorgens J, Ramchuran S (2008a) Multi-copy expression and fed-batch production of *Rhodotorula araucariae* epoxide hydrolase in *Yarrowia lipolytica*. Appl Microbiol Biotechnol 79:235–244

Maharajh D, Lalloo R, Gorgens J (2008b) Effect of an exponential feeding regime on the production of *Rhodotorula araucariae* epoxide hydrolase in *Yarrowia lipolytica*. Lett Appl Microbiol 47:520–525

Malito E, Coda A, Bilyeu KD, Fraaije MW, Mattevi A (2004) Structures of Michaelis and product complexes of plant cytokinin dehydrogenase: implications for flavoenzyme catalysis. J Mol Biol 341:1237–1249

Massonneau A, Houba-Hérin N, Pethe C, Madzak C, Falque M, Mercy M, Kopecny D, Majira A, Rogowsky P, Laloue M (2004) Maize cytokinin oxidase genes: differential expression and cloning of two new cDNAs. J Exp Bot 55(408):2549–2557

Matoba S, Fukayama J, Wing RA, Ogrydziak DM (1988) Intracellular precursors and secretion of alkaline extracellular protease of *Yarrowia lipolytica*. Mol Cell Biol 8(11):4904–4916

Matsuoka M, Matsubara M, Daidoh H, Imanaka T, Uchida K, Aiba S (1993) Analysis of regions essential for the function of chromosomal replicator sequences from *Yarrowia lipolytica*. Mol Gen Genet 237:327–333

Mayer AM, Staples RC (2002) Laccase: new functions for an old enzyme. Phytochemistry 60 (6):551–565

Mekouar M, Blanc-Lenfle I, Ozanne C, Da Silva C, Cruaud C, Wincker P, Gaillardin C, Neuvéglise C (2010) Detection and analysis of alternative splicing in *Yarrowia lipolytica* reveal structural constraints facilitating nonsense-mediated decay of intron-retaining transcripts. Genome Biol 11(6):R65

Menendez JA, Vellon L, Colomer R, Lupu R (2005) Effect of gamma-linolenic acid on the transcriptional activity of the Her-2/neu (erbB-2) oncogene. J Natl Cancer Inst 97 (21):1611–1615

Mougin C, Jolivalt C, Briozzo P, Madzak C (2003) Fungal laccases: from structure-activity studies to environmental applications. Environ Chem Lett 1:145–148

Muesch A, Hartmann E, Rohde K, Rubartelli A, Sitia R, Rapoport TA (1990) A novel pathway for secretory proteins? Trends Biochem Sci 15(3):86–88

Muller S, Sandal T, Kamp-Hansen P, Dalboge H (1998) Comparison of expression systems in the yeasts *Saccharomyces cerevisiae*, *Hansenula polymorpha*, *Kluyveromyces lactis*, *Schizosaccharomyces pombe* and *Yarrowia lipolytica*. Cloning of two novel promoters from Yarrowia lipolytica. Yeast 14:1267–1283

Nakagawa S, Niimura Y, Gojobori T, Tanaka H, Miura K (2008) Diversity of preferred nucleotide sequences around the translation initiation codon in eukaryote genomes. Nucleic Acids Res 36 (3):861–871

Nakajima T, Ballou CE (1975) Yeast manno-protein biosynthesis: solubilization and selective assay of four mannosyltransferases. Proc Natl Acad Sci USA 72:3912–3916

Ni X, Chi Z, Ma C, Madzak C (2008) Cloning, characterization, and expression of the gene encoding alkaline protease in the marine yeast *Aureobasidium pullulans* 10. Mar Biotechnol 10 (3):319–327

Ni X, Yue L, Chi Z, Li J, Wang X, Madzak C (2009) Alkaline protease gene cloning from the marine yeast *Aureobasidium pullulans* HN2-3 and the protease surface display on *Yarrowia lipolytica* for bioactive peptide production. Mar Biotechnol 11:81–89

Nicaud JM, Fabre E, Gaillardin C (1989) Expression of invertase activity in *Yarrowia lipolytica* and its use as a selective marker. Curr Genet 16:253–260

Nicaud JM, Fournier P, La Bonnardiere C, Chasles M, Gaillardin C (1991) Use of ars18 based vectors to increase protein production in *Yarrowia lipolytica*. J Biotechnol 19:259–270

Nicaud JM, Gaillardin C, Seman M, Pignède G (2000) Procédé de transformation non-homologue de *Yarrowia lipolytica*. French Patent Application FR2782733 (A1)

Nicaud JM, Madzak C, van den Broek P, Gysler C, Duboc P, Niederberger P, Gaillardin C (2002) Protein expression and secretion in the yeast *Yarrowia lipolytica*. FEMS Yeast Res 2:371–379

Nicaud JM, Fudalej F, Neuvéglise C, Beckerich JM (2009) Procédé d'intégration ciblée de multicopies d'un gène d'intérêt dans une souche de Yarrowia. French Patent Application FR2927089 (A1)

Nicaud JM, Chardot T, Beopoulos A (2010) New mutant yeast strains capable of accumulating a large quantity of lipids. World Patent Application WO2010004141 (A2)

Niu DK, Yang YF (2011) Why eukaryotic cells use introns to enhance gene expression: splicing reduces transcription-associated mutagenesis by inhibiting topoisomerase I cutting activity. Biol Direct 6:24

Nthangeni MB, Urban P, Pompon D, Smit MS, Nicaud JM (2004) The use of *Yarrowia lipolytica* for the expression of human cytochrome P450 CYP1A1. Yeast 21(7):583–592

Ogrydziak DM, Scharf SJ (1982) Alkaline extracellular protease produced by *Saccharomycopsis lipolytica* CX161-1B. J Gen Microbiol 128:1225–1234

Ogrydziak DM, Demain AL, Tannenbaum SR (1977) Regulation of extracellular protease production in *Candida lipolytica*. Biochim Biophys Acta 497:525–538

Oh DB, Park JS, Kim MW, Cheon SA, Kim EJ, Moon HY, Kwon O, Rhee SK, Kang HA (2008) Glycoengineering of the methylotrophic yeast *Hansenula polymorpha* for the production of glycoproteins with trimannosyl core N-glycan by blocking core oligosaccharide assembly. Biotechnol J 3(5):659–668

Otterbein L, Record E, Longhi S, Asther M, Moukha S (2000) Molecular cloning of the cDNA encoding laccase from *Pycnoporus cinnabarinus* I-937 and expression in *Pichia pastoris*. Eur J Biochem 267:1619–1625

Papanikolaou S, Komaitis M, Aggelis G (2004) Single cell oil (SCO) production by *Mortierella isabellina* grown on high-sugar content media. Bioresour Technol 95:287–291

Park CS, Chang CC, Kim JY, Ogrydziak DM, Ryu DDY (1997) Expression, secretion and processing of rice α-amylase in the yeast *Yarrowia lipolytica*. J Biol Chem 272:6876–6881

Park CS, Chang CC, Ryu DDY (2000) Expression and high-level secretion of *Trichoderma reesei* endoglucanase I in *Yarrowia lipolytica*. Appl Biochem Biotechnol 87(1):1–15

Park JN, Song Y, Kim JY, Oh DB, Kang HA (2008) A novel Yl*MPO1* gene derived from *Yarrowia lipolytica* and a process for preparing a glycoprotein not being mannosylphosphorylated by using a mutated *Yarrowia lipolytica* in which Yl*MPO1* gene is disrupted. World Patent Application WO2008136564 (A1)

Park JN, Song Y, Cheon SA, Kwon O, Oh DB, Jigami Y, Kim JY, Kang HA (2011) Essential role of Yl*MPO1*, a novel *Yarrowia lipolytica* homologue of *Saccharomyces cerevisiae* MNN4, in mannosylphosphorylation of N- and O-linked glycans. Appl Environ Microbiol 77 (4):1187–1195

Paulick MG, Bertozzi CR (2008) The glycosylphosphatidylinositol anchor: a complex membrane-anchoring structure for proteins. Biochemistry 47(27):6991–7000

Peng Y, Liu GL, Yu XJ, Wang XH, Jing L, Chi ZM (2011) Cloning of exo-β-1,3-glucanase gene from a marine yeast *Williopsis saturnus* and its overexpression in *Yarrowia lipolytica*. Mar Biotechnol (NY) 13(2):193–204

Pepper LR, Cho YK, Boder ET, Shusta EV (2008) A decade of yeast surface display technology: where are we now? Comb Chem High Throughput Screen 11(2):127–134

Picataggio SK, Zhu Q (2005) Glyceraldehyde-3-phosphate dehydrogenase and phosphoglycerate mutase promoters for gene expression in oleaginous yeast. World Patent Application WO2005003310 (A2)

Pignède G, Wang H, Fudalej F, Seman M, Gaillardin C, Nicaud JM (2000) Autocloning and amplification of *LIP2* in *Yarrowia lipolytica*. Appl Environ Microbiol 66(8):3283–3289

Poirier Y, Erard N, Petétot JM (2001) Synthesis of polyhydroxyalkanoate in the peroxisome of *Saccharomyces cerevisiae* by using intermediates of fatty acid beta-oxidation. Appl Environ Microbiol 67(11):5254–5260

Pollak DM, Zhu Q (2005) Fructose biphosphate aldolase regulatory sequences for gene expression in oleaginous yeast. World Patent Application WO2005049805 (A2)

Rao A, Pimprikar P, Bendigiri C, Kumar AR, Zinjarde S (2011) Cloning and expression of a tyrosinase from *Aspergillus oryzae* in *Yarrowia lipolytica*: application in L-DOPA biotransformation. Appl Microbiol Biotechnol 92(5):951–959

Record E, Punt PJ, Chamkha M, Labat M, van den Hondel CA, Asther M (2002) Expression of the *Pycnoporus cinnabarinus* laccase gene in *Aspergillus niger* and characterization of the recombinant enzyme. Eur J Biochem 269(2):602–609

Richard M, Quijano RR, Bezzate S, Bordon-Pallier F, Gaillardin C (2001) Tagging morphogenetic genes by insertional mutagenesis in the yeast *Yarrowia lipolytica*. J Bacteriol 183 (10):3098–3107

Romanos MA, Scorer CA, Clare JJ (1992) Foreign gene expression in yeast: a review. Yeast 8:423–488

Roth R, Moodley V, van Zyl P (2009) Heterologous expression and optimized production of an *Aspergillus aculeatus* endo-1,4-beta-mannanase in *Yarrowia lipolytica*. Mol Biotechnol 43 (2):112–120

Rothblatt JA, Meyer DI (1986) Secretion in yeast: translocation and glycosylation of prepro-alpha-factor in vitro can occur via an ATP-dependent post-translational mechanism. EMBO J 5 (5):1031–1036

Ruiz-Herrera J, Sentandreu R (2002) Different effectors of dimorphism in *Yarrowia lipolytica*. Arch Microbiol 178(6):477–483

Schmid-Berger N, Schmid B, Barth G (1994) Ylt1, a highly repetitive retrotransposon in the genome of the dimorphic fungus *Yarrowia lipolytica*. J Bacteriol 176:2477–2482

Setati ME, Ademark P, van Zyl WH, Hahn-Hägerdal B, Stålbrand H (2001) Expression of the *Aspergillus aculeatus* endo-beta-1,4-mannanase encoding gene (man1) in *Saccharomyces cerevisiae* and characterization of the recombinant enzyme. Protein Expr Purif 21(1):105–114

Sharpe P, Ye R, Zhu Q (2008) Production de caroténoïdes dans une levure oléagineuse recombinante. World Patent Application WO2008073367 (A1)

Shibasaki S, Maeda H, Ueda M (2009) Molecular display technology using yeast-arming technology. Anal Sci 25(1):41–49

Shiningavamwe A, Obiero G, Albertyn J, Nicaud JM, Smit M (2006) Heterologous expression of the benzoate para-hydroxylase encoding gene (CYP53B1) from *Rhodotorula minuta* in *Yarrowia lipolytica*. Appl Microbiol Biotechnol 72(2):323–329

Shusta EV, Raines RT, Plückthun A, Wittrup KD (1998) Increasing the secretory capacity of *Saccharomyces cerevisiae* for production of single-chain antibody fragments. Nat Biotechnol 16(8):773–777

Sohn YS, Park CS, Lee SB, Ryu DD (1998) Disruption of PMR1, encoding a Ca2+-ATPase homolog in *Yarrowia lipolytica*, affects secretion and processing of homologous and heterologous proteins. J Bacteriol 180(24):6736–6742

Song Y, Choi MH, Park JN, Kim MW, Kim EJ, Kang HA, Kim JY (2007) Engineering of the yeast *Yarrowia lipolytica* for the production of glycoproteins lacking the outer-chain mannose residues of N-glycans. Appl Environ Microbiol 73(14):4446–4454

Song H, Zhou L, Zhang L, Gao B, Wei D, Shen Y, Wang R, Madzak C, Jiang Z (2011) Construction of a whole-cell catalyst displaying a fungal lipase for effective treatment of oily wastewaters. J Mol Catal B Enzym 71:166–170

Steinborn G, Böer E, Scholz A, Tag K, Kunze G, Gellissen G (2006) Application of a wide-range yeast vector (CoMed) system to recombinant protein production in dimorphic *Arxula adeninivorans*, methylotrophic *Hansenula polymorpha* and other yeasts. Microb Cell Fact 5:33–45

Swennen D, Beckerich JM (2007) *Yarrowia lipolytica* vesicle-mediated protein transport pathways. BMC Evol Biol 7:219–237

Swennen D, Paul MF, Vernis L, Beckerich JM, Fournier A, Gaillardin C (2002) Secretion of active anti-Ras single-chain Fv antibody by the yeasts *Yarrowia lipolytica* and *Kluyveromyces lactis*. Microbiology 148:41–50

Swennen D, Henry C, Beckerich JM (2010) Folding proteome of *Yarrowia lipolytica* targeting with uracil permease mutants. J Proteome Res 9(12):6169–6179

Tharaud C, Ribet AM, Costes C, Gaillardin C (1992) Secretion of human blood coagulation factor XIIIa by the yeast *Yarrowia lipolytica*. Gene 121:111–119

Thompson A, Gasson MJ (2001) Location effects of a reporter gene on expression levels and on native protein synthesis in *Lactococcus lactis* and *Saccharomyces cerevisiae*. Appl Environ Microbiol 67(8):3434–3439

Tobe S, Takami T, Ikeda S, Mitsugi K (1976) Production and some enzymatic properties of alkaline protease of *Candida lipolytica*. Agric Biol Chem 40:1087–1092

Trimble RB, Atkinson PH, Tschopp JF, Townsend RR, Maley F (1991) Structure of oligosaccharide on Saccharomyces SUC2 invertase secreted by the methylotrophic yeast *Pichia pastoris*. J Biol Chem 266:22807–22817

van Uden N, Fell JW (1968) Marine yeasts. Adv Microbiol Sea 1:167–201

van Zyl PJ (2010) A method for producing a polypeptide in *Yarrowia lipolytica*. World Patent Application WO2011148339 (A1)

van Zyl PJ, Moodley V, Rose SH, Roth RL, van Zyl WH (2009) Production of the *Aspergillus aculeatus* endo-1,4-beta-mannanase in *A. niger*. J Ind Microbiol Biotechnol 36(4):611–617

Vernis L, Abbas A, Chasles M, Gaillardin C, Brun C, Huberman JA, Fournier P (1997) An origin of replication and a centromere are both needed to establish a replicative plasmid in the yeast *Yarrowia lipolytica*. Mol Cell Biol 17:1995–2004

Walter P, Blobel G (1980) Purification of a membrane-associated protein complex required for protein translocation across the endoplasmic reticulum. Proc Natl Acad Sci USA 77 (12):7112–7116

Walter P, Blobel G (1983) Disassembly and reconstitution of signal recognition particle. Cell 34 (2):525–533

Wang F, Yue L, Wang L, Madzak C, Li J, Wang X, Chi Z (2009) Genetic modification of the marine-derived yeast *Yarrowia lipolytica* with high-protein content using a GPI-anchor-fusion expression system. Biotechnol Prog 25(5):1297–1303

Wang JH, Hung W, Tsai SH (2011) High efficiency transformation by electroporation of *Yarrowia lipolytica*. J Microbiol 49(3):469–472

Werck-Reichhart D, Feyereisen R (2000) Cytochromes P450: a success story. Genome Biol 1(6): reviews3003.1–reviews3003.9

Werner T, Schmulling T (2009) Cytokinin action in plant development. Curr Opin Plant Biol 12 (5):527–538

Wojtatowicz M, Rymowicz W, Robak M, Zarowska B, Nicaud JM (1997) Kinetics of cell growth and citric acid production by *Yarrowia lipolytica* Suc^+ transformants in sucrose media. Pol J Food Nutr Sci 6/47(4):49–54

Xuan JW, Fournier P, Gaillardin C (1988) Cloning of the *LYS5* gene encoding saccharopine dehydrogenase from the yeast *Yarrowia lipolytica* by target integration. Curr Genet 14:15–21

Xue Z, Zhu Q (2006a) Glycerol-3-phosphate o-acyl transferase promoter for gene expression in oleaginous yeast. World Patent Application WO2006031937 (A2)

Xue Z, Zhu Q (2006b) Ammonium transporter promoter for gene expression in oleaginous yeast. World Patent Application WO2006052754 (A2)

Xue Z, Yadav N, Zhu Q (2007) Optimized strains of Yarrowia lipolytica for high eicosapentaenoic acid production. European Patent Application EP2195415 (A1)

Yang CH, Liu WH (2007) Cloning and characterization of a maltotriose-producing α-amylase gene from *Thermobifida fusca*. J Ind Microbiol Biotechnol 34(4):325–330

Yang XS, Jiang ZB, Song HT, Jiang SJ, Madzak C, Ma LX (2009) Cell-surface display of the active mannanase in *Yarrowia lipolytica* with a novel surface-display system. Biotechnol Appl Biochem 54(3):171–176

Yang CH, Huang YC, Chen CY, Wen CY (2010a) Expression of *Thermobifida fusca* thermostable raw starch digesting alpha-amylase in *Pichia pastoris* and its application in raw sago starch hydrolysis. J Ind Microbiol Biotechnol 37(4):401–406

Yang CH, Huang YC, Chen CY, Wen CY (2010b) Heterologous expression of *Thermobifida fusca* thermostable alpha-amylase in *Yarrowia lipolytica* and its application in boiling stable resistant sago starch preparation. J Ind Microbiol Biotechnol 37(9):953–960

Yaver DS, Matoba S, Ogrydziak DM (1992) A mutation in the signal recognition particle 7S RNA of the yeast *Yarrowia lipolytica* preferentially affects synthesis of the alkaline extracellular protease: in vivo evidence for translational arrest. J Cell Biol 116:605–616

Yu XJ, Madzak C, Li HJ, Chi ZM, Li J (2010) Surface display of acid protease on the cells of *Yarrowia lipolytica* for milk clotting. Appl Microbiol Biotechnol 87(2):669–677

Yue L, Chi Z, Wang L, Liu J, Madzak C, Li J, Wang X (2008) Construction of a new plasmid for surface display on cells of *Yarrowia lipolytica*. J Microbiol Methods 72:116–123

Yuzbashev TV, Yuzbasheva EY, Vibornaya TV, Sobolevskaya TI, Laptev IA, Gavrikov AV, Sineoky SP (2012) Production of recombinant *Rhizopus oryzae* lipase by the yeast *Yarrowia lipolytica* results in increased enzymatic thermostability. Protein Expr Purif 82:83–89

Zhang N, Suen WC, Windsor W, Xiao L, Madison V, Zaks A (2003) Improving tolerance of *Candida antarctica* lipase B towards irreversible thermal inactivation through directed evolution. Protein Eng 16(8):599–605

Zhao CH, Cui W, Liu XY, Chi ZM, Madzak C (2010) Expression of inulinase gene in the oleaginous yeast *Yarrowia lipolytica* and single cell oil production from inulin-containing materials. Metab Eng 12(6):510–517

Zhu KL, Chi ZM, Li J, Zhang FL, Li MJ, Yasoda HN, Wu LF (2006) The surface display of haemolysin from *Vibrio harveyi* on yeast cells and their potential applications as live vaccine in marine fish. Vaccine 24:6046–6052

Acid and Alkaline Extracellular Proteases of *Yarrowia lipolytica*

David Ogrydziak

Abstract *XPR2* and *AXP1*, coding for alkaline (AEP) and acid (AXP) extracellular proteases, have been sequenced for several strains. For *XPR2*, the three sequenced strains are not closely related and produce significantly different levels of AEP, yet the coding sequences are identical, and there is only a single nucleotide difference in one promoter suggesting that host physiology, not promoter differences, determines AEP production. The possibility that pro-mAEP forms a dimer that can inhibit mature AEP (mAEP) proteolytic activity in *trans* is examined. AXP contains a predicted signal sequence and a 44 amino acid prepro-region. Activation involves pH-dependent autoprocessing that occurs extracellularly. *XPR2* UAS1 and UAS2 promoter elements have been identified and their roles in regulation explored. *Cis*-sequences and Rim pathway components involved in pH regulation of the proteases have been discovered and characterized. *YlOPT1* and *YLSSY5* are in a signaling pathway(s) regulating *AXP1* and *XPR2*, perhaps by sensing amino acids. Both pepsin-like (30 potentially secreted members) and subtilisin-like (16 potentially secreted members) gene families have undergone lineage-specific expansion compared to other yeast and filamentous fungi. To determine if expression of secretory pathway components is regulated in response to secretory demand, rapid AEP induction conditions and *XPR2* multicopy strains were developed. Changes in genomic transcription were measured when growth started to slow after AEP induction. For secretory pathway components, mostly repression was found. Possibly, their induction had occurred by the control time point, and the turning off of this short-term response at later time points appeared as repression.

D. Ogrydziak (✉)
Department of Food Science and Technology, University of California, One Shields Avenue, Davis, CA 95616, USA

Institute of Marine Resources, University of California, Davis, CA, USA
e-mail: dmogrydziak@ucdavis.edu

G. Barth (ed.), *Yarrowia lipolytica*, Microbiology Monographs 25,
DOI 10.1007/978-3-642-38583-4_2, © Springer-Verlag Berlin Heidelberg 2013

Contents

1 Introduction 78
2 AEP Processing 79
  2.1 Signal Peptide Cleavage/Dipeptidyl Aminopeptidase Processing 79
  2.2 Is Pro-mAEP a Dimer? 80
3 AXP Processing 82
4 AEP Mode of Translocation 82
5 *XPR2* Regulation 83
6 *AXP1* Regulation 86
7 Heterologous Protein Expression 87
8 Other Extracellular Proteases 87
9 Regulation of Secretory Pathway Components 91
  9.1 Construction and Screening of *XPR2* Multicopy Strains 91
  9.2 AEP Levels 93
  9.3 Transcriptional Response to Secretion Stress 93
References 95

1 Introduction

Y. lipolytica alkaline and acid extracellular proteases were last extensively reviewed in an overall review of yeast extracellular proteases (Ogrydziak 1993).

Studies on the *Y. lipolytica* alkaline protease grew out of a project in which a yeast producing an extracellular protease(s) would solubilize fish proteins in a non-sterile process at acidic pH. *Y. lipolytica* was chosen because of its high extracellular proteolytic activity (Ahearn et al. 1968). When it became clear that the major protease secreted was an alkaline protease, the project changed to regulation and characterization of this protease.

Most likely all *Y. lipolytica* strains can produce an alkaline extracellular protease (AEP) . Various screens revealed substantial variation in AEP production between strains. Extensive inbreeding of *Y. lipolytica* strains has been done to improve their genetic manipulability. Strain CX161-1B, developed in Mortimer's laboratory, produced more AEP than strains developed in Heslot's laboratory, and wild strain W29 produced even more (see below).

XPR2 (coding for AEP) has been cloned and sequenced from three *Y. lipolytica* strains—NRRL Y-1094 (Davidow et al. 1987), CX161-1B (Matoba et al. 1988), and W29 (Nicaud et al. 1989). These strains are not closely related, but the coding sequences are identical. The promoter sequences of NRRL Y-1094 and W29 are identical and CX161-1B has one difference at position -404. Therefore, differences in AEP production probably reflect differences in host cell physiology, not promoter differences.

Surprisingly, *XPR2* from CLIB122, whose genome has been sequenced, has not been sequenced. CLIB122 contains *xpr2-322*, an *Apa*I deletion that removes amino acids 1–34 from the N-terminus of AEP (Barth and Gaillardin 1996). It was constructed with W29 *XPR2* DNA and integrated at the *Mlu*I site located over 900 bp upstream of the *Apa*I site. Therefore, most of the CLIB122 *XPR2* promoter sequence and an unknown amount of the coding sequence, depending on where the crossover occurred for the "pop out," are derived from W29.

2 AEP Processing

AEP is co-translationally translocated into the endoplasmic reticulum (ER), the signal peptide cleaved, and the pro-region N-linked glycosylated to produce a 55-kDa polypeptide, the first precursor detected intracellularly in pulse-chase immunoprecipitation experiments (Ogrydziak 1993). This precursor is transported to the Golgi where nine N terminal X-Ala, X-Pro dipeptides are removed by dipeptidyl aminopeptidase (DPAPase) to produce the 52 kDa precursor. Xpr6p cleavage after Lys.Arg at the end of the propeptide produces a 20 kDa propeptide and 32 kDa mature AEP (mAEP). S397A, an active site serine mutation that eliminates proteolytic activity, produces mAEP indicating that AEP proteolytic activity is not required for AEP processing (Matoba et al. 1997).

2.1 Signal Peptide Cleavage/Dipeptidyl Aminopeptidase Processing

Radiosequencing of the 55 kDa AEP precursor labeled with 3H-Leu and 35S-Met confirmed cleavage between Ala15 and Ala16 (Matoba et al. 1997). If signal peptide cleavage was after Ala15, Leu should be found at position 3 and there should be no Met signal. This is what was found. A 15 amino acid signal peptide was also found for the P17M AEP 55 kDa precursor (Yaver et al. 1992).

A recent review states that the AEP signal peptide is 13 amino acids long (Thevenieau et al. 2009). If signal peptide cleavage was after position 13, Leu should have been found at positions 1 and 5, and it was not. The possibility of an additional cleavage site(s) after Leu18 cannot be eliminated, as there is no Leu or Met for the next 26 amino acids in the pro-region. No 52 kDa AEP precursor was detected for A19V AEP (removal of only one dipeptide should occur), suggesting there is no downstream signal peptide cleavage site in-frame for DPAPase processing.

Analysis by SignalP 3.0 (Bendtsen et al. 2004) yields a prediction of maximum likelihood for the signal peptide cleavage site at Ala15-Ala16 using Neural Networks (NN) with Ala19-Ala20 as second most likely. Using Hidden Markov Models (HMM), maximum cleavage site probabilities were 0.605 between Pro21 and Ala22, 0.15 between Ala15 and Ala16, and 0.10 between Ala19 and Ala20. These predicted cleavage sites are all in the correct frame to allow removal of the downstream dipeptides by DPAPase.

The original review included evidence that DPAPase processing occurred in vivo (Ogrydziak 1993). N-terminal sequencing revealed that the 52 kDa precursor secreted by an *xpr6* mutant contained one species starting at the end of the dipeptide stretch and species starting one or two dipeptides upstream (Matoba and Ogrydziak 1989). Modifications of AEP (A19V and P17M), use of a DPAPase inhibitor proBoro, and expression of *XPR2* in *Saccharomyces cerevisiae* strains lacking both the Golgi and vacuolar DPAPases all demonstrated that mature AEP was still secreted when DPAPase processing was eliminated (Matoba et al. 1997). Therefore, Xpr6p processing was not dependent on DPAPase processing.

Deleting the *Y. lipolytica* Golgi DPAPase (YALI0B2838g) to determine effects on AEP processing is now possible. Unexpectedly, there does not appear to be a vacuolar DPAPase in CLIB 122.

2.2 *Is Pro-mAEP a Dimer?*

Wild-type strains secrete mAEP and propeptide but no AEP precursors (Matoba et al. 1988). The propeptide helps with the folding of AEP and inhibits its proteolytic activity; it was able in *trans* to enhance folding of a premature AEP construct lacking the pro-region (Fabre et al. 1992). If propeptide assistance in mAEP folding is intramolecular, this result suggests that the interaction occurs in the ER with unfolded or partially folded polypeptides. Inhibition of AEP activity by propeptide was demonstrated by the lack of proteolytic activity of the 52 kDa AEP precursor secreted by *xpr6* mutants (Enderlin and Ogrydziak 1994).

Secreted mAEP levels are significantly higher than for propeptide, suggesting propeptide is less stable. Propeptide degradation might start in the secretory pathway after Xpr6p cleavage. Or propeptide and mAEP might be secreted in equimolar amounts, perhaps as a non-covalently bound complex. Propeptide degradation would then begin when the complex dissociates.

Modification of this AEP processing model is proposed based on observations made decades ago. On skim milk (SKM) plates, segregants of crosses between wild-type and *xpr6* strains produced noncircular zones of clearing. If *xpr6* and wild-type segregants were adjacent, the edge of the zone of clearing between them was fairly straight. If the wild-type segregant was surrounded on four sides by *xpr6* segregants, it appears to form a square zone of clearing (Fig. 1).

Square zones strongly suggest that *xpr6* segregants are secreting something that interferes with clearing of SKM plates by mAEP. *xpr6-13* predominately secretes the 52 kDa pro-mAEP, lesser amounts of mAEP, and even less propeptide (Enderlin and Ogrydziak 1994). If only propeptide could inhibit mAEP activity, then any propeptide that diffused into the area normally cleared on SKM plates by an adjacent wild-type strain should be rapidly degraded, and no inhibition of clearing would be expected.

Square zones could be explained if pro-mAEP is more stable than propeptide to AEP degradation and if pro-mAEP can inhibit fully folded mAEP activity.

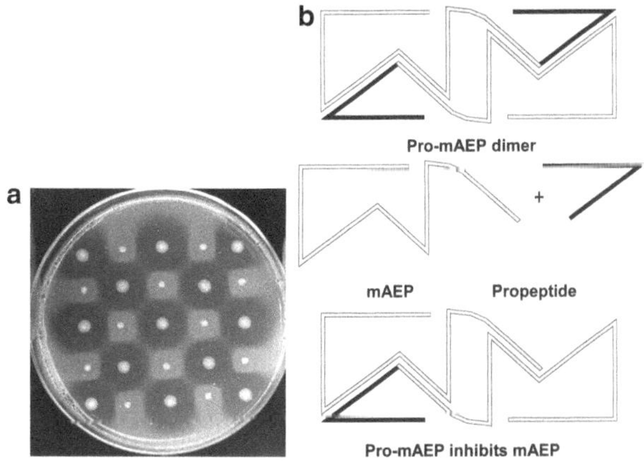

Fig. 1 (**a**) Wild-type and *xpr6-13* strains were toothpicked in an alternating pattern on SKM plates. Cleared zones appear *darker* because of the black background. (**b**) Schematic drawings representing the pro-mAEP dimer, mAEP and propeptide, and the pro-mAEP and mAEP complex. The pro-region is *black* and mature AEP *gray*. Adapted from Hu et al. (1996)

The front of mAEP diffusing from the wild type would meet the front of pro-mAEP diffusing from *xpr6*, and clearing on SKM plates would be inhibited in regions where the ratio of pro-mAEP to mAEP was sufficiently high, transforming the curved front of clearing into a relatively straight line.

For years, pro-mAEP inhibition of fully folded mAEP was not considered because it was assumed that the pro-region would already be involved in inhibiting the covalently attached mAEP region. AEP is a subtilisin-like protease, and Hu et al. (1996) showed that pro-subtilisin forms a dimer and that its assistance in folding and inhibition of the mature region could be intermolecular. This provided a model for how pro-mAEP could inhibit mAEP (Fig. 1). If pro-mAEP forms dimers and enhancement of folding and inhibition of protease activity by the pro-region can occur intermolecularly, then square zones can be explained.

Supernatant mixing experiments showed that pro-mAEP could inhibit mAEP. A 3:1 mixture of *xpr6-13* and wild-type supernatants yielded a 46 % decrease in AEP activity compared to fourfold dilution of the wild-type supernatant (unpublished results). If pro-mAEP was rapidly degraded by mAEP, such a significant inhibition of AEP activity might not be expected.

The *xpr6* mutants are slightly leaky. AEP activity measured by a casein hydrolysis assay was 0.1 % and 0.4 % of wild type for *xpr6-13* and *xpr6-25*, respectively (Ogrydziak and Mortimer 1977). Western blots of extracellular samples revealed no mAEP for a *Δxpr6* strain but what appears to be much greater than 0.1 % of wild-type levels of mAEP from *xpr6-13* (Enderlin and Ogrydziak 1994). Both *xpr6-13* and *xpr6-25* grow much faster on SKM plates than *Δxpr6*, consistent with some mAEP being produced and some proteolysis occurring. Inhibition of mAEP by pro-mAEP might explain why no zones of clearing are detected. There would be far more pro-mAEP than mAEP molecules, and if pro-mAEP were relatively stable, insufficient casein hydrolysis would occur to produce a zone of clearing.

Proving existence of the pro-mAEP dimer by examining samples from a *Δxpr6* strain under nondenaturing conditions would provide support for this new model. All previous samples were examined by SDS-PAGE, and a pro-mAEP dimer and pro-mAEP/mAEP interactions would not have been detected. Reagents needed for more detailed studies of AEP inhibition by pro-mAEP and propeptide could be readily obtained—pro-mAEP from *Δxpr6* and propeptide from S397A.

3 AXP Processing

Since the original review, the N-terminal amino acid sequence of the extracellular acid protease from *Y. lipolytica* 148 was determined and *AXP1* cloned (Young et al. 1996). AXP is a 397 amino acid pepsin-like protein that includes a 44 amino acid prepro-region. The predicted amino acid sequence is identical to that of AXP from CLIB122. There is one silent nucleotide difference in the coding region, six nucleotide differences and a GTA insertion in the 5′ upstream region, and one nucleotide difference in the 3′ downstream region.

Predicted signal peptide cleavage occurs between Ala17 and Ala18. Unlike *XPR2*, the pro-region does not end in Lys.Arg; cleavage occurs between Phe44 and Ala45. This cleavage is due to pH-dependent autoprocessing (McEwen and Young 1998). In pulse-chase immunoprecipitation studies using an AXP-specific antibody, only a 42 kDa precursor but no mature AXP (39 kDa) was detected intracellularly. The 42 kDa precursor was secreted and at lower pHs (4.0 and 4.6) converted to mature AXP. This conversion happened extremely slowly, if at all, at pH 5.6 and 6.0. The 42 kDa precursor maturation was only slightly affected by the inhibitor pepstatin suggesting that autocatalytic activation was primarily intramolecular.

The mature region contains three potential N-linked glycosylation sites and the pro-region none. Mobility of the 42 kDa precursor on SDS-PAGE was unchanged for cells treated with tunicamycin indicating that no sites are used.

4 AEP Mode of Translocation

Yaver et al. (1992) found SRP-dependent co-translational translocation for wild-type AEP and SRP-independent posttranslational translocation for P17M—Pro to Met two amino acids after the signal peptide cleavage site.

For *S. cerevisiae*, proteins using the SRP-dependent pathway tend to have more hydrophobic signal peptides—HB12 (average hydrophobicity for 12 residues after the last positively charged residue of the n-region) values around 3.0 or more (Ng et al. 1996). Values around 2.0 or less were characteristic of proteins not using the SRP-targeting pathway. SRP-independent translocation of AEP would be predicted based on its HB 12 of 2.15. The P17M mutation, not in the signal peptide, did not change the signal peptide cleavage site and actually increased HB 16 slightly. Therefore, P17M translocation becoming SRP-independent posttranslational was unexpected.

SRP-dependent co-translational translocation could be restored by increasing hydrophobicity of the signal peptide (Matoba and Ogrydziak 1998). However, combining the A12P and P17M mutations had the same effect and did not increase hydrophobicity. Kinkiness was proposed as another factor(s) that may be important in translocation pathway choice. Structural models of the extended form of the wild-type N-terminus have a kink; P17M removes the kink caused by Pro17 resulting in a relatively straight structure. Addition of the A12P mutation restores the kink. Possibly the conformation of the signal peptide at the ribosome surface may affect the interaction with SRP. Less hydrophobic signal peptides might depend on the kink to approach SRP more closely more often to enhance interactions. These effects might only have been seen because of AEP signal peptide's short length and low average hydrophobicity. Not all the mutants fit this model; factors besides hydrophobicity and kinkiness can be important.

5 *XPR2* Regulation

Since the 1993 review, progress in understanding *XPR2* regulation has been substantial. The model was that partial derepression of AEP occurred when *Y. lipolytica* did not have access to preferred C-, N-, and S-sources. Protein in the medium had an inductive effect (Ogrydziak et al. 1977).

Deletion analysis of the *XPR2* promoter (p*XPR2*) identified *cis*-acting elements conferring high expression (Blanchin-Roland et al. 1994). A p*XPR2-lacZ* fusion integrated into the genome was used as a reporter for *XPR2* expression. Effects of deletions were measured in three media buffered with phosphate buffer (pH 6.8)—repressing (MMAm) containing glycerol (10 g/L), a good carbon source for *Y. lipolytica*, and ammonium sulfate (2 g/L); noninducing (YEg) with 1 % yeast extract and 0.1 % glucose; and inducing (YPDm) with 0.2 % yeast extract, 0.1 % glucose, and 5 % proteose peptone (PP). Two major UASs were identified—UAS1 between −822 and −769 and UAS2 between −208 and −140. Somewhat surprisingly, given the evolutionary distances, regulatory sequences similar to binding sites for *S. cerevisiae* transcription factors such as *GCN4* and *TUF*/*RAP1* were found in both UASs. In vivo footprinting revealed that proteins bound the UASs constitutively. The authors suggest that other regulatory proteins may modulate the activity of these bound proteins. Finally, *XPR2* expression was greatly reduced by deletion of both UASs, but regulation was retained.

The function of each UAS was defined more precisely using hybrid promoters consisting of p*XPR2* functional elements upstream of a minimal *LEU2* promoter (p*LEU2*) driving expression of the *lacZ* gene integrated into the genome (Madzak et al. 1999). UAS1A, corresponding to position -805 to -776 of p*XPR2*, carries *GCN4*-like and *TUF*/*RAP1*-like sites. UAS1B, corresponding to position -805 to -701, contains UAS1 and an adjacent region containing two imperfect repeats. UAS1B, and especially two copies of UAS1B (UAS1B2), enhanced β-galactosidase production more than UAS1A and UAS1A repeats on inducing medium. UAS1B and UAS1B2 also increased activity on noninducing and repressing media indicating that the UAS1B hybrid promoters were less affected by environmental conditions. Levels were similar in YPDm and YEg suggesting absence of a peptone sensing element.

Three DNA fragments from UAS2 were tested. UAS2A (position -149 to -124) carries a decameric repeat that overlaps two sites identical to *Aspergillus nidulans* PacC site. UAS2C (position -126 to -106) partially overlaps a *GCN4*-like site and contains a site matching the *S. cerevisiae ABF1*-site consensus sequence. UAS2B (position -150 to -106) includes both UAS2A and UAS2C. UAS2A did not activate expression on YPDm, but in one orientation UAS2B caused a fivefold increase. Two copies of UAS2B caused more than an additive increase. UAS2C alone did not activate expression. Mutations in the *GCN4*-like or *ABF1*-like sites of UAS2B revealed that only the *ABF1*-like site is necessary for UAS activity.

Comparison of activity on YPDm at pH 6.8 and 4.0 revealed an effect for UAS2A and UAS2B but not UAS2C, suggesting that the decameric repeat containing the putative binding site for Rim101p/PacC regulator was important for pH control. *YlRIM101* context did not affect UAS1-driven expression but has major effects on UAS2-driven expression.

Madzak et al. (1999) concluded that UAS1 can increase activity in all media, especially as number of copies increase and that UAS1 is "poorly sensitive to repressing conditions." UAS2 seems to be involved in regulation by C- and N-sources and pH regulation. They proposed that at neutral or alkaline pH, YlRim101p is activated by a C-terminal truncation. This form can bind to PacC-like sites and activate p*XPR2* expression through UAS2. However, preferred C- and N-sources prevent this UAS2-driven expression suggesting that binding of another regulatory protein impairs YlRim101p activation.

Deletion of the UASs maintained regulation but at a much lower level, suggesting that they were not involved in regulation. But the hybrid promoters unexpectedly showed regulation similar to that of intact p*XPR2*. The authors attribute this to "compensation effects from other sequences still present in the deleted promoter" (Madzak et al. 1999).

The first search for *trans*-acting factors affecting AEP production concerned pH regulation (Otero and Gaillardin 1996). Strains with p*XPR2:lacZ* and p*XPR2:hph* fusions are white on YPD-X-Gal and hygromycin sensitive at pH 4.0. Nine stable mutants (PH1-9) that were blue and hygromycin resistant at pH 4.0 defined two RPH loci. The PH mutations seemed to affect only pH and not metabolic regulation of *XPR2*.

trans-acting regulatory mutations that prevent either p*XPR2*-driven expression under conditions of C- and N-limitation at pH 6.8, i.e., derepressing medium Y (1 % yeast extract), or force expression under repressing conditions were isolated (Lambert et al. 1997). To avoid *cis*-acting mutations, two reporters were used, production of AEP by *XPR2* and a p*XPR2-lacZ* fusion. Mutants that were Lac- and Xpr- on Y6.8 medium were considered nonderepressible. Seven recessive monogenic mutations defined four complementation groups—*PAL1–PAL4*. The *pal1*, *2*, and *3* mutants were transformed with a replicating vector library to find genes that restore the Lac+ phenotype on Y-X-Gal 6.8. Plasmid pINA935 complemented only *pal2* mutations. pINA937 suppressed all four *pal* mutations and contained a truncated version of a gene designated *YlRIM101* based on homology to *S. cerevisiae* Rim101p and *A. nidulans* PacC.

YlRIM101p was shown to be absolutely required for transcriptional activation of *XPR2* at pH 6.8. *XPR2* expression at pH 4.0 was undetectable in the *YlRIM101* disruptant indicating that YLRim101p was not required for shutdown of expression in these conditions. C-terminal truncations of *YLRIM101* activated *XPR2* transcription.

A search was done for mutants (made using an mTnYl1-transposase library) that affected *AXP1* and/or *XPR2* expression to identify additional components of the pH signaling pathway and pH-independent regulators (Gonzalez-Lopez et al. 2002). To avoid *cis*-acting mutations, AEP and LacZ were assayed under p*XPR2* control and AXP and GusC under p*AXP1* control. In derepressing medium (Y) at pH 7.0, the parental strain made clear zones on SKM plates (Aep+) and turned blue on Y-X-Gal7 medium (Lac+). At pH 4.0, it cleared zones on BSA plates (Axp+) and turned blue on Y-Gluc4 medium (Gus+).

Around 190,000 transformants were screened. The four affecting only *XPR2* expression were *ScSIN3* homologs. Thirty mutations affecting only *AXP1* activity were not extensively characterized. The 89 mutations that affected expression of both proteases lead to identification of ten genes. Five were homologs of genes in the Rim pH signaling pathway. Five other genes were not involved in pH signaling—*OPT1*, *SSY5*, *VPS28*, *NUP85*, and *MED4*.

Possible roles of YlSin3p (component of a histone deacetylase), YlNup85p (nuclear pore protein), and YlMed4p (subunit of RNA polymerase II mediator complex) in extracellular protease regulation are presently unclear. YlVps28p is a component of the ESCRT-I complex, and there is now evidence suggesting that Rim and VPS pathways cooperate in ambient pH signaling possibly explaining its role in extracellular protease regulation (Blanchin-Roland et al. 2005).

Mutations affecting *YlOPT1* and *YLSSY5* nearly abolished *XPR2* and *AXP1* transcription but did not affect transcription of another pH-dependent gene. Gonzalez-Lopez et al. (2002) suggested that these genes may affect amino acid sensing. *OPT1* codes for a proton-coupled oligopeptide transporter in *S. cerevisiae*. Interestingly, the *OPT1*-related family of proteins is highly represented in *Y. lipolytica* (17 genes); the seven other yeast in Genolevures have only two to six genes (Sherman et al. 2009).

Possibly, the large number of potential oligopeptide transporters reflects *Y. lipolytica's* nutritional dependence on products of protein degradation by extracellular proteases. If several of these potential oligopeptide transporters are expressed, finding only *OPT1* mutations suggests that its primary role in protease regulation may not be oligopeptide transport. In *S. cerevisiae*, some of the numerous potential glucose transporters turn out to be glucose sensors (Ozcan et al. 1996). By analogy, perhaps YlOpt1p is an oligopeptide sensor, part of a signaling pathway that senses N-availability and/or indirectly inductive proteins/peptides in the medium.

In *S. cerevisiae*, Ssy5p is associated with Ptr3p and the permease-like Ssy1p in the plasma membrane. This complex detects the presence of extracellular amino acids and activates, by endoproteolytic processing, the membrane-bound transcription factor Stp1 to activate a signaling pathway leading to transcriptional induction of genes like AGP1—an amino acid permease (Abdel-Sater et al. 2004). Ssy5p has been shown to be the endoprotease responsible for Stp1p processing. It has been suggested that a similar mechanism involving an Ssy5p-like protein might operate in *Y. lipolytica* (Gonzalez-Lopez et al. 2002). It will be interesting to determine if YlOpt1p and YlSsy5p are part of the same signaling pathway.

YlSSY5 (YALI0E04400g) has strong homology to *ScSSY5* (6e-88). The His, Asp, and Ser triad characteristic of the serine protease active site and surrounding residues is highly conserved, as is its putative self-endoproteolytic processing site (Abdel-Sater et al. 2004). The best candidate for *YlPTR3* is YALI0D02673g with two non-overlapping homologous regions—(2e-45; 85/218) and (2e-21; 93/190). The best candidates for *YlSSY1*, YALI0B19338g (2e-66) and YALI0C00451g (9e-63), and for *YlSTP1*, YALI0E24937g (1e-36) and YALI0B05478g (8e-20), are not as convincing.

In conclusion, progress has been significant on pH and *XPR2* regulation, including some on protein/peptide induction.

6 *AXP1* Regulation

Compared to *XPR2*, much less is known about *AXP1* regulation. AXP is produced during exponential growth. Production was higher with proteins and peptones than amino acids and ammonium ion (Yamada and Ogrydziak 1983). The effects of C- and S-availability have not been specifically studied. *AXP1* expression is also regulated by pH. *AXP1* mRNA expression was maximum at pH 5.5 and decreased as neutral pH was approached (Glover et al. 1997). There are three copies of the core PacC hexanucleotide in the *AXP1* promoter (Lambert et al. 1997).

pH regulation was studied using continuous cultures and pH-controlled batch cultures (Glover et al. 1997). *AXP1* mRNA was found from pH 4.5–6.5 and *XPR2* mRNA from pH 5.5–7.5. At alkaline pH, a strong band for *XPR2* and no discernible band for *AXP1*, and at acidic pH, a strong band for *AXP1* and no discernible band for *XPR2*, were found on Northern blots for pH 6.8 versus pH 4.5 (Lambert et al. 1997) and pH 7.0 versus 4.0 (Gonzalez-Lopez et al. 2002).

In the batch culture, AEP activity was produced at pH as low as 5.0 where there was little *XPR2* mRNA. The drop off in AXP activity was faster than for *AXP1* mRNA levels at pH 6.5 and pH 7.0. Glover et al. (1997) suggested that AXP may be degraded by AEP at these pHs. Surprisingly, stability of AEP to AXP and vice versa at different pHs has never been directly measured using purified enzymes.

The Rim pathway is involved in *AXP1* regulation. The model, explored in the most detail in *A. nidulans*, is that activated (C-truncated) Rim101p would be predicted to induce alkaline genes such as *XPR2* and to repress acidic genes such as *AXP1* at neutral/alkaline pH. Unexpectedly, if the Rim pathway is disrupted, *AXP1* is not expressed at pH 7.0 (Gonzalez-Lopez et al. 2002). It was proposed that at pH 7.0, where AXP is inactive, induction cannot occur because production of a specific *AXP1* inducer is dependent on AEP activity.

The model predicts that at acidic pH, little or no Rim101p is activated and alkaline genes are not induced and acidic genes not repressed. As predicted, no *XPR2* mRNA was detected at pH 4.0. Surprisingly, when the Rim pathway was disrupted, reduced levels of *AXP1* mRNA were detected at pH 4.0 compared to wild type suggesting that under acidic conditions full induction of *AXP1* requires the Rim pathway (Gonzalez-Lopez et al. 2002). The authors proposed that induction of *AXP1* is not dependent on Rim pathway for pH sensing but for interpretation of other signals, possibly nutrient limitation.

Both *ssy5* and *opt1* mutants make no *AXP1* mRNA at pH 4.0 suggesting that the signaling pathway(s) involving these genes has a role in *AXP1* expression.

7 Heterologous Protein Expression

Extracellular proteases have played a major role in heterologous protein production. The *XPR2* prepro-region has provided localization and processing signals. Heterologous expression using p*XPR2* in a *Δaxp1 Δxpr2* host was commonly done using YPDm medium at pH 6.8. The host cannot degrade the proteins in YPDm. However, 5.0 % PP provides enough low molecular weight C- and N-sources for cell growth. When these become limiting late in log phase, growth slows, p*XPR2* is induced, and heterologous protein expression starts. In contrast, for Xpr2+ strains growing in medium with lower levels of PP, AEP production starts early in log phase and basically stops in stationary phase.

Minimal medium is less expensive than YPDm, and it simplifies product purification from undegraded proteins and peptides. UAS1B hybrid promoters were less affected by environmental conditions, and a promoter consisting of four UAS1B fragments with a minimal downstream region from p*LEU2* made use of minimal media possible (Madzak et al. 2000). Expression was quasi-constitutive in minimal MMAm medium and 60–85 % of that for native p*XPR2* on YPDm medium.

8 Other Extracellular Proteases

The MEROPS Peptidase Database (Release 9.1) was used to predict potential extracellular proteases in *Y. lipolytica* CLIB122 (Rawlings et al. 2010). There are 131 known and putative peptidase entries. The MEROPS definition of peptidases includes broad specificity endoproteinases (emphasized in this chapter) and exopeptidases that remove one or two amino acids at a time. The entries were screened for signal peptides using SignalP 3.0. Proteins predicted to have signal peptides by both NN and HMM (probability ≥0.900) were considered to be potentially secreted.

MEROPS defines 12 peptidase families and reports frequency of family members versus frequency in genomes of the other sequenced fungi. In *Y. lipolytica*, three families had undergone lineage-specific expansion. *Y. lipolytica* has 34 homologs in A1 (pepsin family)—endoproteinases, usually active at acidic pH, with an aspartic active site. The next highest is *Neurospora crassa* with 18, and *S. cerevisiae* has only 11. *Y. lipolytica* has 19 homologs in the S8 (subtilisin family)—peptidases homologous to the serine endoproteinase subtilisin. *N. crassa* has 10 and *S. cerevisiae* 11. The third expanded family is the serine carboxypeptidase family S10.

Family A1 includes *AXP1* and has 30 potentially secreted members (Table 1). Two had strongest homology to *ScPEP4* and are probably vacuolar. Fourteen had strongest homology to *S. cerevisiae* yapsin and may be GPI anchored at the cell surface. And one had strongest homology to *ScBAR1*.

YALI0E33363g (blastp score of 1e-92) and YALI0D10967g (2e-88) were most similar to *AXP1* (YALI0B05654g) in sequence and predicted protein size. For all three, the predicted signal peptide cleavage site was identical, the next two amino acids were Ala.Pro, and a Phe.Ala autoprocessing site was similarly located. Based on microarray analysis of AEP induction in strains carrying multiple copies of *XPR2* (see below), *AXP1* and YALI0E33363g provided sufficient signal to determine expression ratios at all time points but YALI0D10967G only at one. It may be poorly expressed under these conditions.

There is a large drop off in sequence homology with YALI0B20526g (3e-31) as next highest. Of the 30 pepsin-like genes coding for potentially secreted proteases, 21 provided sufficient signal at all time points to determine expression ratios. The nine remaining genes might be expressed in other conditions.

Four genes (not included in MEROPS) that code for potentially secreted pepsin-like proteins with the characteristic catalytic active site were identified. Three had strongest homology to *S. cerevisiae* yapsin and all weak homology to *AXP1*.

Yamada and Ogrydziak (1983) found three extracellular acid proteases, but Nelson and Young (1987) found only one. The three proteases were separated by ion exchange chromatography and had similar molecular masses—40.3–40.7 kDa—determined by SDS-PAGE. Carbohydrate contents were 25 %, 12 %, and 1.2 % for proteases I, II, and III, respectively. Isoelectric point was 3.8 for protease III and 4.9 for protease I. Based on molecular mass, isoelectric point, and carbohydrate content, protease III corresponds to AXP. Arguments in favor of proteases YALI0E33363p and YALI0D10967p being proteases I and II are similar predicted masses of the mature molecules (37.4–37.9 kDa) and AXP-like location for potential Phe-Ala autoprocessing sites. Also the predicted isoelectric point for mature YALI0D10967p is 4.9, identical to that of protease I. An argument against identity of these proteases is that YALI0E33363p has no potential N-linked glycosylation site and YALI0D10967p only one. If carbohydrate content differences were due to O-linked glycosylation, then SDS-PAGE mobility should differ significantly. Therefore, identity would require that the reported carbohydrate contents of

Table 1 A1 family—potentially secreted pepsin-like proteases

Gene	**Name**[a] Homolog	MEROPS	blastp versus *AXP1*	Amino acids	Signal peptide cleavage site NN	HMM	Microarray data[b]
YALI0B05654g	***AXP1***	MER003667	0.0	397	17–18	17–18	Yes
YALI0E33363g		MER101133	1e-92	397	17–18	17–18	Yes
YALI0D10967g		MER175599	2e-88	403	17–18	17–18	No
YALI0B20526g	*Ct ASP*	MER101142	3e-31	392	20–21	20–21	No
YALI0C20273g	*CtASP2*	MER101138	2e-28	385	17–18	26–27	No
YALI0D14300g	*ScYPS3*	MER101135	8e-28	384	17–18	17–18	Yes
YALI0A02002g	*ScYPS3*	MER101143	5e-27	443	22–23	22–23	No
YALI0B00374g	*S. fib.* ASP	MER101140	7e-27	389	18–19	23–24	Yes
YALI0F10549g	*ScYPS3*	MER101130	2e-24	450	17–18	17–18	Yes
YALI0E25784g	*CtASP4*	MER177669	1e-23	393	17–18	17–18	Yes
YALI0B00132g	*S. fib.* ASP		8e-23	476	17–18	23–24	Yes
YALI0C08547g	*CaASP3*	MER101137	2e-22	385	17–18	17–18	No
YALI0E10175g	*ScYPS3*	MER071463	2e-21	534	17–18	17–18	Yes
YALI0D01331g	*ScYPS3*	MER184679	4e-20	457	17–18	17–18	Yes
YALI0B20174g	*ScYPS3*	MER101141	6e-19	393	17–18	17–18	Yes
YALI0D10835g	*A. ory* ASP	MER175569	3e-17	778	18–19	18–19	Yes
YALI0F27071g	*ScPEP4*	MER107343	8e-17	396	18–19	18–19	Yes
YALI0C14938g	*ScYPS3*	MER184465	1e-16	447	17–18	17–18	Yes
YALI0E22374g	*ScYPS3*	MER071464	2e-15	727	15–16	15–16	Yes
YALI0D17270g	*ScYPS3*		8e-14	391	14–15	18–19	No
YALI0C08899g	*CtASP4*	MER184463	1e-13	359	18–19	18–19	Yes
YALI0E13860g	*ScYPS3*	MER101131	3e-13	378	20–21	20–21	No
YALI0D22957g	*ScYPS3*	MER184467	7e-13	455	17–18	22–23	No
YALI0A16819g	*CaASP4*	MER172235	1e-11	457	18–19	18–19	Yes
YALI0F09163g	*ScYPS3*		3e-11	388	23–24	23–24	Yes
YALI0C10923g	*ScYPS3*		1e-09	416	14–15	25–26	Yes
YALI0D02024g	*ScPEP4*	MER101134	6e-09	625	17–18	17–18	Yes
YALI0C10135g	*ScYPS3*	MER184405	4e-08	411	13–14	16–17	Yes
YALI0C08283g	*CtASP4*	MER101136	9e-07	374	15–16	20–21	No
YALI0E11715g	*ScBAR1*	MER184468	>7.7	394	18–19	18–19	Yes

[a]Gene name is in bold. *Ct*, *Candida tropicalis*; *Sc*, *S. cerevisiae*; *S. fib.*, *Saccharomycopsis fibuligera*; *Ca*, *Candida albicans*; *A. ory.*, *Aspergillus oryzae*; ASP, acid secreted protease
[b]Expressed in AEP induction conditions based on microarray data

proteases I and II be incorrect. Also *Δaxp1* strains made no zone of clearing on BSA (pH 4.0) plates (Otero and Gaillardin 1996). YALI0D10967p and YALI0E33363p would have to be expressed at sufficiently low levels on these plates that no zones of clearing could be detected.

Family S8 includes *XPR2* (YALI0F31889g) and has 16 potentially secreted members (Table 2). Of these, two had strongest homology with *ScPRB1* and with each other (e-170) and may be vacuolar, and one was Xpr6p thought to be Golgi located. YALI0A08360g (not in MEROPS) has strong sequence similarity (5e-98) with *XPR2* including the Asp, His, and Ser catalytic triad.

Table 2 S8 family—potentially secreted subtilisin-like proteases

Gene	Name[a] Homolog	MEROPS	blastp versus *XPR2*	Amino acids	Signal peptide cleavage site NN	HMM	*XPR6* site	Microarray data[b]
YALI0F31889g	***XPR2***	MER000340	0.0	454	15–16	21–22	Yes	Yes
YALI0D02981g		MER191803	1e-167	450	15–16	21–22	Yes	Yes
YALI0A09262g	***AEP2***	MER055295	1e-102	544	15–16	15–16	No	Yes
YALI0A08360g			5e-98	533	15–16	15–16	No	Yes
YALI0B22880g		MER191800	1e-80	467	15–16	20–21	Yes	Yes
YALI0B19316g		MER191799	3e-74	453	23–24	15–16	No	Yes
YALI0C15532g		MER191801	3e-71	892	19–20	19–20	No	No
YALI0E28875g		MER191804	9e-66	475	16–17	16–17	Yes	Yes
YALI0C20691g		MER191802	4e-61	512	19–20	19–20	No	Yes
YALI0A10208g		MER167710	3e-58	419	17–18	17–18	No	No
YALI0B16500g	*ScPRB1*	MER191798	1e-55	516	18–19	18–19	No	Yes
YALI0A06435g	*ScPRB1*	MER167615	6e-48	471	16–17	18–17	No	Yes
YALI0F19646g		MER191806	1e-46	415	14–15	19–20	No	No
YALI0B02794g	*Dh* SP	MER191797	1e-39	379	19–20	19–20	No	Yes
YALI0F24453g	*Ca* SP	MER191807	7e-37	440	15–16	15–16	No	No
YALI0F13189g	***XPR6***	MER000366	5e-01	976	17–18	17–18	–	Yes

[a]Gene name is in bold. *Sc*, *S. cerevisiae*; *Dh*, *Debaryomyces hansenii*; *Ca*, *Candida albicans*; *SP*, serine protease
[b]Expressed in AEP induction conditions based on microarray data

YALI0D02981g has strongest homology to *XPR2* (e-167). The sizes are very similar, the predicted signal peptide cleavage sites are identical, there are four X-Ala, X-Pro dipeptides versus nine for AEP, and N-linked glycosylation and dibasic processing (Lys.Arg) sites appear in similar locations.

YALI0A09262g (AEP2) (e-102) and YALI0A08360g (5e-98) are the next closest homologs to *XPR2*. They are more closely related in size and sequence (1e-0.0) to each other than to *XPR2*. Like *XPR2*, the predicted signal peptide cleavage sites are 15–16. Neither has an N-linked glycosylation or Lys.Arg site in appropriate locations to mimic AEP processing.

YALI0B22880g has similar size and the next strongest homology (1e-80). It has identical predicted signal peptide cleavage site (15–16) using NN and near identical (20–21 versus 21–22) using HMM. Both a potential N-linked glycosylation and Lys.Arg sites are in similar locations compared to AEP.

Except for YALI0C15532g, the remaining nine potentially secreted S8 family proteins are similar in size to AEP. Blastp scores range from 7e-37 to 3e-74. Only one had the 15–16 signal peptide cleavage site. None had the appropriately located N-linked glycosylation and Lys.Arg processing sites.

It is not known how many of this S8 family are actually expressed and secreted. In microarray experiments (see below), only four did not provide sufficient signal to determine expression ratios at all time points. Starting with only the sequencing information, it would have been a daunting task to knock out all the potentially secreted S8 family member genes to stabilize secreted heterologous proteins. However, *Δxpr2* strains make no zone of clearing on SKM plates (pH 6.8), and including *Δxpr2* in host strains seems sufficient to largely stabilize secreted heterologous proteins. *AEP2* has been disrupted in a *Δxpr2* background, and no phenotypic difference for heterologous protein production was observed (J.Y. Kim, personal communication).

The remaining potentially secreted peptidases were primarily amino- and carboxypeptidases but also included homologs to peptidases involved in mating factor processing.

It would be interesting to determine how expression of S8 family members is regulated, where they are located, and their biological roles.

9 Regulation of Secretory Pathway Components

Synonymous codon usage was more biased for *Y. lipolytica* than for *S. cerevisiae* for genes involved in protein targeting and secretion, suggesting that *Y. lipolytica* might produce these components at much higher levels. The hypothesis that *Y. lipolytica* may regulate components of the secretory pathway in response to secretory demand was tested by overexpressing AEP and measuring changes in genomic transcription.

9.1 Construction and Screening of XPR2 *Multicopy Strains*

Multiple copies of *XPR2* can be integrated by targeting plasmid pINA773 (containing *XPR2*, *URA3d* with a greatly truncated promoter, a fragment of *Y. lipolytica* rDNA, and a pBR322 fragment) to rDNA (Le Dall et al. 1994). Because of the truncated promoter, only multicopy integrants make enough Ura3p to support growth. Strains with 25–60 copies of *XPR2* were transferred from a rich medium at pH 4.0 (*XPR2* is repressed) to a glucose/proteose peptone (GPP) medium at pH 6.8. The strains grew rapidly for 10–11 h, but then AEP was induced and growth slowed for several hours before increasing again. By this time *XPR2* copy numbers had decreased.

To avoid loss of *XPR2* gene copies, a method for more rapid induction was developed. Major modifications were to induce at 23 °C where AEP differential productivity ((ΔUnits/ml)/(Δcell dry weight/ml)) is highest (Dedeoglu, E., unpublished results) and to replace PP with a 90 % dialyzed PP/10 % PP mixture. This removed most of the low molecular weight N-compounds that repress *XPR2* expression. In the new induction conditions, AEP production for strain 773–2 (containing 40–50 copies of *XPR2*) had started an hour after transfer and growth slowed by 3 h. Considering the short time and that cell mass had increased less than 50 % at 3 h, little decrease in *XPR2* copy number should have been possible.

Strain CX161-1B grew more rapidly, produced more AEP, and had higher cell yields than 20–12, parent of 773–2 (unpublished data). Also growing inoculum in a defined synthetic complete (SC) medium at pH 6.8 with ammonia as the N-source would more reliably repress AEP synthesis and avoid the large pH change (4.0–6.8) on induction. Both 20–12 and 773–2, related to inbred E129 strains, utilize ammonia very poorly. Therefore, *XPR2* multicopy strains were constructed with a CX161-1B-related strain as parent. These strains grew more slowly than expected in SC medium. W29 (isolated from the sewers of Paris) produced slightly more AEP and grew much better in SC medium than CX161-1B (unpublished results). Therefore, W29-related *XPR2* multicopy strains and additional CX161-1B *XPR2* multicopy strains were constructed.

About 200 transformants were prescreened by streaking on YLT-ura plates (SC supplemented with amino acids and nucleotides). AEP production of 64 isolates was estimated by measuring the diameters of colonies and zones of clearing for cells spotted on SKM plates. Growth on YLT-ura plates was estimated by colony diameter. Plates were incubated at 23 °C and measurements done every 24 h.

The diameters of zones of clearing and colonies on SKM plates were slightly larger for W29 compared to CX161-1B. For *XPR2* multicopy W29 transformants, zones of clearing were comparable to those for parental W29, but colony diameters were always smaller, consistent with the *XPR2* multicopy strains producing more AEP per cell. Colony sizes on YLT-ura were slightly larger for W29 than CX161-1B. For most of the *XPR2* multicopy transformants, colonies on YLT-ura were noticeably smaller than for the parents.

AEP activity can be estimated by measuring the zones of clearing on SKM plates around wells filled with liquid samples. The logarithm of AEP concentration is linearly related to the diameter of the zone of clearing (Ogrydziak and Scharf 1982). AEP production was estimated using the diameter of the zone of clearing around the colony and the log conversion. Colonies were disc shaped, and cell numbers were estimated from the area. Fifteen transformants were screened several times. The eight producing the highest levels of AEP were screened further in liquid culture. All were derived from W29.

Cells were grown overnight in defined medium at pH 6.8 and resuspended at ~1 g/L dry weight cells in induction medium in flasks incubated at 23 °C. Klett readings, AEP well assays on SKM plates, and pH measurements were done. Strains 3 and 44 were chosen for further study. Compared to W29, AEP production by strain 3 was 6-fold higher at 12 h and 12.4-fold higher at 24 h. These were among the highest values. The increase in AEP production of strain 44 compared to W29 (6.2-fold) was similar to that for strain 3 at 12 h and somewhat lower (9.2-fold) at 24 h. However, growth after transfer of strain 44 slowed the least of any of the high-producing transformants. Significant AEP overproduction combined with more wild type-like growth rates would be of interest because some changes in gene expression after AEP induction may not be specific for secretion stress but due to decreases in growth rate.

9.2 AEP Levels

Ten hours after transfer, AEP production (AEP-related protein/dry cell weight) for W29 is about 2.3 % of total protein. This is calculated using 9,000 U = 1 mg of AEP (Ogrydziak and Scharf 1982), a 1.486 mass ratio of pro-mAEP (the propeptide is also secreted) to mAEP, and the assumption that cells are 50 % protein. The differential rate of AEP production ((ΔAEP-related protein/ml)/(Δcell protein/ml)) varied from 3.8 to 3.2 % over 10 h.

For strain 3 at 10 h, average pro-mAEP production was 5.6 %. At 24 h, it was 11.5 %. Values were much higher for differential rates of pro-mAEP production—16.4 % for 0–3 h and as high as 29.3 % for 16–28 h.

Estimating secretory stress on a cell is difficult. Even when cells are not secreting proteins into the extracellular medium, there is substantial traffic through the secretory pathway to support cell growth. The percentage of total protein synthesized that enters the secretory pathway is unknown. If it were 30 % and if AEP is 10–15 % of total protein synthesized, then AEP induction would increase secretory demand by 33–50 %. Heterogeneity of AEP production from cell to cell would result in heterogeneity of secretion stress. In any case, for strain 3, AEP induction leads to slowing of cell growth (td of 8.5 h from 3 to 7 h versus 3.6 h for W29 over the same time period) and a decrease of the differential rate of AEP production after 3 h.

For strain 44 at 10 h, average pro-mAEP production was 9.4 %. Differential rates were 11.4 % for 0–3 h, 12.1 % for 3–6 h, and as high as 20.6 % for 8–10 h. Differential rates were lower than for strain 3 from 0 to 3 h, actually slightly increased after 3 h, and growth did not slow as much (td of 4.1 h from 3 to 7 h).

9.3 Transcriptional Response to Secretion Stress

A DNA microarray experiment was done to determine what genes responded to secretion stress. Cells were transferred to AEP induction medium and samples taken at various times. For strain 3, based on quality/quantity of the RNA preparations, growth rates, and AEP levels, the 1.5, 4, 6, 12, 16, and 24 h samples were used. For strain 44, 2, 4, 6, 7, and 9 h samples were used. The earliest time point (control) was not taken right after transfer but when AEP was already induced and before growth had slowed. Comparison of gene expression levels when growth starts to slow with this time point was expected to more specifically address how cells adapt to secretion stress.

Genes were considered to be differentially expressed based on SAM analysis (Tusher et al. 2001), EDGE analysis (Leek et al. 2006), or if at least three time points had a threefold difference in expression. There was data for at least three times points for 5,915 genes for strain 3 and 5,216 genes for strain 44. Fifty genes were induced in both strains. For strains 3 and 44, 84 and 62 genes were uniquely induced, respectively.

For both strains, none of the ten most strongly induced genes at 4 h were secretory pathway genes. Some make biological sense—an ammonium transporter and two vacuolar proteases. Most presently do not.

The only induced secretory pathway-related genes are homologs of *VPS13*, *SSA4*, and *SSO1*, and induction was strongest at later time points. Overexpression in *S. cerevisiae* of *S. cerevisiae* and *K. lactis SSO1* homologs, plasma membrane t-SNAREs involved in secretory vesicle fusion, enhanced production of secreted proteins, suggesting they may be a rate-limiting components for protein secretion (Toikkanen et al. 2004). Four potential transcription factors were induced—*HOY1*, *RIM101*, and homologs of *ScCAT8* and ScGZF3—but again induction was strongest at later time points.

Besides *XPR2*, several potentially secreted proteases were induced—the *ScPRB1* homolog (YALI0B16500g) in both strains and the *ScPEP4* homolog (YALI0F27071g) in strain 44. The *ScBAR1* homolog (YALI0B00374g) was induced in strain 3 and YALI0A08360g (more closely related to *AEP2* than *XPR2*) in strain 44.

The differentially expressed genes were analyzed by GO Term Finder (http://go.princeton.edu/cgi-bin/GOTermFinder); it identifies significantly enriched genes in a list of genes. This is not possible directly in *Y. lipolytica*. Therefore, *S. cerevisiae* genes with blastp score $\leq$1e-20 against their *Y. lipolytica* homologs were included in the gene list and analyzed against overall GO term frequencies for *S. cerevisiae*. As discussed above, gene frequencies for certain classes of genes can differ significantly between the two yeasts. Of the 50 genes induced in both strains, 31 met the criterion. Cation transport was the only process ontology term enriched. Six genes were annotated to the term including two ammonium permeases.

For strain 44, of the 112 genes induced, 71 made the list. Transmembrane transport was the only process ontology term enriched. Fourteen genes were annotated to the term including nine transporters of low molecular weight N-sources.

Eighty of 134 genes induced in strain 3 made the list. Fifteen process ontology terms were enriched. The same 18 genes were annotated to four terms including carboxylic acid, oxoacid, cellular ketone, and organic acid metabolic processes. Eleven genes were annotated to ion transport including zinc, copper, and lactate transporters and three involved with ammonium, also annotated to nitrogen utilization and ammonium transport. Thirteen genes were annotated to response to abiotic stress including *ATG8*, *PRB1*, *CTT1*, and *HSP31*.

Far more genes (250) were repressed in both strains. For strains 3 and 44, 229 and 74 genes were uniquely repressed, respectively. For both strains, the ten most repressed genes at 4 h had no obvious connection to protein secretion. No potentially secreted proteases were repressed, and three transcription factors (*MHR1*, *YDR296W* and *ZPR1*) were repressed in both strains. Several heat shock genes, cochaperones, and *EGD1* and *EGD2* (nascent polypeptide-associated complex (NAC) subunits involved in protein targeting) were repressed in both strains. In strain 3, unexpectedly *SRP21* and *SRP14*, components of the signal recognition particle, were also repressed. In strain 44, *SEC61*, a component of the translocon was repressed.

Many of the repressed genes corresponded to genes whose expression level was positively correlated with growth rate in *S. cerevisiae* (Brauer et al. 2008), suggesting that their repression is a response to the slower growth rate and not specific for secretion stress. GO analysis terms were enriched in genes involved in ribosomal (both mitochondrial and cytosolic) biogenesis and assembly, ribosomal constituents, rRNA processing, translation, and mitochondrial protein import.

More genes are repressed than induced after AEP induction in *XPR2* multicopy strains. Unfortunately, the time points chosen did not allow an unambiguous answer to the question of whether or not the unfolded protein response (Ron and Walter 2007) is induced. Surprisingly, for strain 3 the increase in *XPR2* expression was only two- to threefold at earlier time points, and for strain 44 there was little increase in expression. The ten genes with the highest average fluorescent intensity at the earliest time point (control) were identified. Although several factors may affect intensity, high intensity should reflect a high mRNA level. *XPR2* ranked first for strain 44 and first or second (depending on the chip) for strain 3, suggesting that *XPR2* was already highly expressed and almost fully induced in the control samples.

KAR2 and *PDI1* are induced in UPR. Except for the 4 h sample for strain 44, there was no evidence of induction of these two genes. For strain 3, there was evidence of repression for *KAR2* at later time points. However, for strain 3 *KAR2* ranked seventh for average fluorescent intensity at 1.5 h, again suggesting it was already almost fully induced. *PDI1* appears to be expressed at more moderate levels. Unfortunately, much of the evidence for UPR-like response may have been missed if it had already occurred when the control sample was taken. It is possible that UPR and other responses to secretion stress such as increases in ER and translocon components had already happened. The repression of *SEC61*, *SRP14*, and *SRP21* and induction of the autophagy gene *ATG8* and vacuolar proteases *PRB1* and *PEP4* might reflect a turning down of this response. It would be interesting to (1) use an earlier control sample and examine more closely spaced time points, (2) follow AEP induction in a strain with a single copy of *XPR2*, and (3) determine if ER had increased as part of the response.

References

Abdel-Sater F, El Bakkoury M, Urrestarazu A, Vissers S, Andre B (2004) Amino acid signaling in yeast: casein kinase I and the *Ssy5* endoprotease are key determinants of endoproteolytic activation of the membrane-bound Stp1 transcription factor. Mol Cell Biol 24:9771–9785

Ahearn DG, Meyers SP, Nichols RA (1968) Extracellular proteinases of yeasts and yeastlike fungi. Appl Microbiol 16:1370–1374

Barth G, Gaillardin C (1996) *Yarrowia lipolytica*. In: Wolf K (ed) Non-conventional yeasts in biotechnology a handbook. Springer, Berlin, pp 313–388

Bendtsen JD, Nielsen H, von Heijne G, Brunak S (2004) Improved prediction of signal peptides: SignalP 3.0. J Mol Biol 340:783–795

Blanchin-Roland S, Cordero Otero RR, Gaillardin C (1994) Two upstream activation sequences control the expression of the *XPR2* gene in the yeast *Yarrowia lipolytica*. Mol Cell Biol 14:327–338

Blanchin-Roland S, Da Costa G, Gaillardin C (2005) ESCRT-I components of the endocytic machinery are required for Rim101-dependent ambient pH regulation in the yeast *Yarrowia lipolytica*. Microbiology 151:3627–3637

Brauer MJ, Huttenhower C, Airoldi EM, Rosenstein R, Matese JC, Gresham D, Boer VM, Troyanskaya OG, Botstein D (2008) Coordination of growth rate, cell cycle, stress response, and metabolic activity in yeast. Mol Biol Cell 19:352–367

Davidow LS, O'Donnell MM, Kaczmarek FS, Pereira DA, DeZeeuw JR, Franke AE (1987) Cloning and sequencing of the alkaline extracellular protease gene of *Yarrowia lipolytica*. J Bacteriol 169:4621–4629

Enderlin CS, Ogrydziak DM (1994) Cloning, nucleotide sequence and functions of *XPR6*, which codes for a dibasic processing endoprotease from the yeast *Yarrowia lipolytica*. Yeast 10:67–79

Fabre E, Tharaud C, Gaillardin C (1992) Intracellular transit of a yeast protease is rescued by trans-complementation with its prodomain. J Biol Chem 267:15049–15055

Glover DJ, McEwen RK, Thomas CR, Young TW (1997) pH-regulated expression of the acid and alkaline extracellular proteases of *Yarrowia lipolytica*. Microbiology 143(Pt 9):3045–3054

Gonzalez-Lopez CI, Szabo R, Blanchin-Roland S, Gaillardin C (2002) Genetic control of extracellular protease synthesis in the yeast *Yarrowia lipolytica*. Genetics 160:417–427

Hu Z, Haghjoo K, Jordan F (1996) Further evidence for the structure of the subtilisin propeptide and for its interactions with mature subtilisin. J Biol Chem 271:3375–3384

Lambert M, Blanchin-Roland S, Le Louedec F, Lepingle A, Gaillardin C (1997) Genetic analysis of regulatory mutants affecting synthesis of extracellular proteinases in the yeast *Yarrowia lipolytica*: identification of a RIM101/pacC homolog. Mol Cell Biol 17:3966–3976

Le Dall MT, Nicaud JM, Gaillardin C (1994) Multiple-copy integration in the yeast *Yarrowia lipolytica*. Curr Genet 26:38–44

Leek JT, Monsen E, Dabney AR, Storey JD (2006) EDGE: extraction and analysis of differential gene expression. Bioinformatics 22:507–508

Madzak C, Blanchin-Roland S, Cordero Otero RR, Gaillardin C (1999) Functional analysis of upstream regulating regions from the *Yarrowia lipolytica XPR2* promoter. Microbiology 145 (Pt 1):75–87

Madzak C, Treton B, Blanchin-Roland S (2000) Strong hybrid promoters and integrative expression/secretion vectors for quasi-constitutive expression of heterologous proteins in the yeast *Yarrowia lipolytica*. J Mol Microbiol Biotechnol 2:207–216

Matoba S, Ogrydziak DM (1989) A novel location for dipeptidyl aminopeptidase processing sites in the alkaline extracellular protease of *Yarrowia lipolytica*. J Biol Chem 264:6037–6043

Matoba S, Ogrydziak DM (1998) Another factor besides hydrophobicity can affect signal peptide interaction with signal recognition particle. J Biol Chem 273:18841–18847

Matoba S, Fukayama J, Wing RA, Ogrydziak DM (1988) Intracellular precursors and secretion of alkaline extracellular protease of *Yarrowia lipolytica*. Mol Cell Biol 8:4904–4916

Matoba S, Morano KA, Klionsky DJ, Kim K, Ogrydziak DM (1997) Dipeptidyl aminopeptidase processing and biosynthesis of alkaline extracellular protease from *Yarrowia lipolytica*. Microbiology 143(Pt 10):3263–3272

McEwen RK, Young TW (1998) Secretion and pH-dependent self-processing of the pro-form of the *Yarrowia lipolytica* acid extracellular protease. Yeast 14:1115–1125

Nelson G, Young TW (1987) Extracellular acid and alkaline proteases from *Candida olea*. J Gen Microbiol 133:1461–1469

Ng DT, Brown JD, Walter P (1996) Signal sequences specify the targeting route to the endoplasmic reticulum membrane. J Cell Biol 134:269–278

Nicaud J-M, Fabre E, Beckerich JM, Fournier P, Gaillardin C (1989) Cloning, sequencing and amplification of the alkaline extracellular protease (*XPR2*) gene of the yeast *Yarrowia lipolytica*. J Biotechnol 12:285–297

Ogrydziak DM (1993) Yeast extracellular proteases. Crit Rev Biotechnol 13:1–55

Ogrydziak DM, Mortimer RK (1977) Genetics of extracellular protease production in *Saccharomycopsis lipolytica*. Genetics 87:621–632

Ogrydziak DM, Scharf SJ (1982) Alkaline extracellular protease produced by *Saccharomycopsis lipolytica* CX161-1B. J Gen Microbiol 128:1225–1234

Ogrydziak DM, Demain AL, Tannenbaum SR (1977) Regulation of extracellular protease production in *Candida lipolytica*. Biochim Biophys Acta 497:525–538

Otero RC, Gaillardin C (1996) Dominant mutations affecting expression of pH-regulated genes in *Yarrowia lipolytica*. Mol Gen Genet 252:311–319

Ozcan S, Dover J, Rosenwald AG, Wolf S, Johnston M (1996) Two glucose transporters in *Saccharomyces cerevisiae* are glucose sensors that generate a signal for induction of gene expression. Proc Natl Acad Sci USA 93:12428–12432

Rawlings ND, Barrett AJ, Bateman A (2010) MEROPS: the peptidase database. Nucleic Acids Res 38:D227–D233

Ron D, Walter P (2007) Signal integration in the endoplasmic reticulum unfolded protein response. Nat Rev Mol Cell Biol 8:519–529

Sherman DJ, Martin T, Nikolski M, Cayla C, Souciet JL, Durrens P (2009) Genolevures: protein families and synteny among complete hemiascomycetous yeast proteomes and genomes. Nucleic Acids Res 37:D550–D554

Thevenieau F, Nicaud J-M, Gaillardin C (2009) Applications of non-conventional yeast *Yarrowia lipolytica*. In: Satyanarayana T, Kunze G (eds) Yeast biotechnology: diversity and applications. Springer, Dordrecht, pp 580–613

Toikkanen JH, Sundqvist L, Keranen S (2004) Kluyveromyces lactis *SSO1* and *SEB1* genes are functional in *Saccharomyces cerevisiae* and enhance production of secreted proteins when overexpressed. Yeast 21:1045–1055

Tusher VG, Tibshirani R, Chu G (2001) Significance analysis of microarrays applied to the ionizing radiation response. Proc Natl Acad Sci USA 98:5116–5121

Yamada T, Ogrydziak DM (1983) Extracellular acid proteases produced by *Saccharomycopsis lipolytica*. J Bacteriol 154:23–31

Yaver DS, Matoba S, Ogrydziak DM (1992) A mutation in the signal recognition particle 7S RNA of the yeast *Yarrowia lipolytica* preferentially affects synthesis of the alkaline extracellular protease: in vivo evidence for translational arrest. J Cell Biol 116:605–616

Young TW, Wadeson A, Glover DJ, Quincey RV, Butlin MJ, Kamei EA (1996) The extracellular acid protease gene of *Yarrowia lipolytica*: sequence and pH-regulated transcription. Microbiology 142(Pt 10):2913–2921

The Lipases from *Y. lipolytica*: Genetics, Production, Regulation, and Biochemical Characterization

Patrick Fickers, Vincent Sauveplane, and Jean-Marc Nicaud

Abstract The yeast *Yarrowia lipolytica* possesses multiple paralogues of genes dedicated to hydrophobic substrates metabolization. Among them, 16 lipase encoding genes, involved in lipid or grease breakdown, were highlighted in the yeast genome. However, little information on all those paralogues has been yet obtained. Microarray data suggest that only a few of them could be expressed. Lipase synthesis seems to be dependent on the fatty acid or oil used as carbon source confirming the high adaptation of *Y. lipolytica* to different hydrophobic substrate. This review focuses on the biochemical characterization of those enzymes with special emphasis on the Lip2p lipase which is the isoenzyme mainly synthesized by *Y. lipolytica*. The 3D structure of this lipase established by homology modeling confirms that Lip2p is a lipase sensu stricto with a lid covering the active site of the enzyme in its closed conformation. Recent findings on enzyme conditioning in dehydrated or liquid formulation and in enzyme immobilization by entrapment in natural polymers from either organic or mineral origins are also discussed together with long-term storage strategies.

Contents

1 Lipases Are Atypical Enzymes 100
2 Genetic Determinants of Lipase Enzymes 100
3 Regulation of Lipase Biosynthesis 104

P. Fickers (✉)
Biotechnologies et de Bioprocédés, Université Libre de Bruxelles, Avenue F.-D. Roosevelt 50, C.P. 165/61, 1050 Bruxelles, Belgium
e-mail: pfickers@ulb.ac.be

V. Sauveplane
INRA, UMR1319, Institut MICALIS, Domaine de Vilvert, 78352 Jouy-en-Josas, France
e-mail: vincent.sauveplane@free.fr

J.-M. Nicaud
CNRS, Institut MICALIS, Domaine de Vilvert, 78352 Jouy-en-Josas, France
e-mail: jean-marc.nicaud@grignon.inra.fr

G. Barth (ed.), *Yarrowia lipolytica*, Microbiology Monographs 25,
DOI 10.1007/978-3-642-38583-4_3, © Springer-Verlag Berlin Heidelberg 2013

4 Cloning, Overexpression, Production in Bioreactor, Bioprocess 105
5 Enzyme Purification 107
5.1 Purification of Lip2 107
5.2 Large-Scale Purification of Lip2p Lipase 108
5.3 Purification of Lip7p and Lip8p Lipases 109
6 3D Structure of Lip2p 109
7 Conditioning and Long-Term Storage 111
8 Catalytic Properties 112
8.1 Optimal pH and Temperature, Stability in Organic Solvents, and Effect of Metal Ions on Lip2p 113
8.2 Substrate Specificity 113
8.3 Extracellular Lipase Activators and Inhibitors 115
9 Lipase Immobilization 115
10 Conclusions 116
References 117

1 Lipases Are Atypical Enzymes

Lipase (EC 3.1.1.3) constitutes a group of enzymes that catalyze both the hydrolysis and the ester formation from glycerol and long-chain fatty acids. They differ from esterase (EC 3.1.1.1) due to their ability to hydrolyze triglyceride at lipid–water interface (Sarda and Desnuelle 1958). Lipases belong to the structural super family of α/β-hydrolases whose activities rely mainly on a catalytic triad usually formed by Ser, His, and Asp residues (Ollis et al. 1992). The serine residue usually appears embedded in the conserved pentapeptide Gly-X-Ser-X-Gly forming a characteristic β-turn-α motif named the nucleophilic elbow. Substrate hydrolysis starts with a nucleophilic attack by the catalytic serine oxygen on the carbonyl carbon atom of the ester bond, yielding a tetrahedral intermediate stabilized by hydrogen bonding to nitrogen atoms of main-chain residues belonging to the so-called oxyanion hole. The alcohol molecule is then liberated, leaving an acyl-enzyme complex, which is finally hydrolyzed with liberation of the fatty acid and regeneration of the enzyme (Fig. 1). Besides their hydrolytic activity, lipases remain enzymatically active in organic solvents where they are able, thus, to catalyze a wide range of chemioselective, regioselective, and stereoselective reactions.

2 Genetic Determinants of Lipase Enzymes

The extracellular lipase from *Y. lipolytica* CBS6303 strain was first characterized by Thonart and collaborators. They obtained overexpressing mutants that allowed enzyme purification to homogeneity and N-terminal sequencing of the 49 amino acid of the secreted form (Destain 1998).

The first attempt to identify a genetic determinant coding for a lipase in *Y. lipolytica* E150 strain was performed by reverse genetics. Due to the absence of large conserved blocks in the alignment of protein sequences of different yeast

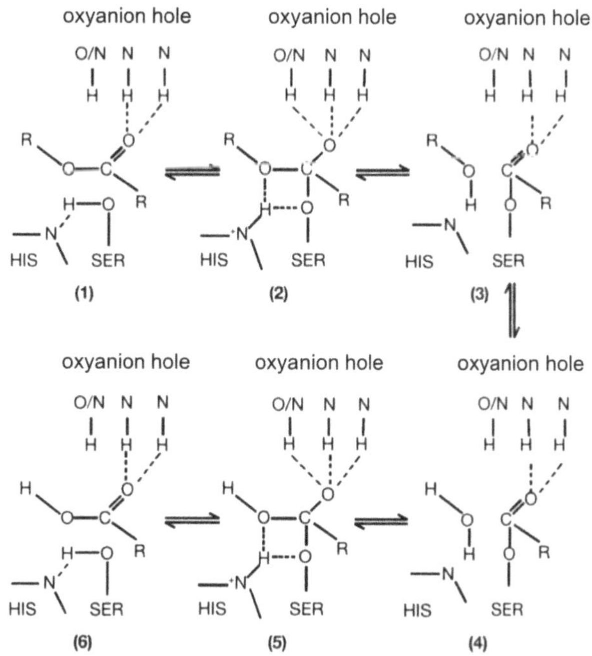

Fig. 1 Schematic representation of ester-bond hydrolysis catalyzed by lipase. (1) Non-covalent Michaelis complex; (2) first tetrahedral intermediate; (3) acyl enzyme and release of the alcohol molecule; (4) nucleophile attack from a water molecule; (5) second tetrahedral intermediate; (6) release of fatty acid

and fungal lipases, direct PCR on genomic DNA could not be used. Indeed, only the highly conserved motif GHSLGG/AA, which contains the catalytic serine of the active site, could be identified. Therefore, Pignede et al. (2000a) used the 5′ rapid amplification of cDNA ends (RACE-PCR) technique with degenerated antisense primers based on the conserved core sequence and abiding to the *Y. lipolytica* codon usage. cDNA libraries from strain PO1d grown in induction media containing either olive oil or oleic acid were used as templates for PCR amplification. This led to the determination of the complete cDNA sequence of the *LIP2* gene. This latter was found to encode a prepro-enzyme containing a 13 amino acid signal sequence, a stretch of four dipeptides (X-Ala or X-Pro) which are substrates of a diaminopeptidase, a 12 amino acid pro-region ending in the KR (Lys-Arg) site, substrate for the endoprotease encoded by the *XPR6* gene, and finally the mature lipase sequence of 301 amino acids (Pignede et al. 2000a) (Fig. 2). The significant reduction in lipase production obtained for a *Δlip2* mutant confirmed that *LIP2* encoded the main extracellular lipase activity, Lip2p. In addition, comparison of the amino acid sequence deduced from *LIP2* and the N-terminal sequence experimentally determined previously confirmed this hypothesis. Enzyme localization by Western blot analysis using a histidine-tagged lipase showed that at the beginning of the growth phase, the extracellular lipase protein is mainly associated with the cell membrane, whereas at the end of the growth phase, lipase is released in the culture supernatant (Fickers et al. 2004). Analysis of the *Lip2p* sequence highlights the presence of two potential N-glycosylation sites (N^{113}IS and N^{134}NT). Jolivet et al. (2007) demonstrated that both sites were glycosylated with mainly two

a

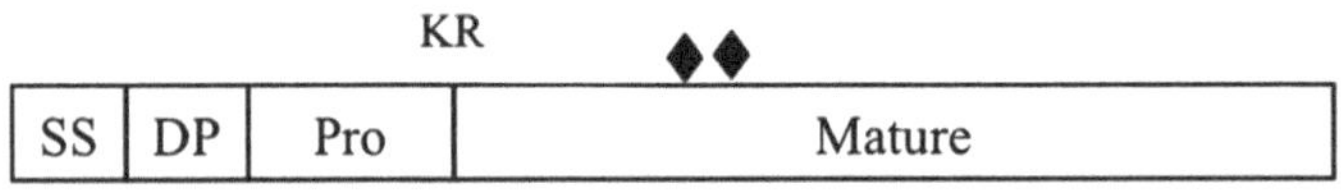

b

```
                                                     X-Ala
ATG AAG CTT TCC ACC ATC CTT TTC ACA GCC TGC GCT ACC |CTG GCT
Met Lys Leu Ser Thr Ile Leu Phe Thr Ala Cys Ala Thr |Leu Ala 15

 X-Pro    Dipeptide
|GCC GCC|CTC CCT|TCC CCC|ATC ACT CCT TCT GAG GCC GCA GTT CTC
|Ala Ala|Leu Pro|Ser Pro|Ile Thr Pro Ser Glu Ala Ala Val Leu 30

       Pro↓Mature
CAG|AAG CGA|GTG TAC ACC TCT ACC GAG ACC TCT CAC ATT GAC CAG
Gln|Lys Arg|Val Tyr Thr Ser Thr Glu Thr Ser His Ile Asp Gln 45
```

Fig. 2 Targeting sequence of the lipase Lip2p. (**a**) Schematic representation of the *LIP2* gene: the pre-signal sequence (SS), the four dipeptide motifs X-Ala, X-Pro (DP), and the pro-region (PRO) with the Lys-Arg site (KR) cleaved by the endoprotease encoded by the *XPR6* gene and the mature region (mature) containing the two potential glycosylation sites N113IS and N134NT (*diamond*). (**b**) Nucleotide and amino acid sequences for the 33 amino acids of the prepro sequence together with the first 12 amino acids of the mature form (taken from Pignede et al. 2000a, b)

different mannose structures $Man_8GlcNAc_2$ and $Man_9GlcNAc_2$. Single mutants with S115V or N134Q mutations led to lipase enzyme with similar or event increased specific activity compared to WT lipase depending on the substrates. On the contrary, the T136V mutation drastically reduced the specific activity toward p-NPB and tributyrin (<1 %) but conserved a reduced activity for triolein (>20 %). For the double mutant S115V-N134Q, activity was either increased or decreased depending on the substrate and the pH. This demonstrated that glycosylation was important for lipase activity but not for lipase secretion (Jolivet et al. 2007).

The second approach to isolate genes coding for lipases in a *Y. lipolytica*-related yeast was obtained by complementation of a non-lipolytic yeast (*S. cerevisiae*) with a genomic DNA library of a lipolytic yeast (*Candida deformans*). This was performed by Bigey and coworkers who isolated triacylglycerol lipase genes from *C. deformans*, a closely related yeast which belongs to the *Y. lipolytica* clade. Screening for lipase expression in *S. cerevisiae*, they identified a gene family of three lipases: cd*LIP1*, cd*LIP2*, and cd*LIP3* (Bigey et al. 2003). cd*LIP1* was shown to be highly similar to *LIP2* (91.6 % identity). Using cd*LIP2* and cd*LIP3* sequences, the *Y. lipolytica* homologues were cloned and sequenced (Fickers et al. 2005b). The deduced amino acid sequences of Lip7p (336 aa) and Lip8p (341 aa) proteins presented 86.9 % and 89.2 % identity to cdLip2 and cdLip3, respectively. SignalP analysis pointed out a potential 25 aa signal sequence for Lip7p, giving rise to a 341 (38.3 kDa) mature protein. Similarly, a potential 28 aa signal sequence was detected for the 371 aa sequence deduced from the *LIP8* ORF, although these putative signal sequence do not fulfill all signal sequence characteristics such as the presence of charged amino acids at position +1. By contrast to Lip2p, which is secreted at the end of the growth phase, it was suggested that Lip7p and Lip8p remain mainly associated to the cell wall. However, they could easily be released by washing the cells with phosphate buffer (Fickers et al. 2004, 2005b). The N-terminal sequence of Lip7p and Lip8p were found to correspond to the 8 aa of the putative signal sequence, indicating that they are present as precursor when associated to the cell. These results suggest that Lip7p and Lip8p could correspond to cell-bound lipases described previously by Ota et al. (1982). Both Lip7p and Lip8p proteins presented a potential glycosylation site at aa position 140. Deglycosylation with endoglycosidase H under denaturing condition resulted in a 5 and 5.2 kDa decrease in the apparent molecular mass of both Lip7p and Lip8p, respectively, confirming that both proteins are glycosylated. Sequence alignment between Lip7p and Lip8p and other fungal lipases led to the identification of the GHSLGAA motif at amino acid position 220. Both proteins also contained eight-conserved cysteine residues, which may form disulfide bridges. In all aligned sequences, the aspartic acid, histidine, and serine residues, which belong to the catalytic triad of lipase, were found at perfectly conserved position (Fickers et al. 2005b).

Successive gene disruption analysis, resulting in a triple-deleted strain Δ*lip2*Δ*lip7*Δ*lip8* unable to produce lipase, suggested that all lipases genes had been identified. However, the recent determination of the complete genome sequence of the haploid *Y. lipolytica* strain E150 (CLIB99) by the Génolevures consortium (Casaregola et al. 2000; Dujon et al. 2004) highlights the presence of 16 lipase encoding genes as well as four esterase encoding genes. The lipase family GL3R0084 (http://www.genolevures.org/fam/GL3R0084, formerly family GLS.94) contained genes *LIP2* (YALI0A20350g), *LIP4* (YALI0E08492g), *LIP5* (YALI0E02640g) and *LIP7* (YALI0D19184g), *LIP8* (YALI0B09361g), *LIP9* (YALI0E34507g), *LIP10* (YALI0F11429g), *LIP11* (YALI0D09064g), *LIP12* (YALI0D15906g), *LIP13* (YALI0E00286g), *LIP14* (YALI0B11858g), *LIP15* (YALI0E11561g), *LIP16* (YALI0D18480g), *LIP17* (YALI0F32131g), *LIP18* (YALI0B20350g), and *LIP19* (YALI0A10439g), whereas the esterase family GL3C3695 (http://www.genolevures.org/fam/GL3C3695, formerly GLS.95) presents four members: *LIP1* (YALI0E10659g), *LIP3* (YALI0B08030g), *LIP6* (YALI0C00231g), and *LIP20* (YALI0E05995g) (Theveniau et al. 2009). The presence of those multiple gene paralogues in the *Y. lipolytica* genome is not restricted to the *LIP* family and could be extended to other gene families involved in hydrophobic substrate utilization. For example, the *POX* family clusters 6 genes coding for acyl-CoA oxidase and the *ALK* family 12 genes coding for cytochrome P450 monooxygenase of the CYP52 family. This feature nicely highlights the *Y. lipolytica* addiction to hydrophobic substrates (Fickers et al. 2005a–d; Theveniau et al. 2009). Little information on all those paralogues has been yet obtained. Comparison of microarray data obtained on glucose and oleic acid media revealed that the only lipase gene expressed were *LIP11* in glucose and *LIP2* in oleic acid.

Recently, transcriptomic analysis was performed by creating three cDNA libraries (glucose exponential phase, glucose stationary phase, and oleic acid exponential phase). From each library, the sequencing of 10,000 clones allowed to determine the cDNA copy number for each genes. For the *LIP* genes, only the *LIP2* cDNA was found 15 times in oleic acid media and *LIP13* was found 1 time on glucose media in stationary phase (Fickers et al. 2005a; Mekouar et al. 2010). More detailed transcriptomic analyses are in progress to determine the conditions were these different lipases are induced. This may be dependent on the fatty acid or oil used for growth confirming the high adaptation of *Y. lipolytica* to different hydrophobic substrates.

3 Regulation of Lipase Biosynthesis

Lipase production could be greatly modulated depending on carbon and nitrogen sources (Novotny et al. 1988; Corzo and Revah 1999; Peireira-Meirelles et al. 1997; Fickers et al. 2004; Turki et al. 2009). There is also a high variability of lipase production depending on *Y. lipolytica* strains and environmental conditions (Guerzoni et al. 2001). In fact, media containing the *Y. lipolytica* preferred carbon sources (glucose, glycerol), or easily assimilable nitrogen compounds, did not trigger lipase nor protease secretion although they allow rapid cell growth. Glycerol was shown to be the more repressive carbon source for lipase production (Fickers et al. 2003). By contrast, in conditions of carbon starvation or in the presence of a less assimilable nitrogen compounds, lipase secretion is triggered. Lipase production is also significantly increased in the presence of hydrophobic substrates such as fatty acids, methyl esters, and oils (Pignede et al. 2000a; Fickers et al. 2004; Destain et al. 2005). The regulation of lipase expression and secretion is therefore under the control of multiple signaling pathways. This was evidenced by analyzing mutants affected in lipases production (Nga et al. 1989; Destain 1998; Mauersberger et al. 2001; Fickers et al. 2003; Thevenieau et al. 2007). Comparison of the lipase productivity obtained in the presence of different carbon sources (phase I) and after their depletion in the medium (phase II) was performed for several lipase overproducing mutants obtained by chemical mutagenesis (Fickers et al. 2003). Some of them presented similar productivities in phases I and II in glucose medium but not in the presence of glycerol. Other mutants such as Lgx86 showed an increased productivity in phase II compared to phase I whatever the carbon sources used. However, most of the mutants showed a derepression of lipase production upon glycerol exhaustion in the medium. Particularly, the second-generation mutant LgX64.81 presented the highest lipase productivity of 174.4 U/ml/h/A_{600} (pU) in phase II compared to the 0.02 pU obtained for the wild type (Fickers et al. 2003).

Reduced lipase production level was observed in the presence of mineral nitrogen. Very low level of lipase production was obtained in the presence of ammonium (about 2 pU). A slightly increased lipase production was obtained with casamino acid (6.6 pU), while an intermediate productivity level was obtained with urea, yeast extract, and peptone (about 100 pU). However, utilization of tryptone as nitrogen source significantly increased lipase production (about 500 pU) in both glucose and oleic media (Fickers et al. 2004). Induction by tryptone seems to be due to bioactive peptides present in this particular casein hydrolysate (Turki et al. 2009). The slowly metabolized nitrogen source, *N*-acetylglucosamine, also induced lipase production, but promoted at the same time intensive dimorphic growth (Novotny et al. 1994).

Regulatory pathways involved in lipase induction and regulation at the gene level are poorly understood. Involvement of glucose catabolite repression was demonstrated with the analysis of the LgX64.81 mutant. This strain presented a higher double time and a reduced sugar consumption rate for both glucose and fructose which was correlated with a decrease of hexokinase activity (Hxt). Overexpression of *HXK1* in this strain led to an increase in kinase activity but to a reduction of *LIP2* expression (followed with a β-galactosidase fusion as reporter). This confirmed the involvement of glucose catabolic repression in the regulation of lipase production (Fickers et al. 2005a–d). Recently, Desfougères and collaborators demonstrated that induction by fatty acid and oil is mediated by two different pathways. Indeed, they discovered two new genes coding for small proteins induced on oleic acid media called *SOA1* and *SOA2* (specific oleic acid). The double mutant *Δsoa1 Δsoa2* exhibited a growth defect on oil (tributyrin and triolein), but not on fatty acid (oleic acid). It was unable to secrete lipase on media containing tributyrin or triolein. Analysis of *LIP2* and *POX2* expression showed that *LIP2* expression was reduced in the single mutants and abolished in the double mutants, while *POX2* expression was not affected by these deletions. This suggested that two pathways regulated *LIP2* expression: one sensitive to fatty acid, and also regulating the *POX2* (β-oxidation) pathway, and one specific of the triglyceride pathway, which requires the *SOA* genes (Desfougeres et al. 2010).

4 Cloning, Overexpression, Production in Bioreactor, Bioprocess

Overexpression of lipase in *Y. lipolytica* was initially achieved by two approaches. In the Nicaud's group, the *LIP2* gene was cloned in the vector JMP3 under the control of the strong *POX2* promoter inducible by oleic acid. The JMP3 vector carries the defective *ura3d4* marker that allows gene amplification (Pignede et al. 2000b). Eight multicopy transformants were isolated and their copy number determined. They contained from 6 to 16 copies of the pPOX2-*LIP2* expression cassette. The multicopy strain JMY184 was able to produce 1,500 U/ml in flask which represents a 30-fold increased lipase production over the wild-type level (50 U/ml) (Pignede et al. 2000a). These strains were stable at least over 120 generations. In the Thonart's group, overproducing mutants were isolated from strain CBS6303 by successive rounds of chemical mutagenesis using *N*-methyl-*N*′-nitro-*N*-nitrosoguanidine. After the first round of mutagenesis, several overproducing mutants were isolated including LgX64 which produced about 15-fold more lipase than the wild-type strain. A second round of mutagenesis was performed on mutant LgX64 and led to the isolation of second-generation mutants among which mutant LgX64.81 producing about 25-fold more lipase than the wild type (i.e., 1,500 U/ml) (Destain et al. 1997). No further significant increase of lipase production could be obtained after a third round of mutagenesis (Destain et al. 1997; Fickers et al. 2003). Later, Fickers and collaborators combined both approaches by amplification of the *LIP2* gene into the LgX64.81 mutant. Strain JMY1105 was able to produce 26,450 U/ml of lipase in a medium containing olive oil and tryptone as carbon and nitrogen sources. Moreover, feeding with a combination of tryptone and olive oil at the end of the exponential growth phase further enhanced lipase production to 158,246 U/ml after 80 h of cultivation in a 20 l fermentor (Fickers et al. 2005a–d).

Mutant strain LgX64.81 was the focus of a number of investigations for the development of a lipase production process based on a non-genetically modified strain. Batch culture in 15 l bioreactor led to a lipase production of 1,000 U/ml (Destain 1998). Culture feeding with the complete medium led to a twofold increased lipase production (2,000 U/ml), whereas addition of a combination of glucose and olive oil used as inducer led to a threefold increase (Fickers et al. 2009). Recently, Turki and collaborators set up a stepwise strategy based on methyl oleate and tryptone feeding. In this process, cell growth and lipase production phases were decoupled. This strategy permitted a significant increase in lipase production, (i.e., 10,000 U/ml) after 80 h of culture (Turki et al. 2010). Besides this, a lipase production process compatible with industrial and commercial needs was set up in large-scale fermentor (2,000 l). The medium used was composed of whey powder, corn steep liquor together with a mixture of glucose and olive oil as carbon sources and inducer. This led to a lipase activity of approximately 1,100 U/ml after 53 h of fermentation (Fickers et al. 2006).

Other strategies for lipase production were also investigated. One of them consisted in solid-state fermentation. In those types of process, microorganisms not only grow at the surface of a solid support but can penetrate deeply into it. These supports can also supply nutriment such as sugars and oils for the growing microorganism. In such a context, Dominguez and collaborators used barley bran as solid support for production of extracellular lipase by *Y. lipolytica* strain ATCC1240 in the presence of several triacylglycerol-rich wastes such as triturated nut. They obtained a lipase production of 23 U/ml after 11 days of cultivation (Dominguez et al. 2003). Even if these values are low, they are comparable to those reported for similar processes developed for *Rhizopus* or *Rhizomucor* lipase production (Gombert et al. 1999).

Lipases from *Y. lipolytica* were also produced heterologously in the methylotrophic yeast *Pichia pastoris*. To this end, *LIP2* was cloned into the pPICZaA and integrated in *P. pastoris* X-33. The lipase was successfully expressed and secreted with an apparent molecular weight of 39 kDa using a *Saccharomyces cerevisiae* secretion signal peptide (α-factor) and the methanol-inducible promoter of the alcohol oxidase 1 gene (*AOX1*). A lipase activity level of 12,500 U/ml was obtained in a fed-batch cultivation process (Yu et al. 2007a). Similarly, Song and collaborators cloned, in frame and downstream from the α-factor signal peptide of *S. cerevisiae*, the *LIP7* and *LIP8* genes from *Y. lipolytica* strain AS2.1216 into vector pPIC9K. The vector was integrated into the *P. pastoris* KM71 genome and led to the production of 18.7 and 19.6 U/ml for Lip7p and Lip8p, respectively, after 4 days of methanol induction (Song et al. 2006).

5 Enzyme Purification

This section emphasize on the different methods and schemes developed for lipase purification from the very beginning, when lipase was first studied in *Y. lipolytica*, to nowadays techniques that take into account the physicochemical properties of the enzyme. A large-scale purification process adapted for industrial production is also presented.

5.1 *Purification of Lip2*

The first report on the purification of an extracellular lipase from *Candida paralipolytica* (now known as *Y. lipolytica*) was by Ota et al. (1970). The proposed scheme for lipase purification entailed five steps. The two first consisted of protein precipitation using acetone and ammonium sulfate. They reported that the solvent precipitation was very time-consuming because it took several days to achieve lipase precipitation. However, these two first steps led to a recovery of 63 % of the enzyme present in the crude sample. The three following steps consisted of two gel filtrations and one ion exchange chromatography. The two gel filtration steps were performed on CM Sephadex C-50 in acidic condition (pH 5). In between, the ion exchange chromatography was performed on a DEAE Sephadex A-50 matrix at pH 7 using an increasing NaCl gradient for elution. The overall purification yield was about 132-fold, based on protein content, with a recovery of 32 % from the acetone precipitate. This first report established the basis for the lipase purification procedure that is still partially applied nowadays for the purification of the extracellular lipase from *Y. lipolytica*.

Most of the time, the first steps of the purification scheme consist in the concentration of the lipase enzyme by precipitation or by ultrafiltration. For ultrafiltration, hydrophilic regenerated cellulose membranes with a cutoff of 10 kDa are usually used (Yu et al. 2007a, b; Fickers et al. unpublished). It allows both the concentration of the lipase enzyme and the elimination of small contaminant peptides used as nitrogenous sources in the production medium (peptone, tryptone, yeast extract) (Yu et al. 2007a, b; Aloulou et al. 2007). Lupescu and collaborators investigated the optimal conditions for ethanol precipitation of lipase using various enzymatic solution/ethanol ratios, solvent–enzyme contact durations, precipitation temperatures, and concentrations of the calcium ions used as stabilizer-activator. They found that an enzyme/ethanol ratio of 1:3 led to the highest lipase recovery. Although higher proportion of ethanol did not affect the lipase recovery, it led to lower specific activity. They pointed out that calcium ions had a positive effect on Lip2p precipitation and determined that a complete precipitation was obtained within 2 h at 4 °C in the presence of 0.2 % of $CaCl_2$ (Lupescu et al. 2007). Addition of ice-cold acetone to culture supernatant at a 3:1 ratio was also found valuable to precipitate and concentrate lipase (Yu et al. 2007a, b). Compared to ammonium sulfate precipitation, the utilization of solvent such as ethanol or acetone permitted extraction from the lipases samples of residual oil or triglyceride used as inducers for lipase production. Alternatively, Aloulou and coworker used freeze-drying of culture supernatant as a first step to concentrate lipase, followed by gel filtration on Superdex 200 HR. Only one peak eluted with the dead volume of the column presented lipase activity. However, this peak was found to correspond to high molecular weight lipase aggregates with tightly bound lipid corresponding to residual oleic acid used for inducing lipase production (Aloulou et al. 2007).

The second step in *Y. lipolytica* lipase purification consists usually of an ion exchange chromatography. Strong anion exchangers such as Mono Q or Q Sepharose are used in combination with a linear NaCl gradient for protein elution (Yu et al. 2007a, b; Aloulou et al. 2007; Turki et al. 2010). This step led to a purification factor varying from 1.96 to 17.5. The utilization of a mono Q 5/50 GL anion exchange column led to the separation of four enzymatically active proteins with the same specific activities. These isoforms were found to differ in their molecular mass due to differences in glycosylation patterns. These Lip2p isoforms were denominated Lip2A to Lip2D (Aloulou et al. 2007).

When a third step of purification is required to purify *Y. lipolytica* lipase to homogeneity, gel filtration or hydrophobic chromatography is preferred. For gel filtration, utilization of Sephacryl S-100 matrix allowed obtention of a purification factor of 3.4 and a recovery yield of 72 % (Turki et al. 2010). In the case of hydrophobic interaction, chromatography on butyl Sepharose FF was successfully applied. However, due to the relatively high hydrophobicity of the extracellular lipase from *Y. lipolytica*, addition of detergent was required for protein elution (Yu et al. 2007a, b).

When *Y. lipolytica* lipase Lip2 is heterologously expressed in *P. pastoris*, a similar procedure of purification, based on a concentration steps by ultrafiltration and further purification by ion exchange chromatography, was applied with success (Yu et al. 2007a, b).

5.2 Large-Scale Purification of Lip2p Lipase

For industrial purposes, the purification strategies employed should be inexpensive, rapid, high yielding, and amenable to large scale. In that context, Fickers et al.

(2006) developed a purification scheme of Lip2p compatible with large-scale production. This latter consisted on the clarification of the culture supernatant (950 l) on a plate filter (0.13 m^2, 0.2 μm porosity). This first step allowed elimination of cell fragments and of insoluble compounds (olive oil used as an inducer, antifoam). It led to the reduction by a one-third of the dry matter content without any loss of enzymatic activity. The clarified culture supernatant was then concentrated and further purified by ultrafiltration on a 10 m^2 polysulfone membrane presenting a 10 kDa cutoff. This second step led to a 12-fold reduction in volume and an 8-fold increased lipase activity. SDS-PAGE analysis of the purified samples clearly demonstrated that Lip2p was the main protein present in the sample.

5.3 Purification of Lip7p and Lip8p Lipases

By contrast to Lip2p, which is secreted at the end of the growth phase (Fickers et al. 2004), Lip7p and Lip8p were found mainly associated to the cell membrane. For their semi-purification at small scale, the cell pellet, obtained by centrifugation of the culture broth, was extracted three times vigorously with phosphate buffer at pH 7. Then, two volumes of ice-cold acetone were slowly added to cell washes. Finally, the protein pellet was resuspended and further concentrated by ultrafiltration on a Hydrosart membrane with a 10 kDa cutoff (Fickers et al. 2005b).

Lip7p and Lip8p from *Y. lipolytica* strain AS2.1216 have been heterologously produced in *P. Pastoris* as protein fusions to a histidine-tag at the carboxy termini of the proteins. Both lipases were purified from the culture supernatant by Ni-NTA agarose chromatography. Elution was performed with a linear gradient of imidazole (Song et al. 2006).

6 3D Structure of Lip2p

Given the absence of crystallographic data on Lip2p lipase, Bordes and collaborators established a three-dimensional model by using homology modeling on the X-ray structures of highly similar lipases as templates (Fig. 3) (Bordes et al. 2009). Based on sequence homology, the selected template were the lipases from *Rhizomucor miehei*, *Rhizopus niveus*, and *Thermomyces lanuginosa* as well as the feruloyl esterase from *Aspergillus niger*. The secondary structure pattern is very well conserved among all five enzymes. Five catalytically important amino acid residues that form the conserved catalytic triad and the oxyanion hole were identified. The catalytic Ser of Lip2p, located in the nucleophilic elbow after the β-strand 5, was identified as Ser162 by the well-known GxSxG lipase signature. In the case of Lip2p, this signature is GHSLG, like for other lipase from the filamentous fungi superfamily. The two other amino acids of the catalytic triad were found

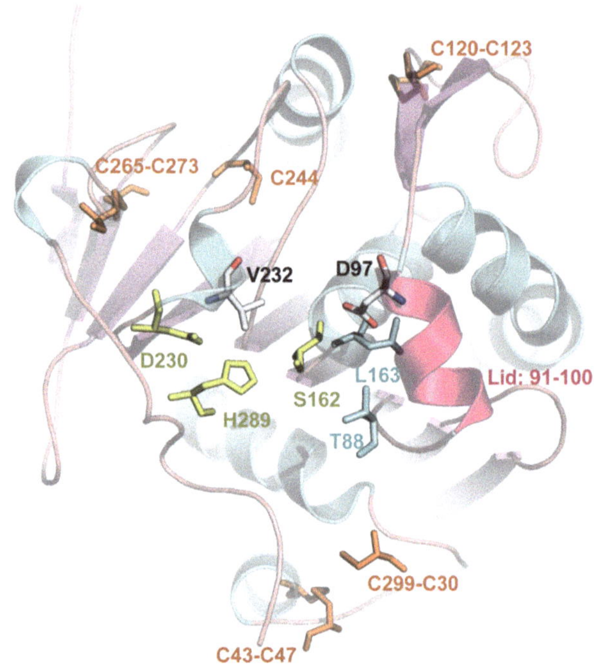

Fig. 3 Overall representation of the Lip2p homology model proposed by Bordes et al. (2009) which highlights the catalytic residues S162, D230, H289, and the four disulfide bonds (Cys30-Cys299, Cys43-Cys47, Cys120-Cys123, and Cys265-Cys273) (taken from Bordes et al. 2009)

perfectly aligned with the catalytic residues of other proteins, namely, Asp230 and His289, located after β-strand 7 and 8, respectively. Another important region of lipase is the oxyanion hole, which consists in two residues that give their backbone amide protons to stabilize the tetrahedral intermediate formed during the reaction mechanism. One of these residues is positioned identically in all lipases, next to the catalytic serine. Like in mucoral lipase, this residue is a Leu (Leu163) in Lip2p. The second residue of the oxyanion hole is located in a loop after the β-strand 3 and next to a Gly residue. As for filamentous fungi, this second residue which forms the oxyanion hole in Lip2p is the rather hydrophilic residue, Thr88, corresponding to Ser82, Ser83, and Thr82 in *R. miehei*, *T. lanuginosa*, and *R. niveus* lipases, respectively. Like other mucoral lipases, Lip2p belongs to the "GX"-type lipase, which presents specificities for medium- and long-chain fatty acid (Pleiss et al. 2000). In the open form of mucoral lipases, the side chain of the hydrophilic residue of the oxyanion hole interacts through hydrogen bonding with a hydrophilic residue, named the anchor residue, located at the end of the α-helix B′1 that forms the lid of the lipase. In Lip2p, the role of anchor residue could be played by Asp97 according to Bordes et al. 2009.

The *Y. lipolytica* lipase Lip2p contains nine Cys residues as deduced from its nucleic acid sequence. Structural alignment with the related lipase sequences suggested the possible formation of four disulfide bonds between Cys30-Cys299, Cys43-Cys47, Cys120-Cys123, and Cys265-Cys273. This also suggests that Lip2p possesses only one free Cys residue (Cys244). In most lipases, the lid covering the active site is composed of one or two short α-helix. Its role is to block the active site in the absence of substrate (closed conformation), while hydrophobic substrate induces a conformational modification of the lid, rendering the active site of the lipase accessible to the substrate (open conformation). On the basis of sequence alignments with the three lipase templates, the α-helix formed by residues Leu91 to Asp97 was assumed to play the role of the lid in Lip2p from *Y. lipolytica* (Bordes et al. 2009).

7 Conditioning and Long-Term Storage

Spray-drying has been used for the dehydration of many enzymes, such as proteases or cellulases, and offers cost advantage compared to freeze-drying. However, dehydration by spray-drying causes a stress to proteins and their unfolding by thermal denaturation, leading to loss of enzymatic activity. Protein unfolding is usually minimized by using additives such as polysaccharide, proteins, or salts. Additives such as lactoses from skim milk stabilized lipase due to a readily attainable amorphous form which reportedly enhances enzyme preservation upon spray-drying (Suzuki et al. 1997). Maltodextrins are low molecular weight products of starch hydrolysis that are non-sweet and soluble in cold water. Consequently, they have found wide application in the food industry (Dokic et al. 1998). Gum arabic is also a versatile compound for most encapsulation methods due to its high solubility, low viscosity, and emulsifying properties.

Without additive, it was not possible to obtain a fluent powder by spray-drying from the semi-purified and concentrated Lip2p lipase preparation. The dry matter content from this preparation was too low and residues of the nutriment medium (corn steep liquor, yeast extract) generate a sticky powder. Fickers et al. (2006) reported that addition of 12 % milk powder (w/v) led to highest lipase recovery, while adding 12 % of milk powder and 3 % (w/v) of gum arabic produced a fluent powder more compatible with commercial products standards. Milk powder is a source of both casein and calcium ions. Both of them were found to increase the enzymatic activity of the extracellular lipase from *Y. lipolytica*. Casein, which forms hydrogen bonds with some lipase residues, could facilitate the interaction of the enzyme with hydrophobic substrate at the lipid–water interface (Brockerhoff 1971). It has been suggested that the role of calcium ions is to reduce the electrical charge of the fatty droplets and remove fatty acids formed during the hydrolysis of triglycerides as insoluble calcium soaps. In addition, crystallographic data suggest that many lipases have a calcium-binding motif around the catalytic site and that the presence of calcium increases the thermostability of the enzyme (Tanaka et al. 2003). The catalytic properties of the dehydrated lipase were investigated by comparing the enzymatic activity in the enzymatic powder and in the non-dehydrated lipase preparation. Optimal activity of the lipase powder was observed at 37 °C confirming previous observation (Destain et al. 1997). In contrast, significant differences were observed in the pH stability of the enzyme.

Both spray-dried and non-dehydrated lipase presented an optimal catalytic activity at pH 7, but surprisingly, the spray-dried lipase seemed less sensitive to acidic pH than the non-dehydrated enzyme. This increased stability of the spray-dried lipase under acidic conditions was related to the stabilization and neutralization of the lipase enzyme by casein peptides of the milk powder (Fickers et al. 2006). This lipase formulation was found effective for long-term conservation. Indeed, loss of lipolytic activity in enzyme powder was less than 3 % after 12 months of storage at 4 °C (Fickers et al. 2006). Other formulations based on gum arabic, maltodextrins, and $CaCl_2$ were also investigated for lipase dehydration by spray-drying of culture supernatant (i.e., without any concentration). Addition of 12 % maltodextrin, 6 % gum arabic, and 3 % $CaCl_2$ had a positive effect on the enzyme and led to an increase in lipase activity by 1.46-fold (Alloue et al. 2007). This formulation was also found effective for the long-term storage of the lipase enzyme. After 30 weeks of self-storage at 20 °C, the loss of the lipolytic activity in the powder was less than 5 % (Alloue et al. 2007). The authors suggested that gum arabic would form a protective coating that preserves the surface of the product from a possible oxidation.

For some applications, liquid formulation could be preferred for their ease of handling. Besides enzyme stabilization by genetic engineering, chemical modification or immobilization and utilization of stabilizing agents were the simplest and cheapest methods to achieve enhanced enzyme stability. Polyhydric alcohols are well known for their high ability to form hydrogen bonds, to increase the organization of water molecules, and to preserve the protein hydration status (Noel et al. 2005). For the extracellular lipase Lip2p, the liquid formulation consisted in the utilization of monopropylene glycol at a concentration of 50 % (v/v) and of protease inhibitor cocktail (PI 2714, 0.1 % v/v; Sigma-Aldrich). When stored at 4 °C, the formulated lipase solution was stable for more than 25 weeks. After this period, more than 80 % of the initial lipase activity could be recovered. By contrast, significant lipase loss was obtained for a temperature storage of 20 °C, due to significant bacterial contaminations. To avoid any microorganism contamination that could lead to loss of lipase enzyme, liquid formulations were gamma irradiated at 10 kGy. This was found sufficient to avoid any microbial growth at least for 24 weeks (Alloue et al. 2008).

8 Catalytic Properties

The particular structure and mode of action of the lipolytic enzyme, i.e., the presence of a lid covering the catalytic triad, the substrate hydrolysis at the lipid–water interface, are at the basis of their catalytic properties. Those from lip2p, lip7p, and lip8p are discussed in this section.

8.1 Optimal pH and Temperature, Stability in Organic Solvents, and Effect of Metal Ions on Lip2p

Aloulou et al. (2007) determined the effect of pH on Lip2p activity using as substrates an emulsion of short chain (tributyrin), medium-chain (trioctanoin), and long-chain (triolein) triglyceride. The maximum specific activities were observed at pH 6.0 with all those substrates. The Lip2p specific activity on emulsion of long-chain triglycerides such as olive oil or triolein was still high at pH 4 but decreased significantly at pH above 8 (Aloulou et al. 2007). Similar results were obtained by several other authors (Destain et al. 1997; Corzo and Revah 1999; Peireira-Meirelles et al. 1997). The effect of pH on the stability of Lip2p was determined by Ota et al. (1970) by incubating the enzyme at 4 °C for 22 h at different pH and by determining the remaining enzymatic activity. The enzyme was found stable in a pH range of 3.5–9.0. Besides this, inactivation of Lip2p under extreme pH conditions was shown to be irreversible (Aloulou et al. 2007). Different authors determined that the optimal temperature was 37 °C and that the stability of the enzyme rapidly decreased at higher temperature (Ota et al. 1970; Destain et al. 1997; Corzo and Revah 1999). However, Yu et al. (2007a, b) reported an optimal temperature of 40 °C, whereas Peireira-Meirelles et al. (1997) observed the maximal activity at 55 °C for the extracellular lipase produced by *Y. lipolytica* strain isolated in an estuary in the vicinity of Rio de Janeiro, Brazil.

The extracellular lipase from *Y. lipolytica* is also relatively stable in the presence of solvent. The lipase retained almost 90 % activity after exposure for 30 min to 10 % acetone, methanol, ethanol, isopropanol, and dimethyl sulfoxide. The enzyme was only found sensitive to acetonitrile. When the organic solvent proportion was increased to 20 %, there was no activity left in acetone, ethanol, and isopropanol. Only, 60 % of Lip2p activity remained in methanol and 95 % in dimethyl sulfoxide (Yu et al. 2007a, b). To explain this observation, Yu and collaborator hypothesized that in aqueous solution, a thin layer of water molecules remains bound to the enzyme allowing the retention of the native conformation of the enzyme. When water-miscible organic solvents are added, they remove the bound water from the enzyme destabilizing its structure and led to its inactivation.

The catalytic properties of the extracellular lipase are also affected by metal ions. The lipase activity was found slightly increased in the presence of Ca^{+} and Mg^{2+} and inhibited in the presence of Zn^{2+}, Ni^{2+}, and Cu^{2+}. Chelating agent such as EDTA had no effect on the lipase activity, suggesting that Lip2p is not a metalloenzyme (Yu et al. 2007a, b; Fickers et al. 2006).

8.2 Substrate Specificity

The substrate specificity of the extracellular lipase Lip2p was determined by the utilization of various triglycerides and fatty acid methyl ester with different chain length of the acyl moiety. For triglycerides, Lip2p showed high activity toward saturated triglyceride-tricaprylin (C8:0) and unsaturated triglycerides-olive oil, trioleine (C18:1). For fatty acid methyl ester, Lip2p preferred long-chain fatty acid ester (C12, C14, and C16). These results, obtained by Yu et al. (2007a, b), showed that lipase was more active on triglycerides than on hydrophilic fatty acid methyl esters, demonstrating that Lip2p is a lipase sensu stricto. The authors hypothesized that the difference of the substrate specificity toward triglycerides and fatty acid methyl ester resulted from the poor emulsification of saturated long-chain triglycerides (>C12) due to their insolubility under the reaction conditions. Corzo and Revah (1999) reported that the extracellular lipase from *Y. lipolytica* 681 was more active on synthetic triacylglycerols such as tricaprylin and tributyrin than on natural oils such as soy and corn oil. The stereoselectivity of Lip2p toward the chemically alike, but sterically non-equivalent ester group, at the sn-1 and sn-3 position of the prochiral triolein substrate was investigated by measuring the enantiomeric excess of 1,2-diolein versus 2,3-dioleine. A clear enantiomeric excess of 1,2-dioleine was observed at low rates of hydrolysis, indicating a slight stereopreference for the hydrolysis of the ester bond at the 3-sn position of triolein. However, this apparent stereopreference of Lip2p was found to be modified by the hydrolysis rate. The enantiomeric excess of diglycerides was found slightly reversed at lipolysis rates higher than 25 % (Aloulou et al. 2007). They hypothesized that this change was probably due to the subsequent hydrolysis of diglyceride into monoglycerides by Lip2p, in which a stereospecific hydrolysis of 1,2-diolein might occur, as previously observed for dog gastric and pancreatic lipases (Carriere et al. 1997).

Lipase hydrolysis of triglyceride is a multistep reaction that includes triglyceride, diglyceride, and monoglyceride hydrolysis. The simplified model of Han and Rhee (1986) based on Michaelis-Menten kinetics allowed the determination of apparent catalytic constants. Based on initial rates of olive oil hydrolysis by the Lip2p, and on the linearity of the Lineweaver–Burk plot, Destain et al. 1997 found a K_m value of 807 mM, a V_m value of 11.4 μmol of fatty acid hydrolyzed per minute, and a k_{cat} value of 500/s.

The substrate specificity of *Y. lipolytica* lipase Lip7p and Lip8p was investigated using synthetic substrate *p*-nitrophenyl esters of varying chain length. Activity data obtained with crude lipase preparation showed that Lip7p had an optimal activity for medium-chain ester *p*-nitrophenyl-C6. It was also found able to hydrolyze longer-chain esters (C8–C12), but it hydrolyzed poorly *p*-nitrophenyl-C2 and *p*-nitrophenyl-C14. By contrast, Lip8p showed a maximal activity toward esters of C10 as well as for medium-chain C8. The efficiency of Lip8p to hydrolyze shorter and longer chains decreased rapidly and it hydrolyzed very poorly C2 and C14 fatty acid esters (Fickers et al. 2005b). Therefore, Lip8p should be considered as a lipase according to Verger (1997), whereas Lip7p rather corresponds to an esterase since its preferential substrates are shorter-chain length ester. Therefore, *Y. lipolytica* produces three extracellular lipase isoenzymes presenting a substrate specificity ranging from medium- to long-chain fatty acids.

8.3 Extracellular Lipase Activators and Inhibitors

The catalytic properties of the extracellular lipase from *Y. lipolytica* were strongly affected by oleic acid, the end product of the hydrolysis of their preferred substrate. The effect of oleic acid was studied using the kinetic model developed by (Han and Rhee 1986). Both the apparent V_{max} and the apparent K_s of the lipase varied with the concentration of oleic acid from 0 to 6 mg/ml. However, the same kinetic parameters were not influenced by glycerol, the second end product of triglyceride hydrolysis. By using the model of Han and Levenspiel (1988), a total inhibition of the enzyme was predicted at an oleic concentration of 12 mg/ml for an enzymatic activity of 10 U/ml (Corzo and Revah 1999). By contrast, lip2p activity increased in the presence of bile salt whatever substrate was used (triolein, tributyrin). This behavior is in contradiction with the strong inhibitory effect of bile salts observed with many lipases, including pancreatic lipase in the absence of colipase. Thus, lip2p is able to adsorb to triglyceride interfaces in the presence of bile salts, without any requirement for a specific cofactor (Aloulou et al. 2007). Many authors reported that calcium ions are able to activate Lip2p. A possible explanation of this phenomenon is that Ca^{++} exerts a special enzyme-activating effect by concentrating it at the lipid–water interface. Therefore, calcium ions might carry out three distinct roles in lipase action: (1) removal of fatty acids as insoluble Ca^{++} salts, (2) direct enzyme activation from concentration at the fat–water interface, and (3) stabilizing effect on the enzyme (Yu et al. 2007a, b; Fickers et al. 2006).

9 Lipase Immobilization

Many interesting enzyme applications are limited due to economic reasons. Enzyme immobilization allows to increase enzyme stability and biological activity, but the main advantage of this technique stems from the possibility of recovering and reusing the enzyme after catalysis. Many methods have been developed for lipase immobilization. Three techniques including inclusion, adsorption, and covalent bond were investigated for the extracellular lipase from *Y. lipolytica*. Entrapment of lipase by inclusion was investigated for alginate and chitosan beads. Higher levels of immobilization were obtained with alginate (66 %) compared to chitosan (23 %) (Alloue et al. 2008). By contrast, the residual lipolytic activity obtained after beads drying was higher for chitosan than for alginate. After 4 months of storage at 4 and 20 °C, freeze-dried beads showed a lower residual lipolytic activity compared to fluidized beads, and immobilization yield in the case of alginate beads was found to decline more rapidly than in chitosan. This could be explained by the fact that alginate is more hydrophilic than chitosan, and therefore, water could be drawn in more rapidly, which causes enzyme to leach from the beads. Using this technique, more than 80 % of the initial lipolytic activity could be recovered after 4 months of storage at 4 and 20 °C (Alloue et al. 2008). Scanning electron microscopy highlighted that freeze-drying procedures led to more friable beads, which in turn would promote enzyme leakage and thus limit bead reuse. Immobilization by inclusion did not modify significantly the pH stability of the lipase despite a protective effect at alkaline pH. By contrast, a significant protective effect was observed upon thermal treatment. Freeze-dried and fluidized entrapped lipase conserved more than 60 % and 80 % of their activity, respectively, after 7 h of incubation at 50 °C, while the free enzyme lost 90 % of its activity under the same conditions (Alloue et al. 2008). Moreover, a significant decrease of lipase activity was obtained after repeated batches of utilization. This loss of activity (more than 50 %) was attributed to enzyme leakage from the beads during their washing.

Lipase adsorption on celite and silica gel led to an immobilization yield of 76 % and 43 %, respectively. This type of immobilization was also found to protect the lipase enzyme at alkaline pH and to increase its thermoresistance. This mode of immobilization was found valuable for the protection of the lipase enzyme from solvents such as hexane or heptane. With this technique, more than 80 % of the enzymatic activity could be recovered after five cycle of enzymatic activity. Immobilization by covalent bond formation on HiTrap NHS-activated matrix led to an immobilization yield of 70 %. After repeated use, the immobilized lipase was found stable. After 5 months of utilization, more than 80 % of the initial lipolytic activity could be recovered. All of these suggest that immobilization of the extracellular lipase from *Y. lipolytica* is a valuable method that permits increased reutilization and thus leads to decreased costs of utilization.

10 Conclusions

The yeast *Yarrowia lipolytica* very efficiently uses lipids as carbon source. Genomic exploration revealed that 16 genes coding for lipases were present in this yeast. The major secreted lipase Lip2p is secreted via a prepro targeting sequence. Mutant strains and overexpressing strains were constructed, allowing large-scale production under optimized conditions and purification of this lipase. Lip2p was shown to have interesting potential applications (see Fickers and Nicaud 2013). More recently, Lip7p and Lip8p have been identified and shown to present different substrate specificities. Purification procedures for these three enzymes have been developed allowing in-depth characterization of their enzymatic properties. A 3D model for Lip2p was proposed and allowed rational design of Lip2p variants. Further improvements are foreseen as a result of the recent determination of Lip2p structure by X-ray diffraction (A. Marty, personal communication). Future studies will target the function and specificities of the 13 other lipases present in the *Y. lipolytica* genome, in order to better understand the remarkable adaptation to lipid utilization of this species.

References

Alloue W, Destain J, Amighi K, Thonart P (2007) Storage of *Yarrowia lipolytica* lipase after spray-drying in the presence of additives. Process Biochem 42:1357–1361

Alloue W, Destain J, Ongena M, Blecker C, Thonart P (2008) Effect of monopropylene glycol and gamma irradiation on *Yarrowia lipolytica* lipase stabilization. Prep Biochem Biotechnol 38:217–228

Aloulou A, Rodriguez JA, Puccinelli D, Mouz N, Leclaire J, Leblond Y, Carriere F (2007) Purification and biochemical characterization of the *LIP2* lipase from *Yarrowia lipolytica*. Biochim Biophys Acta 1771:228–237

Bigey F, Tuery K, Bougard D, Nicaud JM, Moulin G (2003) Identification of a triacylglycerol lipase gene family in *Candida deformans*: molecular cloning and functional expression. Yeast 20:233–248

Bordes F, Cambon E, Dossat-Letisse V, Andre I, Croux C, Nicaud JM, Marty A (2009) Improvement of *Yarrowia lipolytica* lipase enantioselectivity by using mutagenesis targeted to the substrate binding site. Chembiochem 10:1705–1713

Brockerhoff H (1971) On the function of bile salts and proteins as cofactors of lipase. J Biol Chem 246:5828–5831

Carriere F, Thirstrup K, Hjorth S, Ferrato F, Nielsen PF, Withers-Martinez C, Cambillau C, Boel E, Thim L, Verger R (1997) Pancreatic lipase structure-function relationships by domain exchange. Biochemistry 36:239–248

Casaregola S, Neuveglise C, Lepingle A, Bon E, Feynerol C, Artiguenave F, Wincker P, Gaillardin C (2000) Genomic exploration of the hemiascomycetous yeasts: 17. *Yarrowia lipolytica*. FEBS Lett 487:95–100

Corzo G, Revah S (1999) Production and characteristics of lipase from *Yarrowia lipolytica* 681. Bioresour Technol 70:173–180

Desfougeres T, Haddouche R, Fudalej F, Neuveglise C, Nicaud JM (2010) *SOA* genes encode proteins controlling lipase expression in response to triacylglycerol utilization in the yeast *Yarrowia lipolytica*. FEMS Yeast Res 10:93–103

Destain J (1998) Production, purification et caractérisation de la lipase de *Yarrowia lipolytica*. PhD thesis, Faculté Universitaire des Sciences Agronomiques de Gembloux, Gembloux, Belgium

Destain J, Roblain D, Thonart P (1997) Improvement of lipase production from *Yarrowia lipolytica*. Biotechnol Lett 19:105–107

Destain J, Fickers P, Weekers F, Moreau B, Thonart P (2005) Utilization of methyl oleate in production of microbial lipase. Appl Biochem Biotechnol 121–124:269–277

Dokic P, Jakovljevic J, Dokic-Baucal L (1998) Molecular characteristics of maltodextrin and rheological behaviour of diluted and concentrated solutions. Colloids Surf A 141:435–440

Dominguez A, Costas M, Longo MA, Sanroman A (2003) A novel application of solid state culture: production of lipases by *Yarrowia lipolytica*. Biotechnol Lett 25:1225–1229

Dujon B, Sherman D, Fischer G, Durrens P, Casaregola S, Lafontaine I, De Montigny J, Marck C, Neuveglise C, Talla E, Goffard N, Frangeul L, Aigle M, Anthouard V, Babour A, Barbe V, Barnay S, Blanchin S, Beckerich JM, Beyne E, Bleykasten C, Boisrame A, Boyer J, Cattolico L, Confanioleri F, De Daruvar A, Despons L, Fabre E, Fairhead C, Ferry-Dumazet H, Groppi A, Hantraye F, Hennequin C, Jauniaux N, Joyet P, Kachouri R, Kerrest A, Koszul R, Lemaire M, Lesur I, Ma L, Muller H, Nicaud JM, Nikolski M, Oztas S, Ozier-Kalogeropoulos O, Pellenz S, Potier S, Richard GF, Straub ML, Suleau A, Swennen D, Tekaia F, Wesolowski-Louvel M, Westhof E, Wirth B, Zeniou-Meyer M, Zivanovic I, Bolotin-Fukuhara M, Thierry A, Bouchier C, Caudron B, Scarpelli C, Gaillardin C, Weissenbach J, Wincker P, Souciet JL (2004) Genome evolution in yeasts. Nature 430:35–44

Fickers P, Nicaud JM (2013) Biotechnological applications of *Yarrowia lipolytica* lipases: an overview. In: Barth G (ed) *Yarrowia lipolytica*. Springer, Heidelberg

Fickers P, Nicaud JM, Destain J, Thonart P (2003) Overproduction of lipase by *Yarrowia lipolytica* mutants. Appl Microbiol Biotechnol 63:136–142

Fickers P, Nicaud JM, Gaillardin C, Destain J, Thonart P (2004) Carbon and nitrogen sources modulate lipase production in the yeast *Yarrowia lipolytica*. J Appl Microbiol 96:742–749

Fickers P, Benetti PH, Wache Y, Marty A, Mauersberger S, Smit MS, Nicaud JM (2005a) Hydrophobic substrate utilisation by the yeast *Yarrowia lipolytica*, and its potential applications. FEMS Yeast Res 5:527–543

Fickers P, Fudalej F, Le Dall MT, Casaregola S, Gaillardin C, Thonart P, Nicaud JM (2005b) Identification and characterisation of *LIP7* and *LIP8* genes encoding two extracellular triacylglycerol lipases in the yeast *Yarrowia lipolytica*. Fungal Genet Biol 42:264–274

Fickers P, Fudalej F, Nicaud JM, Destain J, Thonart P (2005c) Selection of new over-producing derivatives for the improvement of extracellular lipase production by the non-conventional yeast *Yarrowia lipolytica*. J Biotechnol 115:379–386

Fickers P, Nicaud JM, Destain J, Thonart P (2005d) Involvement of hexokinase Hxk1 in glucose catabolite repression of *LIP2* encoding extracellular lipase in the yeast *Yarrowia lipolytica*. Curr Microbiol 50:133–137

Fickers P, Ongena M, Destain J, Weekers F, Thonart P (2006) Production and down-stream processing of an extracellular lipase from the yeast *Yarrowia lipolytica*. Enzyme Microb Technol 38:756–759

Fickers P, Destain J, Thonart P (2009) Improvement of *Yarrowia lipolytica* lipase production by fed-batch fermentation. J Basic Microbiol 49:212–215

Gombert A, Pinto A, Castilho L, Freire D (1999) Lipase production by Penicillium restrictum in solid state fermentation. Process Biochem 35:85–90

Guerzoni ME, Lanciotti R, Vannini L, Galgano F, Favati F, Gardini F, Suzzi G (2001) Variability of the lipolytic activity in *Yarrowia lipolytica* and its dependence on environmental conditions. Int J Food Microbiol 69:79–89

Han D, Rhee JS (1986) Characteristics of lipase-catalyzed hydrolysis of olive oil in AOT-isooctane reversed micelles. Biotechnol Bioeng 28:1250–1255

Han K, Levenspiel O (1988) Extended monod kinetics for substrate, product, and cell inhibition. Biotechnol Bioeng 32:430–447

Jolivet P, Bordes F, Fudalej F, Cancino M, Vignaud C, Dossat V, Burghoffer C, Marty A, Chardot T, Nicaud JM (2007) Analysis of *Yarrowia lipolytica* extracellular lipase Lip2p glycosylation. FEMS Yeast Res 7:1317–1327

Lupescu I, Groposila-Constantinescu D, Jurcoane S, Diguta C, Cozea A, Tcacenco L (2007) Production of lipase by a strain of the non-conventional yeast *Yarrowia lipolytica* and isolation of crude enzyme. Rom Biotechnol Lett 12:27–37

Mauersberger S, Wang HJ, Gaillardin C, Barth G, Nicaud JM (2001) Insertional mutagenesis in the n-alkane-assimilating yeast *Yarrowia lipolytica*: generation of tagged mutations in genes involved in hydrophobic substrate utilization. J Bacteriol 183:5102–5109

Mekouar M, Blanc-Lenfle I, Ozanne C, Da Silva C, Cruaud C, Wincker P (2010) Detection and analysis of alternative splicing in Yarrowia lipolytica reveal structural constraints facilitating nonsense-mediated decay of intron-retaining transcripts. Genome Biol 11:R65

Nga B, Gaillardin C, Fournier P, Heslot H (1989) Genetic analysis of lipase low producing mutant of *Yarrowia lipolytica*. J Gen Microbiol 135:2439–2443

Noel M, Lozano P, Combes D (2005) Polyhedric alcohol protective effect on *Rhizomucor miehei* lipase deactivation enhanced by pressure and temperature treatment. Bioprocess Biosyst Eng 27:375–380

Novotny C, Dolezalova L, Musil P, Novak M (1988) The production of lipase by some *Candida* and *Yarrowia* yeast. J Basic Microbiol 4:221–227

Novotny C, Dolezalova L, Lieblova J (1994) Dimorphic growth and lipase production in lipolytic yeasts. Folia Microbiol (Praha) 39:71–73

Ollis DL, Cheah E, Cygler M, Dijkstra B, Frolow F, Franken SM, Harel M, Remington SJ, Silman I, Schrag J (1992) The alpha/beta hydrolase fold. Protein Eng 5:197–211

Ota Y, Nakamiya T, Yamada K (1970) Lipase from *Candida paralipolytica*: purification, some properties and modification of the purified enzyme with concentrated solution of sodium chloride. Agric Biol Chem 34:1368–1374

Ota Y, Gomi S, Morimoto Y, Minod Y (1982) Purification and some properties of cell-bound lipase from Saccharomycopsis lipolytica. Agric Biol Chem 46:2885–2893

Peireira-Meirelles F, Rocha L, Sant'Anna G (1997) A stable lipase from *Candida lipolytica* : cultivation conditions and crude enzyme characteristics. Appl Biochem Biotechnol 63–65:73–85

Pignede G, Wang H, Fudalej F, Gaillardin C, Seman M, Nicaud JM (2000a) Characterization of an extracellular lipase encoded by *LIP2* in *Yarrowia lipolytica*. J Bacteriol 182:2802–2810

Pignede G, Wang HJ, Fudalej F, Seman M, Gaillardin C, Nicaud JM (2000b) Autocloning and amplification of *LIP2* in *Yarrowia lipolytica*. Appl Environ Microbiol 66:3283–3289

Pleiss J, Scheib H, Schmid RD (2000) The His gap motif in microbial lipases: a determinant of stereoselectivity toward triacylglycerols and analogs. Biochimie 82:1043–1052

Sarda L, Desnuelle P (1958) Actions of pancreatic lipase on esters in emulsions. Biochim Biophys Acta 30:513–521

Song H, Jiang Z, Ma L (2006) Expression and purification of two lipase from *Yarrowia lipolytica* AS 2.1216. Protein Expr Purif 47:393–397

Suzuki T, Imamura K, Yamamoto K, Satoh T, Okazaki M (1997) Thermal stabilization of freeze-dried enzymes by sugars. J Chem Eng Jpn 30:609–613

Tanaka A, Sugimoto H, Muta Y, Mizuno T, Senoo K, Obata H (2003) Differential scanning calorimetry of the effects of calcium on the thermal unfolding of *Pseudomonas cepacia* lipase. Biosci Biotechnol Biochem 67:207–210

Theveniau F, Gaillardin C, Nicaud JM (2009) Application of the non-conventional yeast *Yarrowia lipolytica*. In: Satyanarayana T, Kunze G (eds) Yeast biotechnology: diversity and applications. Springer, Dordrecht, pp 589–613

Thevenieau F, Le Dall MT, Nthangeni B, Mauersberger S, Marchal R, Nicaud JM (2007) Characterization of *Yarrowia lipolytica* mutants affected in hydrophobic substrate utilization. Fungal Genet Biol 44:531–542

Turki S, Kraeim IB, Weeckers F, Thonart P, Kallel H (2009) Isolation of bioactive peptides from tryptone that modulate lipase production in *Yarrowia lipolytica*. Bioresour Technol 100:2724–2731

Turki S, Ayed A, Chalghoumi N, Weekers F, Thonart P, Kallel H (2010) An enhanced process for the production of a highly purified extracellular lipase in the non-conventional yeast *Yarrowia lipolytica*. Appl Biochem Biotechnol 160:1371–1385

Verger R (1997) Interfacial activation of lipase: facts and artifacts. Trends Biotechnol 15:32–38

Yu M, Lange S, Richter S, Tan T, Schmid RD (2007a) High-level expression of extracellular lipase Lip2 from *Yarrowia lipolytica* in *Pichia pastoris* and its purification and characterization. Protein Expr Purif 53:255–263

Yu M, Qin S, Tan T (2007b) Purification and characterisation of the extracellular lipase Lip2 from *Yarrowia lipolytica*. Process Biochem 42:384–391

Biotechnological Applications of *Yarrowia lipolytica* Lipases: An Overview

Patrick Fickers and Jean-Marc Nicaud

Abstract The yeast *Yarrowia lipolytica* has the ability to grow on substrates such as lipid or grease. The first step in the breakdown and metabolism of those hydrophobic substrates requires the actions of lipolytic enzymes. Lipases from *Y. lipolytica*, and particularly the lip2 enzyme, possess amazing properties in both aqueous and nonaqueous media. Those catalytic properties have enabled the development of many applications. Oil mill wastewater represents a serious risk of pollution due to its high contents of organic compounds, mainly fatty acids and triglycerides. Besides the physicochemical processes developed to treat those wastewaters, the Lip2p lipase was found as an attractive and low-cost biological alternative. In that field, they were also used successfully to degrade sludge for grease trap from the food industries. Applications were also developed in fine chemistry due to the abilities of Lip2p as an enantioselective catalyzer in organic medium for the synthesis of pharmaceutical drug used as a single enantiomer. Besides this, Lip2p was also found effective in the polymerization and modification processes of different fatty compounds of industrial interest. Microorganisms and their produced enzymes are also known for their role in traditional food making. The role of the lipase enzyme from *Y. lipolytica* in flavor development in cheese and fermented sausage is detailed in this review. Their involvement in the production of organic acids such as citric acid from hydrophobic compounds is also presented.

P. Fickers (✉)
Unité de Biotechnologies et de Bioprocédés, Université Libre de Bruxelles, Av. F.-D. Roosevelt 50, C.P. 165/61, 1050 Bruxelles, Belgium
e-mail: pfickers@ulb.ac.be

J.-M. Nicaud
CNRS, Institut MICALIS, Domaine de Vilvert, 78352 Jouy-en-Josas, France
e-mail: jean-marc.nicaud@grignon.inra.fr

G. Barth (ed.), *Yarrowia lipolytica*, Microbiology Monographs 25,
DOI 10.1007/978-3-642-38583-4_4, © Springer-Verlag Berlin Heidelberg 2013

Contents

1 Introduction 122
2 Waste Treatment 122
2.1 Treatment of Oil Mill Effluents 122
2.2 Treatment of Sewage Sludge from the Food Industry 126
3 Fine Chemistry 127
3.1 Resolution of Racemic Mixture 127
3.2 Polymerization Reaction and Modification of Fat 129
4 Applications in Traditional Food Making 130
4.1 Cheese Ripening and Maturation 131
4.2 Fermented Sausage 132
5 Production of Citric Acid 133
6 Conclusions 134
References 134

1 Introduction

The yeast *Yarrowia lipolytica* is often associated with proteinaceous or hydrophobic substrates such as lipid or grease. To assimilate these hydrophobic substrates, *Y. lipolytica* has developed an adaptative strategy in which lipolytic enzymes play a crucial role. Besides their physiological functions, the amazing catalytic properties of lipases have enabled the development of many industrial and/or environmental processes. The main applications developed with *Y. lipolytica* lipase will be discussed in this review. Of these, the first corresponds to the treatment of waste and sludge from the agro-food industry. The second leads to fine chemical production and enantioseparation of pharmaceutical compounds by lipase in aqueous or aqueous-free medium. The third area of application refers to traditional food manufacturing, while the last one is linked to the production of metabolites such as citric acid from hydrophobic substrates.

2 Waste Treatment

Many food industries generate in their process large amount of by-products with a high content of organic compounds that must be treated before their release. Nowadays, more ecological solutions are developed to substitute physicochemical process to degrade those by-products. Alternative applications involving the lipase from *Y. lipolytica* for the treatment of those wastes are presented in this section.

2.1 Treatment of Oil Mill Effluents

Wastewaters from the olive oil processing industry may cause severe pollution especially in the Mediterranean countries which account for about 95 % of the world olive production (Al-Malah et al. 2000). Olive mill wastewater (OMW) represents an annual volume of 3×10^7 m^3 which is produced in a short period of time (from November to late February) (D'Annibale et al. 1998, 2004). The quality and the quantity of the different constituents of OMW are dependent on several factors: the type of olives, the cultivation system, or the production process (Lopes et al. 2009). OMW forms a dark acidic stable emulsion emitting a strong oily smell. It is composed of olive "vegetation waters," waters from processing, olive pulp, and oil. OMW contains, in addition to fat and triglycerides, sugars, phosphate, polyphenols, polyalcohols, pectins, and metals. Another negative property of OMW is its extremely high organic content. Generally, OMW has a biochemical oxygen demand (BOD) ranging between 12,000 and 63,000 mg/l and a chemical oxygen demand (COD) value between 80,000 and 200,000 mg/l (Al-Malah et al. 2000). These values are around 200 to 400 times higher than typical municipal sewage (Cossu et al. 1993). Hence, OMW represents a serious source of water pollution if released into local rivers without any treatment. Several physicochemical processes, including simple evaporation, reverse osmosis, or ultrafiltration, have been proposed to reduce the polluting effect of OMW (Mameri et al. 2000; Al-Malah et al. 2000). Besides these, biological degradation is considered as a safe, effective, and low-cost process of removing harmful pollutants. For OMW, the conventional anaerobic systems of waste treatment have been experimented with. Although it yields to a methane production of 57 l/l of OMW, anaerobic cultures require an adaptation period of 15–25 days, which increase consequently the costs of storage (Andreoni et al. 1993). Therefore, several researches aimed to select microorganisms that are able to grow more rapidly using OMW as primary carbon source. In the early 1960s, Ros de Ursinos (1961) selected a strain of *Torulopsis utilis* that is able to grow on OMW that produced proteins and vitamins. Pilot experiments with this yeast yielded a 50 % in pollution. More recently, Scioli and De Felice (1993) tested the ability of different yeast species to grow on oil mill wastewaters and selected *Y. lipolytica* as the most adapted organism to grow in these conditions (Table 1). This yeast is particularly well suited to these applications since it has a GRAS status, limited nutritional requirements, and a capacity of adaption to stringent environmental conditions. In addition, *Y. lipolytica* is known to produce several different lipases with specific catalytic properties, indispensable for lipid and fat metabolization. The *Y. lipolytica* strain ATCC20255 isolated by Scioli and De Felice (1993) was found capable of reducing the initial COD value (146 g/l) by 80 % within 24 h in a 3.5-l tank fermenter with an increase of its biomass to 23 g/l despite the presence of large amount of phenols (200 mg/l). The production of lipase (770 U/l after 24 h), free and cell bound, could be detected and correlated to the fat degradation. At the same time, Scioli and Vollaro (1997), who obtained similar results with strain ATCC20255, also reported that after processing, the waters had a pleasant smell and did not exhibit the initial oily aspect and intense smell. With the aim to be more exhaustive, Lanciotti et al. (2005) tested the ability of 62 different *Y. lipolytica* isolates to grow in non-diluted and non-supplemented OMW and to reduce its COD level (Table 1). PO1 strain was found the most effective to significantly reduce both

Table 1 Application of *Y. lipolytica* in the treatment of oil mill effluent

Strain	Type of waste	Reduction	Lipase (U/l)	Other product	Refs
ATCC 20255	OMW + $(NH_4)_2SO_4$ + YE	80 % COD	770 cell free 980 cell bound	Biomass 22 g/l	Scioli and Vollaro (1997); de Felice et al. (1997)
62 different strains	OMW (diluted or not)	1.5–41 % COD 0–18 % polyphenol	35–2,315	0–5.2 g/l CA	Lanciotti et al. (2005)
ACA-DC 50109	OMW + glucose	15 % polyphenol		28.5 g/l CA	Papanikolaou et al. (2008)
W29	OMW + $(NH_4)_2SO_4$ + YE	61–79 % COD 57–72 % polyphenol	49–78	n.d.	Lopes et al. (2009)
IMUFRT 50682	OMW + $(NH_4)_2SO_4$ + YE	75–80 % COD 39–68 % polyphenol	16–27	n.d.	Lopes et al. (2009)
W29	Different crude OMW	21–36 % COD 30 % polyphenol	320–451	n.d.	Goncalves et al. (2009)
CBS 2073	Different crude OMW	23–51 % COD 25 % polyphenol	828–1,041	n.d.	Goncalves et al. (2009)
IMUFRJ 50682	Different crude OMW	23–50 % COD 20 % polyphenol	317–533	n.d.	Goncalves et al. (2009)
W29 immobilized	Oil wastewater	82 % COD	n.d.	n.d.	Wu et al. (2009)
W29	Oil wastewater	67 % COD	n.d.	n.d.	Wu et al. (2009)
NCIM 3589	POME	97 % COD 80 % BOD	n.d.	n.d.	Oswal et al. (2002)

n.d. not determined, *YE* yeast extract, *CA* citric acid, *OMW* olive mill wastewater, *POME* palm oil mill effluent

the COD (43 %) and the polyphenol (18 %) content. Other strains were found to produce citric acid or lipase at a high level in these conditions. From this work, it appears that among the different strains tested, a great variability in lipase activity occurred both in the yield of production and in the enzyme specificity. Besides this, *Y. lipolytica* ACA-DC-50109 cultivated on OMW-based media enriched with commercial-industrial glucose was found to present a citric acid production yield of 28.9 g/l (Papanikolaou et al. 2008) (Table 1). All of these suggest that the lipid content of OMW could be metabolized, upon the action of lipase, to synthesize valuable metabolites. Lopes et al. (2009) investigated OMW-based medium supplementation with either nitrogen (ammonium) or surfactant (Tween 80) on cell growth, COD, and phenol reduction as well as lipase synthesis. They showed that the lipase productivity was improved in the presence of ammonium sulfate (6 g/l), whereas an addition of Tween 80 had a negative effect on lipase activity and a significant positive effect on both cell growth and COD reduction. With the same *Y. lipolytica* strain (W29), Goncalves et al. (2009) reported a lipase production of 3,500 U/l in the same supplemented medium, showing that this particular strain could be used for the scale up of lipase production from OMW (Table 1).

All those applications for the treatment of OMW were first developed exclusively with free cells. However, Wu et al. (2009) recently demonstrated that the immobilization of *Y. lipolytica* cells by calcium alginate could present several advantages in the treatment of wastewater from the oil industry (Table 1). These authors reported a COD degradation rate significantly higher for immobilized cells than for free cells. They hypothesized to explain this phenomenon, that for a certain period, the alginate matrix acted as oil and COD adsorbent as well as immobilized carrier. The immobilization of cells also permitted a wider temperature range for COD reduction. Indeed, the suitable temperature for oil degradation by free and immobilized cells range from 25 to 30 °C and 25 to 35 °C, respectively, which is close to the optimal catalytic temperature of the main extracellular lipase Lip2p from *Y. lipolytica* CBS6303 (Destain et al. 1997). From a technological viewpoint, reuse of the immobilized cells is of great advantage. Indeed, this practice can decrease waste of cells, save time, and cut down cultivation costs. For an artificial oil wastewater containing 20 % of oil, immobilized cells could be reused for a maximum of 12 cycles for a total oil degradation of 2,351 mg. Even after the 12 utilizations, a 77 % degradation of the oil content could be obtained. However, for the COD reduction, the immobilized cells could only be reused for six cycles with a total COD degradation of 1,745 mg without any loss of the degradation capacity (Wu and Wan 2008). These authors also investigated the storage stability of oil degradation activity. They observed that the oil degradation was not significantly altered upon storage of both immobilized and free cells in distilled water at 4 °C.

Palm oil mill effluent (POME) is another possible source of inland water pollution. POME composition is somewhat different to OMW. It contains mainly lignocellulosic wastes with a mixture of carbohydrates and oil. POME also presents very high BOD and COD values of 246,000 and 11,000 mg/l, respectively (Oswal et al. 2002). In addition, incomplete extraction of palm oil from the palm nut could also lead to a substantial increase of COD values. POME treatment using the *Y. lipolytica* marine strain NCIM 3589 was investigated without any addition of nutriment or dilution (Table 1). This strain, known for its efficient lipase production, yielded to a 97 % and 80 % reduction of the COD and BOD, respectively, within 48 h. These results are comparable to those obtained previously with a pond treatment system (Chin et al. 1996). Oswal et al. (2002) reported that the initial acidic pH of POME became alkaline after treatment with *Y. lipolytica* probably due to the utilization of fatty acids present in the raw POME by this yeast.

2.2 Treatment of Sewage Sludge from the Food Industry

The food industries, especially those involved in ready-to-eat meal manufacturing, generate sewage sludge with high lipid contents and high COD values. Similarly to oil mill effluent, this waste must be treated before its release into nature. This is usually carried out through a stepwise process. Of these, one is constituted by a grease trap where lipids and grease could accumulate based on their relative low density (Ansenne et al. 1992). Sludge from this grease trap is acidic (pH 4.5–5.5) and contains a dry weight content ranging from 25 to 75 % and a COD value close to 4,000 mg/l. The dry matter is composed of up to 90 % of lipid and grease, mainly triglyceride, free fatty acid, and sterol ester (Thonart et al. 1997). Such effluents were used to grow different yeast and bacteria strains selected for their ability to grow on hydrophobic substrates and to secrete large amount of hydrolases, including lipases. From those experiments, a *Y. lipolytica* strain producing a large amount of the extracellular Lip2p lipase was selected. This strain was tested for its ability to reduce the lipid contents in a 6-m^3 grease trap fed continuously with fresh sewage at a flow rate of 6 m^3 every 24 h. This led to a significant reduction of the lipid content and the maintaining of the COD at a value of 3,000 mg/l during 33 weeks of treatments (Thonart et al. 1997).

In Asia and especially in Japan, where seafood is consumed at a high level per capita, the amount of the fish waste was estimated to be more than two million tons per year, and half is discarded as industrial waste (Hirai 2001). Fish waste mainly consists of viscera, heads, and fish bones as well as fish that are too small to be processed. Some of these wastes have been utilized for the production of condiments and valuable materials (Saito 2004). However, such kinds of processing often produce new waste in addition to the product. Fishmeal is widely used as a supplementary protein source for livestock and culturing fish. Processing of fish waste to fishmeal appeared as the best way to utilize this waste. However, as fish waste contains viscera, it has a relatively high lipid and a low protein content that confers on it a poor nutritive quality. Thus, the reduction of this lipid content could permit the utilization of fish waste for the production of fishmeal. In that context, different microorganisms were screened for their ability to metabolize lipids in fish samples without any additional materials. Yano et al. (2008) isolated the *Y. lipolytica* strain NRBC-10073 that was able to reduce the lipid content of minced anchovy samples by 30 % without any modification of the proteic composition. The lipid reduction by this strain was especially affected by the ratio of the surface area to the weight in the fermented minced samples and by the water content, suggesting the importance of the oxygen supply in the process. Hence, this work demonstrates that the fermentation by *Y. lipolytica* can improve the quality of fishmeal from fish waste.

Besides this, Dominguez et al. (2005) reported the utilization of waste cooking oil for lipase production by a *Y. lipolytica* strain. After 3 days of treatment, a biodegradation degree of 80 % was obtained, as indicated by the COD value, with an important lipase production. The high hydrolytic activity towards medium-chain-length ester hinted at the occurrence of both lipases and esterases in this process.

3 Fine Chemistry

The demand for compounds synthesized as a single enantiomer has increased during the last years. In that context, the utilization of enzymes, which enable stereoselective reaction, as catalyzers was found as an attractive alternative to the classical chiral chemistry. Applications involving the lipase from *Y. lipolytica* were developed recently for the resolution of racemic mixture and in the polymerization and modification of fat.

3.1 Resolution of Racemic Mixture

In recent years, the synthesis or resolution of optically active drugs and their intermediates has been intensively investigated due to the medical benefit of using single enantiomer. There are numerous examples in which the desired biological activity solely resides in one enantiomer of a chiral drug, with the other isomer being less potent, inactive, or even acting with cross-purpose effects (Tsai et al. 1997). Hence, the need for enantiomerically pure molecules, especially in the pharmaceutical industry, has grown since the legislation required investigations into the pharmacological effect of both enantiomers. The mark of drugs sold as single enantiomer has more than a 10 % growth per year and represented around $225 billion worldwide in 2005 (Bordes et al. 2009). Since classical ways to obtain pure enantiomer, i.e., chemical asymmetric synthesis, stereoselective crystallization, or chiral chromatography, are often expensive, the use of enzymes as catalyzers becomes very attractive. In that context, the extracellular lipase Lip2p from *Y. lipolytica* was found to be very effective for the resolution of 2-halogeno-carboxylic acids which are important intermediates in the synthesis pathways of a number of drugs (analgesics, prostaglandin, prostacyclin, semisynthetic penicillin) (Fig. 1). The resolution of 2-bromo-*p*-tolyacetic acid ethyl ester catalyzed by *Y. lipolytica* lipase Lip2p showed an (S)-enantiopreference of 28, which is similar to the best results obtained with the lipase from *Burkholderia cepacia* ($E = 30$). Moreover, *Y. lipolytica* lipase presents a higher catalytic activity and an (S)-enantiopreference, while *B. cepacia* lipase is (R)-enantiomer selective (Guieysse et al. 2004). For the 2-bromo-*o*-tolylacetic acid, a precursor of analgesics and non-peptide angiotensin II receptor antagonists, none of the commercial lipase was able to resolve the racemic mixture (Guieysse et al. 2003a). Only the *Y. lipolytica* lipase Lip2p was able to perform this resolution (Guieysse et al. 2003b). Despite enantioselectivity values obtained with Lip2p being promising, they might not be sufficient for pharmaceutical applications that require a high purity level. However, Lip2p is a good candidate to develop enantioselective catalysts through site-directed mutagenesis or directed evolution. Bordes et al. (2009) reported the improvement by site-directed mutagenesis of the enantioselectivity of the *Y. lipase* Lip2p for the resolution of 2-bromo-arylacetic ester.

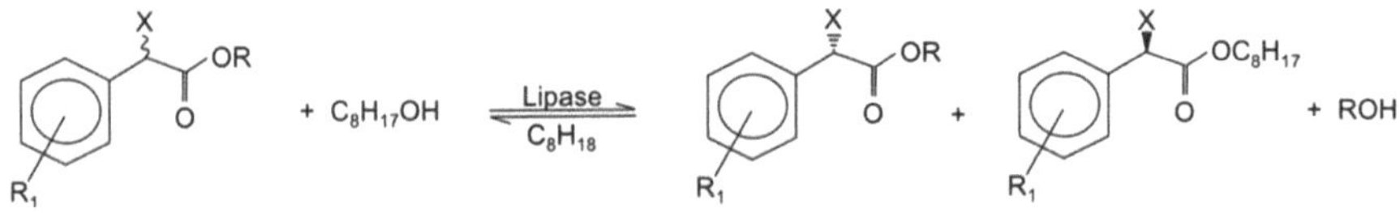

Fig. 1 Transesterification reaction between 2-halogeno-carboxylic acids and 1-octanol catalyzed by the extracellular lipase Lip2p from *Y. lipolytica* in *n*-octane

On the basis of a Lip2p structural model obtained by modeling techniques, five amino acid residues (T88, V94, D97, V232, V285) that form the hydrophobic substrate-binding site of the lipase were selected for site-directed mutagenesis. Position 232 was identified as crucial for the discrimination between enantiomers. Variant V232A displayed an enhanced enantioselectivity by one order of magnitude, whereas variant V232L exhibited a selectivity inversion. To further explore the diversity, position 232 was mutated by 19 other amino acids. Analysis of the obtained mutants led to the selection of the V232S variant, which has a tremendously increased *E* value compared to the wild-type enzyme for the resolution of 2-bromo-phenylacetic acid ethyl ester (58-fold) and 2-bromo-*o*-tolylacetic acid ester (16-fold). In addition to the gain in enantioselectivity, an eightfold increased velocity was observed for both substances (Bordes et al. 2009).

Ibuprofen is an arylpropionic acid related to the class of nonsteroidal, anti-inflammatory drugs. It is well known that the anti-inflammatory activity of this class of compounds is mainly due to the active (S)-enantiomer (Hutt and Caldwell 1984), and (S)-ibuprofen is 160-fold more active than its antipode in the in vitro synthesis of prostaglandin (Adams et al. 1976). In addition, (R)-ibuprofen displays toxicity due to its storage in fatty tissue as a hybrid glycerol ester, whose long-term effects are not known. In that context, Liu et al. (2009) reported the cyclic resolution of ibuprofen using coupled acid-base and lipase catalysis. They showed that the Lip2p lipase from *Y. lipolytica* had a higher affinity for the (S)-enantiomer. In their process, the unreacted (R)-enantiomer is extracted, racemized in a basic solvent–water mixture before being re-resolved. The (S)-ester was separated and hydrolyzed to (S)-ibuprofen in acidic dimethyl sulfoxide–water mixture. The high purity (S)-ibuprofen (ee $= 0.98$) was obtained using Lip2p lipase at pH $= 8$.

Optically pure amines can be used in the fine chemical industry as resolving agents, chiral auxiliaries, and chiral synthetic building blocks for pharmaceuticals as well as agrochemical compounds (Breuer et al. 2004; O'Donnell 2001). A variety of methodologies have been developed for the production of enantiopure amines. However, the latter are time-consuming, expensive, and still present a high risk of racemization. However, hydrolytic enzyme and especially lipase have been widely exploited to solve these problems. ($\pm$)α-Phenylethyl amine is widely used as a powerful intermediate in industrial asymmetric synthesis or chiral adjuvant (Juaristi et al. 1999). Wen et al. (2008) studied the enantioselective aminolysis of immobilized extracellular lipase from *Y. lipolytica* by catalyzing enantioselective acylation of ($\pm$) α-phenylethyl amine with acetic ester in a cosolvent medium. In their optimized condition, i.e., in hexane containing 3 % DMSO at 45 °C, the

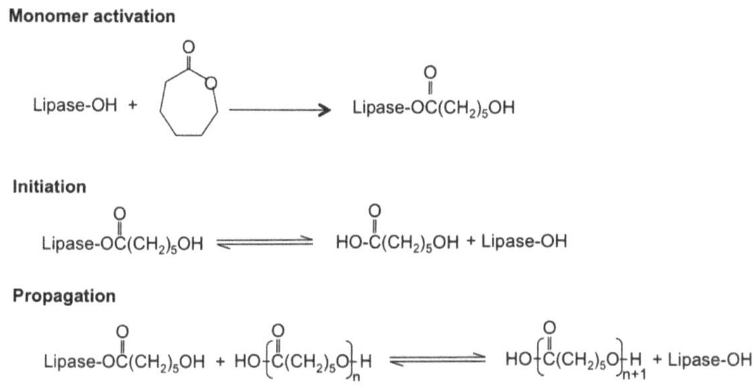

Fig. 2 Scheme for the polyether synthesis by the so-called ring-opening polymerization (ROP) (inspired from Barrera-Rivera et al. 2008)

enantiomeric excess of the product markedly increased from 0.35 to 0.96 after 6 days of reaction with an *E* value close to 190. They also showed that the immobilized lipase could be reused in at least five consecutive batches with a high *E* value.

3.2 Polymerization Reaction and Modification of Fat

Polymerization reactions catalyzed by enzymes proceed generally through quimioselective, regiospecific, and stereoselective pathways. However, lipase-catalyzed reactions are quimioselective and proceed via the formation of an acyl-enzyme intermediate. For polyether synthesis by the so-called ring-opening polymerization (ROP), the key step is the reaction of the lactone ring with the lipase to provoke the ring opening and the formation of an acyl-enzyme intermediate (Fig. 2). The initiation step involves a nucleophilic attack of a water molecule onto the acyl carbon of the intermediate to produce a ω−hydroxycarboxylic acid ($n = 1$), the shortest propagating entities. In the propagation stage, the intermediate is nucleophilically attacked by the terminal hydroxyl group of a propagating polymer to produce a one-unit-more elongated polymer chain. The kinetics of the polymerization showed that the rate-determining step of the overall polymerization is the formation of the enzyme activated monomer. Therefore, the polymerization probably proceeds via an "activated monomer mechanism." Based on that scheme, Barrera-Rivera et al. (2008) investigated the ring-opening polymerization reaction of the ε-caprolactone in the presence of *n*-heptane by the extracellular lipase from *Y. lipolytica*. After 360 h of reaction at 50 °C in the presence of 3 mmol of ε-caprolactone and 100 mg of enzyme, a 100 % conversion rate was obtained and a polyester molecule presented an average mass of 970 Da. The final polymers were found to correspond to an asymmetric α-hydroxy-ω-carboxylic acid poly (ε-caprolactones).

Besides these polymerization reactions, there has been considerable interest in the recent years in lipase-catalyzed reaction for the production of fatty acids, modification of oils and fats, and synthesis of various esters. The numerous substances obtained have many applications in the food, chemical, pharmaceutical, and medical sectors. Despite the fact that lipases are generally regarded as ester hydrolase, they can also catalyze the synthesis and transfer of ester with a regio- and/or stereospecificity. Ester transfer reaction (i.e., alcoholysis) does not directly involve water; they are usually carried out at a low water activity in order to hinder ester hydrolysis. Briand et al. (1994) investigated the alcoholysis of rapeseed oil by methanol using the extracellular lipase from *Y. lipolytica*. The transesterification reaction leads to an equilibrium state after 8 h with a reaction yield (methyl esters formed/total fatty acid initially present in the acylglycerols) of 73, 45, and 65 % for linolenic, linoleic, and oleic acid, respectively. The presence of methanol was found to favor transesterification reaction to the detriment of ester hydrolysis due to methanol inhibition of hydrolysis. This work demonstrated the ability of methyl-ester synthesis using the lipase from *Y. lipolytica*.

Monoacylglycerols (MAGs) are nonionic surfactant emulsifiers with their hydrophilic and hydrophobic parts. They are widely used in the food industry, with applications in dairy products, margarines, and bakery products. In addition, they present lubricant and plasticizing properties, and they are used in textile processing, production of plastics, and formulation of oil for different machineries (Esmelindro et al. 2008). Currently, MAG production is performed by chemical glycerolysis of fats and oils at high temperature in the presence of inorganic alkaline catalysts. Because this process is energy consuming, provided low reaction yield (30–40 %), and required product post-purification by molecular distillation, new techniques for MAG production were developed. These are mainly based on lipase-catalyzed reaction and proceed in organic medium, in solvent-free medium, in ionic liquids, or using compressed fluids as reaction media. In that context, Esmelindro et al. (2008) reported the production of MAG from olive oil in pressurized propane as solvent medium using different type of lipase, including the extracellular lipase from *Y. lipolytica*. Their results showed that lipase-catalyzed glycerolysis in compressed propane might be a potential alternative to conventional methods, as a high content of reaction products was obtained at mild temperatures (30 °C) and pressure conditions (30 bars) with a low solvent to substrate mass ratio (4:1) and in a short reaction time (3 h).

4 Applications in Traditional Food Making

It is well established that *Y. lipolytica* is naturally present in different kinds of food, underlining the importance of this yeast in the agro-food industry. The appearance of different yeasts, including *Y. lipolytica*, and their role in the production, ripening, or spoilage of traditional dairy products and meat products such as sausage have been studied for many years.

4.1 Cheese Ripening and Maturation

The production of mold-ripened cheeses, such as the Camembert and blue-veined cheese varieties, involves a maturation stage that is characterized by the growth of a complex ecology of yeast, bacteria, and filamentous fungi. The microbiological interactions and associated biochemical activities that occur during this stage determine product acceptability and value through their impact on sensory quality, shelf life, and safety (Addis et al. 1998). The role of yeast depends on the type of cheese: in some varieties they are responsible for spoilage, gassiness, slime formation, and discoloration, while in others they are involved in the ripening process and contribute to microbial interactions, texture changes, and biosynthesis of flavor compounds (Suzzi et al. 2001). Due to features such as proteolytic and lipolytic activities, yeast species including *Y. lipolytica* play an important role in the production of aroma precursors, especially amino acids, fatty acids, and ester. In particular, the dynamics of the free fatty acids (FFAs) release, which largely depends on the milk fat composition and microbial lipase selectivity and activity, determines the flavor of many dairy products (Suzzi et al. 2001). In order to better characterize this phenomenon, Addis et al. (1998) monitored the growth of yeast and bacteria during maturation of retail Camembert and blue-veined cheeses. Yeasts were found predominantly throughout the maturation process with *Debaryomyces hansenii* and *Y. lipolytica*, however, to a lesser extent, being the main constituent of the microbial flora. Among the 25 *Y. lipolytica* strains isolated from different Camembert and blue-veined cheese samples, 23 of them were found to present a lipolytic activity (varying from weak to strong) on tributyrin and butterfat agar. Suzzi et al. (2001), who performed a similar experiment, observed two different behaviors in terms of lipolytic activity. Some *Y. lipolytica* strains showed very high lipolytic activities over the first 3 days of maturation, producing the highest amounts of total FFAs. However, extended incubation of these strains resulted in a significant decrease of total FFA concentration. For other strains, characterized by a lower lipolytic activity after those 3 days, the total FFA content increased until the end of the maturation period. However, the authors reported that the specificity of the individual fatty acids released, and presumably their subsequent metabolisms, does not necessarily rely on the level of lipase production during the three first days of maturation. When the released FFAs are analyzed in more detail, it appears that short-chain FFAs (C4–C10) were produced by all the strains at low levels (1–2 % of the total FFAs) during the first days of maturation. After 6 days, longer-chain FFAs, such as palmitic (C16:0), palmitoleic (C16:1), stearic (C18:0), oleic (C18:1), and linoleic (C18:2), decrease when strains producing high level of lipase are present. The linolenic acid (C18:3) released during the first 3 days tended to disappear whatever the level of lipase production. In regard to lipase specificity, all the isolated strains hydrolyzed both saturated and unsaturated fatty acids from milk fat with the liberation of high proportion of even-numbered carbon FFAs. The latter are then subsequently transformed by microbial enzymes such as lipoxygenases, epoxydases, and hydratases in the corresponding hydroxy acids which are then oxidized in shorter molecules including lactones, by means of α-, β-, and δ-oxidations (Fickers et al. 2005). The diminution of total FFAs occurring in the presence of certain *Y. lipolytica* strains suggests that these molecules are transformed into the corresponding alcohols, ketones, and lactones. It was also highlighted that the levels of the observed lipolytic activities did not correlate with genetic variations observed in the different producing strains (Suzzi et al. 2001).

By contrast to Camembert and blue-veined cheese, raclette is a semihard cheese. In this type of cheese, short-chain FFAs are important in the flavor development during ripening. *Y. lipolytica* was found responsible for the increase of *n*-butyric acid as a result of milk fat hydrolysis (Wyder et al. 1999). The lack of modification in *n*-butyric and *n*-caproic ratios during the maturation process suggests a lipase action rather than an esterase activity. Surprisingly, the authors found that *Y. lipolytica* did not seem to contribute to the breakdown of proteins and peptides despite its strong extracellular proteolytic capacity. Since practically no yeast cells were detected in the mature cheese, the action of yeasts in the maturation process could be attributed to enzymes released after cell lysis.

4.2 Fermented Sausage

In Europe, dry fermented sausages have a long tradition originating from Mediterranean countries since Roman times. Many types of fermented sausages have been developed together with processing conditions, additives, and ingredient formulation. Among the flavor products identified in dry sausages, the oxidation products of lipids, released upon meat fat hydrolysis by lipase, account for about 60 % of the total compounds which influence the flavor (Berdague et al. 1993). Lactic acid bacteria *Micrococci* and coagulase-negative *Staphylococci* have the most relevant role in the fermentative process and ripening. However, yeast and mold were also found associated with this process notably in the development of specific organoleptic characteristics (Samelis et al. 1993). Gardini et al. (2001) followed the yeast population during manufacturing and ripening of "salsiccia sotto sugna," a typical salami of the Lucania region (Italy). Four different batches, from four different farms in Luciana, were studied. Although each batch showed a specific yeast population, *Y. lipolytica* strains were isolated in all of them and they were found to hydrolyze pork fat at different levels. Some of them produced high amounts of free fatty acids (FFAs) after 6 days of growth, while others produced a relatively low amount of FFAs, without significant differences in relation to the incubation time. A third group was responsible for the slight decrease of total FFAs in the medium after 6 days of incubation probably due to their metabolization or oxidization into flavor compounds. The major products of the lipolytic activity of *Y. lipolytica* strains from the third group were, in decreasing order, oleic, palmitic, stearic, linoleic, and myristic acid. This clearly demonstrates a specificity of the lipolytic enzymes for the positions *sn*1 and *sn*3 of the triglycerides. In fact, these are the positions of triglycerides most frequently occupied by unsaturated fatty acids in pork fat. Gardini et al. (2001) also reported that saturated and unsaturated fatty acids were present at similar concentrations at pH 5.5, while for other pH values, the saturated fatty acids were at lower concentrations. This highlights that *Y. lipolytica* lipase could favor at pH 5.5 the liberation of saturated FFAs rather than unsaturated. This tendency could have a positive effect by reducing the phenomenon of rancidity in which polyunsaturated FFAs are involved.

5 Production of Citric Acid

Citric acid (CA) is an intermediate of the tricarboxylic acid cycle (TCA) that holds a key position in the central metabolic fluxes in cells. Due to its acidulant, flavoring agent, and antioxidant properties, CA is used mainly in the food and beverage industry. In recent years, the consumption of citric acid and its salt, trisodium citrate, has reached worldwide 800,000 tons with an increase of 5 % per year (Kamzolova et al. 2005). Under certain conditions of fermentation, fungi, bacteria, and yeasts can produce CA in large amounts. Traditionally, the yeast strains, mostly belonging to *Aspergillus niger*, have been used for commercial production of CA from molasses, sucrose, or glucose (Kristiansen and Sinclair 1979). However, the production of CA with the use of fungi is associated with the accumulation of significant amounts of solid and liquid waste. As an alternative, there is a great interest in the possibility of CA production by yeast. *Y. lipolytica* is known to produce a wide range of organic acid, including TCA cycle intermediates, such as CA or isocitric acid (ICA). Many studies have been dedicated to this production using different sources of carbon (*n*-alkane, raw glycerol, ethanol). *Y. lipolytica* strains used in those processes are characterized by a greater resistance to high substrates and metal ion concentrations, thus allowing the use of less refined substrate (Rane and Sims 1993). In that context, Kamzolova et al. (2005, 2007) investigated the production of citric acid in a 3-l bioreactor using different *Y. lipolytica* strains grown in the presence of beef fat or olive oil. For *Y. lipolytica* strain 704, lipase production could be observed for the two hydrophobic substrate sources with a very high value of lipase activity (2,760 U/ml) when olive oil was used as carbon source. In the course of cell cultivation, glycerol and free fatty acid concentrations remain constant in the culture medium, suggesting that they were consumed simultaneously upon hydrolysis of triglycerides by lipase enzymes. FFA composition of the culture medium indicated that oleic and linoleic acids were the most representative hydrophobic substrates. However, the strain producing the higher amount of lipase was not the one that produced the larger quantity of citric acid. This clearly demonstrates that the two phenomena are not directly connected. A maximum of 135 g/l of citric acid was obtained at the end of the culture for strain *Y. lipolytica* 187/1. This value was significantly higher than those reported in the literature for other species (Klasson et al. 1989; Wojtatowicz et al. 1991). In addition, the undesired ICA was produced only in low amount (7.8 g/l). With *Y. lipolytica* strain Y-2373, the concentrations of ICA and CA obtained in similar conditions were 34 and 40 g/l, respectively, which correspond to an ICA/CA ratio of 0.85. However, Kamzolova et al. (2007) noticed that the composition of citric acids depends considerably on the pH value. At pH 6, the concentrations of ICA and CA were 55.4 and 21.7 g/l, respectively, with an ICA/CA ratio of 2.55:1, while at pH 4.5, the ratio of ICA to CA was 1:1.18. It is known that CA transport across the membrane is favored by low pH value (Peltsmane et al. 1988), whereas ICA transport is pH independent. This may explain why at pH 4.5, the ratio of ICA to CA changes in favor of CA. All of these suggest that plant oils appear as a promising substrate for citric acid production by *Y. lipolytica*.

6 Conclusions

The ability of *Y. lipolytica* to grow on raw substrate and to produce enzymes and metabolites in large amounts has received all the attention from industrials and academics for more than three decades. However, the availability nowadays of "omic" techniques and sophisticated molecular tools should permit rapid advances in the field.

References

Adams S, Bresloff P, Mason C (1976) Pharmacological difference between the optical isomers of ibuprofen: evidence for metabolic inversion of the (-)isomer. J Pharm Pharmacol 28:256–257

Addis E, Fleet G, Cox JM (1998) Evaluation of methods for the isolation and enumeration of yeast in mould-ripened cheese. In: Jakobsen M, Narvhus J, Viljoen BC (eds) Yeast in the dairy industry: positive and negative aspects. International Dairy Federation, Brussels, pp 109–116

Al-Malah K, Azzam M, Abu-lail N (2000) Olive mills effluent (OME) wastewater post treatment using activated clay. Sep Purif Technol 20:225–234

Andreoni V, Bonfanti P, Daffonchio D, Sorlini C, Villa M (1993) Anaerobic-digestion of olive mill-effluent: microbiological and processing aspects. J Environ Health Part A 28:2041–2059

Ansenne A, Destain J, Godefroid J, Thonart P (1992) La problématique des séparateur de graisse. Trib eau 558:33–40

Barrera-Rivera K, Flores-Carreon A, Martinez-Richa A (2008) Enzymatic ring opening polymerization of ε-caprolactone by a new lipase from *Yarrowia lipolytica*. J Appl Pol Sci 109:708–719

Berdague JL, Monteil P, Montel M, Talon R (1993) Effects of starter cultures on the formation of flavour compounds in dry sausage. Meat Sci 35:275–287

Bordes F, Cambon E, Dossat-Letisse V, Andre I, Croux C, Nicaud JM, Marty A (2009) Improvement of *Yarrowia lipolytica* lipase enantioselectivity by using mutagenesis targeted to the substrate binding site. Chembiochem 10:1705–1713

Breuer M, Ditrich K, Habicher T, Hauer B, Kesseler M, Stuurmer R, Zelinski T (2004) Industrial methods for the production of optically active intermediates. Angew Chem Int Ed 43:788–824

Briand D, Dubreucq E, Galzy P (1994) Enzymatic fatty esters synthesis in aqueous medium with lipase from *Candida parapsilosis* (Ashford) langeron and talice. Biotechnol Lett 16:813–818

Chin K, Lee S, Mohammad H (1996) A study of palm oil mill effluent treatment using a pond system. Water Sci Technol 34:119–123

Cossu R, Blakey N, Cannas P (1993) Influence of codisposal of municipal solid-waste and oil vegetation water on the anaerobic digestion of a sanitary-landfill. Water Sci Technol 27:261–271

D'Annibale A, Crestini C, Vinciguerra V, Sermanni GG (1998) The biodegradation of recalcitrant effluents from an olive mill by a white-rot fungus. J Biotechnol 61:209–218

D'Annibale A, Ricci M, Quaratino D, Federici F, Fenice M (2004) Panus tigrinus efficiently removes phenols, color and organic load from olive-mill wastewater. Res Microbiol 155:596–603

Destain J, Roblain D, Thonart P (1997) Improvement of lipase production from Yarrowia lipolytica. Biotechnol Lett 19:105–107

De Felice B, Pontecorvo G, Carfagna M (1997) Degradation of waste waters from olive oil mills by Yarrowia lipolytica ATCC 20255 and Pseudomonas putida. Acta Biotechnol 17:231–239

Dominguez A, Longo M, Sanroman M (2005) Recycling of waste cooking oil for lipase production by *Yarrowia lipolytica*. J Biotechnol 118S1:S170

Esmelindro A, Fiametti K, Ceni G, Corazza M, Treichel H, de Oliveira D, Oliveira J (2008) Lipase-catalysed production of monoglycerides in compressed propane and AOT surfactant. J Suercrit Fluids 47:64–69

Fickers P, Benetti PH, Wache Y, Marty A, Mauersberger S, Smit MS, Nicaud JM (2005) Hydrophobic substrate utilisation by the yeast *Yarrowia lipolytica*, and its potential applications. FEMS Yeast Res 5:527–543

Gardini F, Suzzi G, Lombardi A, Galgano F, Crudele MA, Andrighetto C, Schirone M, Tofalo R (2001) A survey of yeasts in traditional sausages of southern Italy. FEMS Yeast Res 1:161–167

Goncalves C, Lopes M, Ferreira J, Belo I (2009) Biological treatment of olive mill wastewater by non-conventional yeast. Bioresour Technol 100:3759–3763

Guieysse D, Salagnad C, Monsan P, Remaud-Simeon M (2003a) Lipase-catalyzed enantioselective transesterification toward esters of 2-bromo-tolylacetic acids. Tetrahedron Asymmetry 14:317–323

Guieysse D, Salagnad C, Monsan P, Remaud-Simeon M (2003b) Towards a novel explanation of Pseudomonas cepacia lipase enantioselectivity via molecular modelling of the enantiomer trajectory into the active site. Tetrahedron Asymmetry 14:1807–1817

Guieysse D, Sandoval G, Faure L, Nicaud J, Monsan P, Marty A (2004) New efficient lipase from *Yarrowia lipolytica* for the resolution of 2-bromo-arylacetic acid esters. Tetrahedron Asymmetry 15:3539–3543

Hirai M (2001) New advanced technology for fish waste. Food Dev 35:11–13

Hutt A, Caldwell J (1984) The importance of stereochemistry in the clinical pharmacokinetics of the 2-arylpropionic acid non-steroidal anti-inflammatory drug. Clin Pharmacokinet 9:371–373

Juaristi E, Leon-Romo J, Reyes A, Escalante J (1999) Recent applications of alpha-phenylalanine in the preparation of enantiopure compounds. Tetrahedron Asymmetry 10:2441–2495

Kamzolova S, Morgunov I, Aurich A, Perevoznikova O, Shishkanova N, Stottmeister U, Finogenova T (2005) Lipase secretion and citric acid production in *Yarrowia lipolytica* yeast grown on animal and vegetable fat. Food Technol Biotechnol 43:113–122

Kamzolova S, Finogenova T, Lunina O, Perevoznikova O, Minachova L, Morgunov I (2007) Characteristics of the growth on rapeseed oil and synthesis of citric and isocitric acids by *Yarrowia lipolytica* yeasts. Microbiology 76:20–24

Klasson T, Clausen J, Gaddy J (1989) Continuous fermentation for the production of citric acid from glucose. Appl Biochem Biotechnol 23:491–509

Kristiansen B, Sinclair C (1979) Production of citric acid in continuous culture. Biotechnol Bioeng 21:297–315

Lanciotti R, Gianotti A, Baldi D, Angrisani R, Suzzi G, Mastrocola D, Guerzoni M (2005) Use of Yarrowia lipolytica strains for the treatment of olive mill wastewater. Bioresour Technol 96:317–322

Liu Y, Wanf F, Tan T (2009) Cyclic resolution of racemic ibuprofen via coupled efficient lipase and acid-base catalysis. Chirality 21:349–353

Lopes M, Araujo C, Aguedo M, Gomes N, Goncalves C, Teixeira J, Belo I (2009) The use of olive mill wastewater by wild-type *Yarrowia lipolytica* strains: medium supplementation and surfactant presence effect by increased air pressure. J Chem Technol Biotechnol 84:533–537

Mameri N, Halet F, Drouiche M, Grib H, Lounici H, Pauss A, Piron D, Belhocine D (2000) Treatment of olive mill washing water by ultrafiltration. Can J Chem Eng 78:590–595

O'Donnell M (2001) The preparation of optically active alpha-amino acids from the benzophenone imines of glycine derivatives. Aldrichim Acta 34:3–15

Oswal N, Sarma PM, Zinjarde S, Pant A (2002) Palm oil mill effluent treatment by tropical marine yeast. Bioresour Technol 85:35–37

Papanikolaou S, Galiotou-Panayotou M, Fakas S, Komaitis M, Aggelis G (2008) Citric acid production by *Yarrowia lipolytica* cultivated on olive-mill wastewater-based media. Bioresour Technol 99:2419–2428

Peltsmane I, Raminya L, Finogenova T, Karklin R (1988) Biosynthesis and isolation of isocitric acide from n-alkane solution. Izv An Latv 487:93–95

Rane K, Sims K (1993) Production of citric acid by *Candida lipolytica* Y1905: effect of glucose concentration on the yield and productivity. Enzyme Microb Technol 15:646–651

Ros de Ursinos F (1961) Estudio del alpechin para su aprovechamiento industrial. VI. Aminoacidos presentes en la levardura *Torulopsis utilis*. Gras Aceit 12:161–165

Saito H (2004) Lipid and FA composition of the pearl oyster *Pinctada fucata martensii:* influence of season and maturation. Lipids 39:997–1005

Samelis J, Aggelis G, Metaxopoulos J (1993) Lipolytic and microbial changes during the natural fermentation and ripening of Greek dry sausages. Meat Sci 35:371–385

Scioli C, De Felice B (1993) Impiega de ceppi di lievito nella depurazione dei feflui dell'industria olearia. Ann Microbiol Enzymol 43:61–69

Scioli C, Vollaro L (1997) The use of *Yarrowia lipolytica* to reduce pollution in olive mill wastewaters. Water Res 31:2520–2524

Suzzi G, Larnote M, Galgano F, Andrighetto C, Lombardi A, Lanciotti R, Guerzoni M (2001) Proteolytic, lipolytic and molecular characterisation of Yarrowia lipolytica isolated from cheese. Int J Food Microbiol 69:69–77

Thonart P, Destain J, Zgoulli S, Antoine P, Godefroid J, Evrard P (1997) Les bacs à graisse: une solution aux problèmes des matières grasses pour les PME. Trib eau 586:41–46

Tsai S, Lin J, Chang C, Chen J (1997) Enzymatic synthesis of (S)-ibuprofen ester prodrug from racemic ibuprofen by lipase in organic solvents. Biotechnol Prog 13:82–87

Wen S, Tan T, Yu M (2008) Immobilized lipase YlLip2-catalysed resolution of alpha-phenylethyl amine in a medium with organic cosolvent. Process Biochem 43:1259–1264

Wojtatowicz M, Rymowicz W, Kautola H (1991) Comparison of different strains of the yeast *Yarrowia lipolytica* for citric acid production from glucose hydrol. Appl Biochem Biotechnol 31:165–174

Wu L, Wan J (2008) Investigation on the capability of disposing the grease wastewater with immobilized *Yarrowia lipolytica*. Chin J Environ Eng 4:482–486

Wu L, Ge G, Wan J (2009) Biodegradation of oil wastewater by free and immobilized *Yarrowia lipolytica* W29. J Environ Sci 21:237–242

Wyder M, Bachmann HP, Puhan Z (1999) Role of selected yeasts in cheese ripening: an evaluation in foil wrapped raclette cheese. Lebensm Wiss Technol 32:333–343

Yano Y, Oikawa H, Satomi M (2008) Reduction of lipids in fish meal prepared from fish waste by a yeast *Yarrowia lipolytica*. Int J Food Microbiol 121:302–307

Production of Organic Acids by *Yarrowia lipolytica*

Christina Otto, Martina Holz, and Gerold Barth

Abstract Wild-type strains of *Yarrowia lipolytica* secrete several organic acids like citric acid, isocitric acid, α-ketoglutaric acid, fumaric acid, malic acid or pyruvic acid, under certain circumstances from different carbon sources. An excess of carbon source together with limitation of growth causes the overproduction of these acids in most cases. Very high amounts of secreted acids (up to 200 g L^{-1}) can be reached in some cases like citric acid and α-ketoglutaric acid, but other acids like fumaric acid and succinic acid are produced in low amounts (below 10 g L^{-1}) by wild-type strains of this yeast. Several mutants were selected and recombinant strains have been constructed to improve the amounts of accumulated acid, the productivity or to change the spectrum of secreted acids. This review summarises the present state of knowledge on this field of overproduction of organic acids by the yeast *Y. lipolytica*.

Contents

1 Introduction 138
2 Citric Acid 138
3 Isocitric Acid 140
4 α-Ketoglutaric Acid 141
5 Succinic Acid 145
6 Future Perspectives 146
References 146

C. Otto (✉) • G. Barth
Institute of Microbiology, Dresden University of Technology, Zellescher Weg 20b, 01062 Dresden, Germany
e-mail: christina.otto@tu-dresden.de

M. Holz
Institute of Bioorganic Chemistry, Heinrich-Heine University, Düsseldorf, Germany

G. Barth (ed.), *Yarrowia lipolytica*, Microbiology Monographs 25,
DOI 10.1007/978-3-642-38583-4_5, © Springer-Verlag Berlin Heidelberg 2013

1 Introduction

Microbial produced organic acids, especially carboxylic acids, represent a key group among the building-block chemicals due to their high potential as intermediate feedstock substances for various industrial production processes. Against the background that the majority of petrochemical production processes come along with critical wastes and conditions for the environment, the rising interest in replacing them by bio-based processes is not surprising. Some acids, e.g. citric acid, have been already produced almost exclusively by microbial fermentation processes, but also other organic acids, e.g. succinic, fumaric, malic and isocitric acid and their biotechnological production by bacteria or yeasts are getting more and more attractive.

Due to its ability to secrete high amounts of a broad spectrum of organic acids and to utilise a broad spectrum of carbon sources, the apathogenicity as well as a well-established toolset of molecular biological and fermentation techniques, the yeast *Yarrowia lipolytica* is already an attractive candidate for the development of bio-based production processes. The overproduction of organic acids, including tricarboxylic-acid-cycle intermediates citric acid (CA), isocitric acid (ICA), α-ketoglutaric acid (KGA) but also pyruvic acid (PA), by this ascomycetal yeast was already well examined for the last decades. The overproduction depends on the utilised carbon source as well as on the availability of particular components, e.g. nitrogen, thiamine or mineral salts. An excess of carbon source, e.g. the renewable and low-cost carbon sources glycerol, ethanol, plant oils, fats, molasses and starch hydrolysates, additionally to alkanes and glucose, combined with a growth limitation by nitrogen exhaustion triggers secretion of citric and isocitric acid, while a combination with limitation by thiamine at low pH values causes secretion of mainly α-ketoglutarate and pyruvate (Treton et al. 1978; Stottmeister et al. 1982, 2005; Barth and Gaillardin 1996, 1997; Aurich et al. 2003; Mauersberger et al. 2003; Anastassiadis and Rehm 2005; Fickers et al. 2005; Finogenova et al. 2005; Förster et al. 2007a, b).

In the last years also the overproduction of other organic acids, like succinic acid, got in the focus of research (Yuzbashev et al. 2010). Here we give a short overview over the previous results which might be relevant for organic acid production by *Yarrowia lipolytica*.

2 Citric Acid

Because of its chemical structure, the tricarboxylic-acid-cycle intermediate citric acid (CA) is of interest for many industrial applications, e.g. as flavour, acidifying and preservative additive in the food and pharmaceutical industry, as stabiliser for vegetable oils and fats or as complex-forming and bleaching component in many washing detergents (Karaffa and Kubicek 2003; Förster et al. 2007a, b). At the beginning of this decade, the worldwide production of citric acid (CA) was approximately 1.0–1.5 million tonnes (BACAS-Belgian Academy Council of Applied Science 2004; http://www.sriconsulting.com). The filamentous fungus *Aspergillus niger* is used mainly for large-scale production of CA from beet or cane molasses or glucose syrup (Crolla and Kennedy 2001; Karaffa and Kubicek 2003). Disadvantages of this common used process are heavy metal contaminated waste water and solid waste gypsum (Röhr et al. 1996; Kubicek 2001). An alternative producer of high CA amounts is the yeast *Y. lipolytica*. Due to several advantages (broader substrate spectrum, higher maximal product formation rate, higher substrate concentrations and yield, greater tolerance to metal ions and low oxygen concentrations, simple process control, waste and sewage minimisation), the yeast *Y. lipolytica* and its ability to produce high amounts of citric acids were researched for the last decades (Stottmeister et al. 1982; Barth and Gaillardin 1997; Finogenova et al. 2005). Citric acid production of *Y. lipolytica* is triggered by an excess of carbon source and nitrogen limitation conditions (Lozinov et al. 1974). During CA production also isocitric acid (ICA) is secreted, which has an inferior buffer capacity and chelating ability compared to CA (Stottmeister et al. 1982; Mattey 1992; Barth and Gaillardin 1997). The production of CA and ICA depends on the used carbon source. On media with glucose, CA was the predominant acid, whereas CA and ICA were found in equal amounts in media containing hexadecane or other *n*-alkanes as sole carbon source. The synthesis of CA and ICA started after complete consumption of the deficient component of the medium and the transition of the culture from the logarithmic phase to the retardation phase; it continued until the carbon source was consumed completely (Finogenova 1975, 1982). It was also shown that this yeast was able to use various substrates for CA and ICA production, like glycerol, ethanol and acetic acid, whereby the CA/ICA product ratio is influenced by the used carbon source and the prevailing cultivation conditions. Wild-type strains secrete mainly CA and about 8–16 % ICA on carbohydrates or glycerol as sole carbon source and approximately 50–65 % CA and 35–50 % ICA on the gluconeogenetic substrates alkanes and the renewable triglycerides, ethanol or acetate (Treton et al. 1978; Stottmeister et al. 1982; Finogenova et al. 1991, 2005; Förster et al. 2007a, b; Holz et al. 2009). To improve the CA/ICA product ratio, the mutants N1 and N2 were generated from production strain 704 by chemical mutagenesis. Ermakova et al. (1986) and Finogenova et al. (1986) have shown with the mutants N1 and N2 that the enzymes isocitrate lyase (ICL) and aconitate hydratase (aconitase) play an important role on the specificity of the CA/ICA ratio. The CA/ICA ratio can also be influenced by modifying the cultivation conditions, i.e. by addition of fluoroacetate (Akiyama et al. 1973), changing temperature, pH-value, air saturation and iron concentration (Stottmeister and Weissbrodt 1991; Finogenova et al. 2002; Anastassiadis and Rehm 2005, 2006), or addition of acetate (Venter et al. 2004).

Further approaches for an optimization of the CA production process with *Y. lipolytica* were performed in the group of Prof. Barth by metabolic engineering. These include the expression of the invertase encoding *ScSUC2* gene for the use of the cheap substrate sucrose. Maximal CA amounts of 127–140 g L^{-1} CA with a yield Y_{CA} of 0.75–0.82 g g^{-1} were reached in a fed-batch cultivation process on sucrose with the recombinant *Y. lipolytica* strain H222-S4(p67ICL1) T5, harbouring the invertase encoding *ScSUC2* gene of *Saccharomyces cerevisiae* under the inducible *XPR2* promoter control and multiple *ICL1* copies (10–15) (Förster et al. 2007a). To further examine the influence of *ICL1* on the CA/ICA product ratio, strains with a gene-dose-dependent overexpression or disruption of the isocitrate lyase (ICL)-encoding gene *ICL1* were constructed. An increased *ICL1* copy number resulted in a strong shift of the CA/ICA ratio into direction of CA. A decrease from 10–12 % to 3–6 % was achieved on sucrose, glycerol and glucose, whereas the ICA proportion decreased on sunflower oil from 37–45 % to 4–7 %. In all cases no influence on the total amount of CA+ICA was detectable. The *icl1* deletion strain showed a moderate 2–5 % increase in the ICA proportion compared to ICL wild-type strains on glucose or glycerol (Förster et al. 2007b).

3 Isocitric Acid

D-*threo*-isocitric acid (also called (*2R,3S*)-isocitric acid) is a 1-hydroxy-1,2,3-tricarboxylic acid and occurs in nature as intermediary product of the tricarboxylic acid cycle. This acid is formed by the enzyme aconitase from citric acid via *cis*-aconitic acid. ICA can be used as ingredients of washing powder and detergents, as dietary supplement or as active substance in pharmaceutical products (Heretsch et al. 2008). In contrast to this possible broad scope of application, ICA is only available as expensive fine chemical on the market.

The first description of secretion of allo-isocitric acid by fungi was done by Sakaguchi et al. (1960). Since the seventieth of the last century, it is known that *Y. lipolytica* can secrete high amounts of ICA but always together with CA after exhaustion of nitrogen and excess of carbon source. The proportion of ICA in the amount of secreted acids is dependent on the carbon source used. On glucose, fructose or glycerol mainly CA is secreted and only 10–12 % ICA by wild-type strains. In contrast, on *n*-alkanes, fats or oils, and ethanol, about the same amount of ICA compared with CA is produced (reviewed in Stottmeister et al. 1982; Mattey 1992; Barth and Gaillardin 1996, 1997; Fickers et al. 2005).

Whereas many studies were focused on reduction of the proportion of ICA to increase the proportion of CA, only few activities were directed to improve the production of ICA. First mutants exhibiting increased secretion of ICA were detected among fluoroacetate-resistant and fluoroacetate-sensitive mutants selected after chemical or UV-light mutagenesis (Akiyama et al. 1973; Ikeno et al. 1975; Stottmeister et al. 1982). Highest yields of ICA produced by mutants or transformants of *Y. lipolytica* were reported by Finogenova et al. (1986) and Holz et al. (2009). Finogenova et al. described two mutants which produced up to 84 % ICA of total amount of secreted acids after growth on *n*-alkanes. Studies of these mutants have shown that enzymes of the tricarboxylic acid cycle and the glyoxylate cycle, like aconitase, isocitrate lyase, citrate synthase, NAD- and NADP-dependent isocitrate dehydrogenases, and malate synthase, are most important for the CA/ICA pattern (Ermakova et al. 1986; Finogenova et al. 1986, 1991, 2005).

Holz et al. (2009) have constructed a strain of *Y. lipolytica* containing eight to ten copies of the gene *ACO1* encoding the enzyme aconitase. This transformant exhibits a seven to nine times higher aconitase activity than the original strain even during the production phase and produces more than 70 % ICA on sunflower oil. On glycerol, glucose or sucrose, the ICA proportion increased only moderately from 10–12 % to 13–17 %.

It was also shown that several cultivation conditions influence the production of ICA. Finogenova's group has demonstrated that pH values of 5.5–6, oxygen concentration of 60–95 % and increased ferrum ion concentration increase the proportion of ICA (Finogenova et al. 1991, 2002, 2005).

4 α-Ketoglutaric Acid

α-Ketoglutaric acid has a broad scope of application, e.g. as building-block chemical for the chemical synthesis of heterocycles, dietary supplement, component of infusion solutions and wound healing compounds (Chernyavskaya et al. 2000; Stottmeister et al. 2005; Huang et al. 2006; Verseck et al. 2007, 2009). KGA can also be used for synthesis of elastomers with a wide range of mechanical and chemical properties or for the production of an agent that protects humans or animals from oxidative stress by increasing the antioxidant capacity (Moser et al. 2007; Barrett and Yousaf 2008).

Currently KGA is produced via different chemical pathways. These multi-step synthetic processes have a lot of disadvantages like the use of risky chemicals, e.g. cyanides, generation of toxic waste, presence of a catalyst containing copper and its disposal or a low product selectivity caused by different by-products (Castleman Evans and Wiselogle 1945; Cooper et al. 1983; Stottmeister et al. 2005; Verseck et al. 2007, 2009). Therefore, a bio-based production process is an attractive alternative for the production KGA and *Y. lipolytica* is a suitable organism for this approach.

The ability of *Y. lipolytica* to produce and secrete KGA was first mentioned in the late 1960s by the groups of Tsugawa and Lozinov. Screening of different microorganisms for KGA production grown on *n*-paraffins showed that *Y. lipolytica* (former *Candida lipolytica*) AJ5004 synthesised about 46 g L^{-1} KGA in 72 h from 8 % (w/v) *n*-paraffin as carbon source (Tsugawa and Okumura 1969; Tsugawa et al. 1969a, b). Additionally works on KGA production with alkanes exhibited maximal KGA amounts of 108.7 g L^{-1} with a substrate-related yield of 120 % (Finogenova et al. 1968; Ermakova and Finogenova 1971; Glazunova et al. 1973; Ermakova et al. 1979, early work summarised in Finogenova et al. 2005). Further approaches for an increase of KGA production with *n*-paraffins were optimization of cultivation conditions and the generation of new suitable strains. For the diploid strain D 1805, the production of 185 g L^{-1} KGA was demonstrated with 10 % *n*-paraffin (C_{13}-C_{18}) after 240 h (Maldonado et al. 1976). The highest product amounts up to 195 g L^{-1} KGA and specific productivities of 1.3–1.4 g L^{-1} h^{-1} were detected for the hyper-producing strain H355 with a mixture of *n*-paraffins (C_{12}-C_{18}) (Weissbrodt et al. 1989). Further studies showed that ethanol also serves as a suitable substrate for KGA production. With the mutant *Y. lipolytica* N1, the highest KGA concentration of 49 g L^{-1} with a yield of 42 % was achieved on ethanol in fed-batch fermentation (Chernyavskaya et al. 2000; Finogenova et al. 2002; Il'chenko et al. 2001, 2002, 2003). Further approaches were done for efficient KGA formation with ethanol including determination of the optimal cultivation conditions, such as concentration of thiamine, dissolved oxygen, nitrogen and initial pH. Also environmental effects were analysed, like the detection of key enzyme activities involved in KGA synthesis and ethanol metabolism (Chernyavskaya et al. 2000; Il'chenko et al. 2002, 2003). The data reveal that a low thiamine concentration (3 μg L^{-1}), an acidic pH (4.5) and nitrogen excess together with a low oxygen concentration (5 %) are of fundamental importance for the overproduction of KGA on ethanol-containing medium (Chernyavskaya et al. 2000). Aurich and Stottmeister (2006) looked for alternative substrates for the production of KGA with *Y. lipolytica* and optimal cultivation conditions. Substrate screening implied refined vegetable oils (olive, canola and sunflower oil) or cold-pressed oils (linseed oil). Fed-batch fermentation with *Y. lipolytica* H355 showed that a maximal KGA amount of 104 g L^{-1} was detected in rapeseed oil-containing media after 360 h (Aurich and Stottmeister 2006). Under optimal production conditions, H355 produced a maximal KGA amount of 126–134 g L^{-1} with a space-time yield of 0.48–0.51 g L^{-1} h^{-1} and a substrate-related yield of about 1.2–1.3 g g^{-1} (Förster et al. 2006). Kamzolova and Morgunov (2013) tested 26 strains of *Yarrowia lipolytica* for KGA production on rapeseed oil as a carbon source. *Y. lipolytica* strain VKM Y-2412 reached a maximum KGA concentration of 102.5 g L^{-1} with the mass yield coefficient of 0.95 g g^{-1} and the volumetric KGA productivity (Q_{KGA}) of 0.8 g L^{-1} h^{-1}.

Förster et al. (2006) showed that glycerol is also a suitable carbon source for KGA production. When 3.5 μg L^{-1} thiamine was included in the medium, 38 g L^{-1} KGA were produced from 100 g L^{-1} glycerol in shaking flasks (Förster et al. 2006). Zhou et al. (2010) also described the use of glycerol as carbon source for the production of KGA. They studied different conditions to enhance KGA production including the effects of initial glycerol concentration, nitrogen source, exogenous vitamins and calcium carbonate. In this context thiamine and calcium ion concentration had the greatest effect on KGA accumulation. The addition of 20 g L^{-1} $CaCO_3$ and 0.8 mg L^{-1} biotin led to an increase in productivity from 16.6 to 39.2 g L^{-1} KGA. This maximal KGA concentration was obtained with the strain WSH-Z06 under optimal conditions from 100 g L^{-1} glycerol with 16.8 g L^{-1} pyruvate as a by-product (Zhou et al. 2010; review see Otto et al. 2011). The KGA production with the strain WSH-Z06 was enhanced by an improved fed-batch strategy. The use of a two-stage pH control strategy, in which pH was buffered by $CaCO_3$ in the growth phase and then maintained at 3.0 in the KGA production phase, led to an increased KGA amount of 53.4 g L^{-1}. A further enhancement to 66.2 g L^{-1} KGA was reached through glycerol fed (Yu et al. 2012).

Furthermore, a systematic approach for studying the effects of changing enzyme activities on the production of KGA by *Y. lipolytica* was done in recent years. Holz et al. (2011) examined the influence of overexpression of the α-ketoglutarate dehydrogenase (KGDH) complex encoding genes on the production process of KGA. It was shown that the constructed strain containing multiple copies of all three KGDH genes encoding the three subunits of the enzyme showed a reduced production of KGA and an elevated production of PA under conditions of KGA production (Holz et al. 2011). Recent studies indicate that an increase in KGA formation process is achievable with the change of specific TCC and TCC-related enzyme activities (Otto et al. 2012; Yovkova, personal communication). An increase in KGA formation was reached through the overexpression of the $NADP^+$-dependent isocitrate dehydrogenase (IDP) encoding gene or simultaneous overexpression of IDP and pyruvate carboxylase genes, suggesting that IDP and not the NAD^+-dependent isocitrate dehydrogenase has the greatest impact on KGA overproduction (Yovkova, personal communication). Recently Otto et al. (2012) showed the improvement of KGA production though reduction of by-products amounts. It was examined whether the concentration of secreted organic acids (main product KGA and PA as major by-product, fumarate [FA], malate [MA] and succinate [SA] as minor by-products) can be influenced by a gene-dose-dependent overexpression of fumarase (FUM) or pyruvate carboxylase (PYC) genes (*FUM1*, *PYC1*) under KGA production conditions. Overexpression of the genes *FUM1* and *PYC1* resulted in strongly increased specific enzyme activities during cultivation of these strains on raw glycerol as carbon source in bioreactors. The recombinant *Y. lipolytica* strains showed different product selectivity of the secreted organic acids KGA, PA, FA, MA and SA. Concentrations of the by-products FA, MA, SA and PA decreased significantly by overproduction of FUM (42 % compared to H355) and increased by overproduction of PYC and also of FUM and PYC simultaneously (161 % resp. 151 % compared to H355). In contrast, the production of KGA was in the range of 137–147 g L^{-1} after 93–114 h with the multicopy strains H355A(FUM1) and H355A(FUM1-PYC1) comparable with the wild-type strain H355 or slightly lower in case of H355(PYC1) (Otto et al. 2012). Yin et al. (2012) showed a regulation of the *Y. lipolytica* pyruvate carboxylation pathway by overexpression of heterologous pyruvate carboxylase genes, named *ScPYC1* from *Saccharomyces cerevisiae* and *RoPYC2* from *Rhizopus oryzae* in *Y. lipolytica* WSH-Z06. *Y. lipolytica*-RoPYC2 reached a maximum concentration of 62.5 g L^{-1} KGA in a pH-controlled 3-L fermenter in which an evident decrease in PA yield from 35.2 to 13.5 g L^{-1} (Yin et al. 2012) took place. Another approach to reduce by-product accumulation for a redistribution of the carbon flux from pyruvate to α-KG was to regulate the cofactor metabolism. The acetyl-CoA synthetase gene, *ACS1*, from *Saccharomyces cerevisiae* and the ATP-citrate lyase gene, ACL, from *Mus musculus* were expressed to regulate the acetyl-CoA metabolism in *Y. lipolytica* WSH-Z06. In a 3-L fermenter, the highest

Table 1 Comparison of KGA producing organisms

Y. lipolytica strain	Carbon source	KGA amount (g L^{-1})	Yield (g g^{-1})	Reference
Candida lipolytica ATCC 16617 and 16618	*n*-Alkanes	48	0.60	Tsugawa and Okumura (1969), Tsugawa et al. (1969a, b)
Candida lipolytica D1805	*n*-Alkanes	185	0.80	Maldonado et al. (1976)
Yarrowia lipolytica H355	*n*-Alkanes	195	0.90	Weißbrodt et al. (1989)
Yarrowia lipolytica N1	Ethanol	49	0.42	Chernyavskaya et al. (2000)
Yarrowia lipolytica H355	Rapeseed oil	134	1.30	Förster et al. (2006)
Yarrowia lipolytica H222-S4(JMP6) T5	Rapeseed oil	126	1.20	Förster et al. (2006)
Yarrowia lipolytica WSH-Z06	Glycerol	35	0.40	Zhou et al. (2010)
Yarrowia lipolytica H222	Glycerol	97	n.d.	Holz et al. (2011)
Yarrowia lipolytica H222-MH1	Glycerol	72	n.d.	Holz et al. (2011)
Yarrowia lipolytica VKM Y-2412	Ethanol	172	0.70	Kamzolova et al. (2012)
Yarrowia lipolytica H355	Raw glycerol	139	0.47	Otto et al. (2012)
Yarrowia lipolytica H355A(FUM1-PYC1)	Raw glycerol	146	0.52	Otto et al. (2012)
Yarrowia lipolytica WSH-Z06	Glycerol	66	n.d.	Yu et al. (2012)
Yarrowia lipolytica WSH-Z06-RoPYC2	Glycerol	62	n.d.	Yin et al. (2012)
Yarrowia lipolytica WSH-Z06-ACL	Glycerol	56	n.d.	Zhou et al. (2012)
Yarrowia lipolytica VKM Y-2412	Rapeseed oil	102	0.95	Kamzolova and Morgunov (2013)

n.d. not determined

yield of α-KG in *Y. lipolytica*-ACL was reached up to 56.5 g L^{-1} with an obvious decrease of pyruvate accumulation from 35.1 to 20.2 g L^{-1} (Zhou et al. 2012).

A further examination of the KGA production from ethanol was published by Kamzolova et al. (2012). In particular they found that an increased amount of zinc and iron ions is required for the KGA production from ethanol. The influence of thiamine limitation and excess of carbon and nitrogen source as well as a low pH on the intensive KGA production reveal with all previous work. Under optimal conditions, the *Y. lipolytica* strain VKM Y-2412 produced up to 172 g L^{-1} of KGA with the mass yield coefficient of 0.70 g g^{-1} (Kamzolova et al. 2012). Table 1 summarises KGA production in *Y. lipolytica* of the last decades.

5 Succinic Acid

Succinic acid is already used in a variety of industrial processes. So far succinic acid is majorly utilised as detergent and surfactant, but also as an ion chelator to avoid corrosion of metals. Even in food industry succinic acid can be used for acidification, as a flavouring and antimicrobial agent. Also in production processes for some pharmaceutical products, antibiotics, amino acids and vitamins, succinic acid plays a role (Zeikus et al. 1999). Besides this, biotechnological produced succinic acid has the potential to become a mass chemical as a basis for the synthesis of important intermediates in industrial processes, for polymers and special chemicals, e.g. polyester, nylon, polyurethane, 1,4-butanediol, adipic acid, tetrahydrofuran, γ-butyrolactone, *N*-methylpyrrolidone and linear aliphatic acids. The development of bio-based production processes can lead to a replacement of petrochemical produced bulk chemicals or intermediates, as long as the costs of these biological processes can compete with these of the petrochemical production (Zeikus et al. 1999).

The ability to produce and secrete succinate was first mentioned by Kamzolova et al. (2009a) who screened 32 yeast species for succinic acid production grown in ethanol-containing media under aerobic conditions and nitrogen limitation. Up to 0.5 g succinic acid per g cells was detected for all tested *Yarrowia lipolytica* strains. Additionally a succinic acid production was detected in glycerol-containing (up to 5.1 g L^{-1}) and glucose-containing media (up to 1.4 g L^{-1}) (Kretzschmar 2010; Yuzbashev et al. 2010). Furthermore, first approaches for an increase of succinic acid production were presented in Kamzolova et al. (2009b) and Yuzbashev et al. (2010). In the latter case, *Yarrowia lipolytica* strains with mutations in genes encoding the succinate dehydrogenase (SDH) were constructed to achieve a reduction or loss of SDH enzyme activity. First of all temperature-sensitive mutant strains with impaired *SDH1* gene (YALI0D11374g) were obtained which lost their ability to grow on glucose under non-permissive conditions, but still were able to grow on glycerol (Yuzbashev et al. 2010). Under non-permissive conditions, the SDH activity of mutant strain Y-3374 was undetectable and an increase of succinate production up to 20.8 g L^{-1} in rich YPD medium with 5 % glycerol in the presence of $CaCO_3$ as a neutralising agent was observed (compared to about 1.0 g L^{-1} succinate produced by the control strain). Based on these findings, the gene *SDH2* (YALI0D23397g) was deleted. After chemical mutagenesis to improve viability, a maximal succinate production of 45 g L^{-1} with pH regulation and 17.4 g L^{-1} without pH regulation was detected (Yuzbashev et al. 2010). The assumption that a reduced SDH activity can lead to an increased succinate production was also confirmed in experiments with strains harbouring an exchange of the native *SDH2* upstream regulating region to reduce expression of *SDH2* and consequently SDH activity (Holz 2011). However, these studies proved that *Y. lipolytica* is able to produce succinic acid at low pH values and that an increase in succinate production can be achieved by genetic engineering.

In contrast to these studies, Kamzolova et al. (2009) presented a two-stage process for succinic acid production. *Y. lipolytica* was cultivated under thiamine limitation condition to produce AKG which was converted to succinate by oxidation with hydrogen peroxide. In these studies, up to 63 g L^{-1} succinate could be produced.

6 Future Perspectives

At present, *Y. lipolytica* is clearly the best studied yeast in respect to overproduction and secretion of organic acids. Its high potential for secretion of large amounts of such compounds, its resistance to low pH values and its family of multiple genes encoding transporters of carboxylic acid make this organism to a favourite for development of biotechnological processes for production of organic acids. Due to the finiteness of fossil raw materials, there is an increasing interest in alternative approaches for replacing conventional, nonbiological processes using fossil fuels as basic substrate. Intermediates of the microbial metabolism like the organic acids of the tricarboxylic acid cycle are such compounds capable to substitute petroleum-derived chemicals. The availability of the annotated genome and the development of powerful genetic tools for metabolic engineering of *Y. lipolytica* open the door for new strategies of construction of further efficient strains for biotechnological production of organic acids in next future.

References

Akiyama S, Suzuki T, Sumino Y, Nakao Y, Fukuda H (1973) Induction and citric acid productivity of fluoroacetate-sensitive mutant strains of *Candida lipolytica*. Agric Biol Chem 37:879–888

Anastassiadis S, Rehm H-J (2005) Continuous citric acid secretion by a high specific pH dependent active transport system in yeast *Candida oleophila* ATCC 20177. Electron J Biotechnol 8:146–161

Anastassiadis S, Rehm H-J (2006) Oxygen and temperature effect on continuous citric acid secretion in *Candida oleophila*. Electron J Biotechnol 9:414–423

Aurich A, Förster A, Mauersberger S, Barth G, Stottmeister U (2003) Citric acid production from renewable resources by Yarrowia lipolytica. Biotechnol Adv 21:454–455

Aurich A, Stottmeister U (2006) Verbundvorhaben: Biotechnologische Gewinnung von Carbonsäuren, Teilvorhaben 1: Biotechnologische Gewinnung von Oxocarbonsäuren als Synthesebausteine. Präambel, Endbericht; Projekt der Fachagentur für nachwachsende Rohstoffe e.V., BMVEL 220010701. http://www.fnr-server.de/ftp/pdf/berichte/22010701.pdf

BACAS-Belgian Academy Council of Applied Science (2004) Industrial biotechnology and sustainable chemistry. Brussels, pp 1–29. http://wbt.dechema.de/img/wbt_/Literatur/BACAS-Studie.pdf

Barrett DG, Yousaf MN (2008) Poly(triol α-ketoglutarate) as Biodegradable, Chemoselective, and Mechanically Tunable Elastomers. Macromolecules 41:6347–6352

Barth G, Gaillardin C (1996) Yarrowia lipolytica. In: Wolf K (ed) Nonconventional yeasts in biotechnology. Springer, Berlin, pp 313–388

Barth G, Gaillardin C (1997) Physiology and genetics of the dimorphic fungus *Yarrowia lipolytica*. FEMS Microbiol Rev 19:219–237

Castleman Evans R, Wiselogle FY (1945) Studies in the pyridazine series. The absorption spectrum of pyridazine. J Am Chem Soc 67:60–62

Chernyavskaya OG, Shishkanova NV, Il'chenko AP, Finogenova TV (2000) Synthesis of alpha-ketoglutaric acid by *Yarrowia lipolytica* yeast grown on ethanol. Appl Microbiol Biotechnol 53:152–158

Cooper AJL, Ginos JZ, Meister A (1983) Synthesis and properties of the α-keto acids. Chem Rev 83:321–358

Crolla A, Kennedy KJ (2001) Optimization of citric acid production from *Candida lipolytica* Y-1095 using n-paraffin. J Biotechnol 89:27–40

Ermakova IT, Finogenova TV (1971) Participation of the glyoxylate cycle in metabolism of the alkane-oxidizing yeast *Candida lipolytica* during biosynthesis of alpha-ketoglutaric acid. Microbiologiya 40:223–226

Ermakova IT, Finogenova TV, Losinov AB (1979) Effect of cultivation conditions on growth of *Candida lipolytica* yeasts and alpha-ketoacid biosynthesis in the presence of thiamine deficiency. Microbiologiya 48:1004–1010

Ermakova IT, Shishkanova NV, Melnikova OF, Finogenova TV (1986) Properties of *Candida lipolytica* mutants with the modified glyoxylate cycle and their ability to produce citric and isocitric acid. I. Physiological, biochemical and cytological characteristics of mutants grown on glucose or hexadecane. Appl Microbiol Biotechnol 23:372–377

Fickers P, Benetti PH, Waché Y, Marty A, Mauersberger S, Smit MS, Nicaud J (2005) Hydrophobic substrate utilization by the yeast Yarrowia lipolytica and its potential applications. FEMS Yeast Res 5:527–543

Finogenova TV (1975) Biokhimiya i fiziologiya mikroorganizmov (Biochemistry and Physiology of Microorganisms). NTsBI AN SSSR, Pushchino, pp 21–24

Finogenova TV (1982) Biosynthesis of organic acids by yeast and its regulation. Extended Abstract of Doctoral (Biol.) Dissertation, IBFM AN SSSR, Pushchino

Finogenova TV, Losinov AB, Belikov VM, Ermakova IT, Muntjan LN, Safonova EN (1968) Keto-acid production by paraffin-oxidizing yeasts. Mikrobiologiya 37:38–43

Finogenova TV, Shishkanova NV, Ermakova IT, Kataeva IA (1986) Properties of *Candida lipolytica* mutants with the modified glyoxylate cycle and their ability to produce citric and isocitric acid. II. Synthesis of citric and isocitric acid by *Candida lipolytica* mutants and peculiarities of their enzyme systems. Appl Microbiol Biotechnol 23:378–383

Finogenova TV, Shishkanova NV, Fausek EA, Eremina CC (1991) Appl Microbiol Biotechnol 36:231–235

Finogenova TV, Kamzolova SV, Dedyukhina EG, Shishkanova NV, Il'chenko AP, Morgunov IG, Chernyavskaya OG, Sokolov AP (2002) Biosynthesis of citric and isocitric acids from ethanol by mutant *Yarrowia lipolytica* N1 under continuous cultivation. Appl Microbiol Biotechnol 59:493–500

Finogenova TV, Morgunov IG, Kamzolova SV, Chernyavskaya OG (2005) Organic acid production by the yeast *Yarrowia lipolytica*: a review of prospects. Appl Biochem Microbiol 41:418–425

Förster A, Mauersberger S, Köllner K, Barth G (2006) Verbundvorhaben: Biotechnologische Gewinnung von Carbonsäuren, Teilvorhaben 3: Genetic engineering oxocarbonsäurebildender Hefen. Endbericht; Projekt der Fachagentur für nachwachsende Rohstoffe e.V., FZ:22002505

Förster A, Aurich A, Mauersberger S, Barth G (2007a) Citric acid production from sucrose using a recombinant strain of the yeast *Yarrowia lipolytica*. Appl Microbiol Biotechnol 75:1409–1417

Förster A, Jacobs K, Juretzek T, Mauersberger S, Barth G (2007b) Overexpression of the *ICL1* gene changes the product ratio of citric acid production by *Yarrowia lipolytica*. Appl Microbiol Biotechnol 77:861–869

Glazunova LM, Finogenova TV, Losinov AB (1973) Mechanism of the formation of alpha-ketoglutaric acid by *Candida lipolytica* yeasts. Mikrobiologiia 42:627–631

Heretsch P, Thomas F, Aurich A, Krautscheid H, Sicker D, Giannis A (2008) Synthesis with a chiral building block from the citric acid cycle: (2R,3S)-isocitric acid by fermentation of sunflower oil. Angew Chem Int Ed Engl 47:1958–1960

Holz M, Förster A, Mauersberger S, Barth G (2009) Aconitase overexpression changes the product ratio of citric acid production by *Yarrowia lipolytica*. Appl Microbiol Biotechnol 81 (6):1087–1096

Holz M, Otto C, Kretzschmar A, Yovkova V, Aurich A, Potter M, Marx A, Barth G (2011) Overexpression of alpha-ketoglutarate dehydrogenase in *Yarrowia lipolytica* and its effect on production of organic acids. Appl Microbiol Biotechnol 89:1519–1526

Holz M (2011) Gentechnische Optimierung der Hefe Yarrowia lipolytica zur biotechnologischen Produktion von Succinat. Ph.D. thesis, Technische Universität Dresden

Huang HJ, Liu LM, Li Y, Du GC, Chen J (2006) Redirecting carbon flux in *Torulopsis glabrata* from pyruvate to alpha-ketoglutaric acid by changing metabolic co-factors. Biotechnol Lett 28:95–98

Ikeno Y, Masuda M, Tanno K, Omori J, Takahashi N (1975) Citric acid production from various raw materials by yeast. J Ferment Technol 53:752–756

Il'chenko AP, Chernyavskaya OG, Shishkanova NV, Finogenova TV (2001) Metabolic characteristics of the mutant *Yarrowia lipolytica* strain 1 producing alpha-ketoglutaric and citric acids from ethanol and the effect of [NH4+] and [O2] on yeast respiration and biosynthesis. Mikrobiologiya 70:189–195

Il'chenko AP, Chernyavskaya OG, Shishkanova NV, Finogenova TV (2002) Metabolism of *Yarrowia lipolytica* grown on ethanol under conditions promoting the production of a-ketoglutaric and citric acids: a comparative study of the central metabolism enzymes. Microbiology 71:269–274

Il'chenko AP, Chernyavskaya OG, Shishkanova NV, Finogenova TV (2003) Biochemical characterization of the yeast *Yarrowia lipolytica* overproducing carboxylic acids from ethanol: nitrogen metabolism enzymes. Microbiologiia 72:470–475

Kamzolova SV, Morgunov IG (2013) α-Ketoglutaric acid production from rapeseed oil by *Yarrowia lipolytica* yeast. Appl Microbiol Biotechnol 97(12):5517–5525

Kamzolova SV, Yusupova AI, Vinokurova NG, Fedotcheva NI, Kondrashova MN, Finogenova TV, Morgunov IG (2009) Chemically assisted microbial production of succinic acid by the yeast *Yarrowia lipolytica* grown on ethanol. Appl Microbiol Biotechnol 83:1027–1034

Kamzolova SV, Chiglintseva MN, Lunina JN, Morgunov IG (2012) α-Ketoglutaric acid production by Yarrowia lipolytica and its regulation. Appl Microbiol Biotechnol 96:783–791

Karaffa L, Kubicek CP (2003) *Aspergillus niger* citric acid accumulation: do we understand this well working black box? Appl Microbiol Biotechnol 61:189–196

Kretzschmar A (2010) Die Beeinflussung der Succinatproduktion durch die veränderte Aktivität der Succinyl-CoA Synthetase und der Pyruvat-Carboxylase in Yarrowia lipolytica. Ph.D. thesis, Technische Universität Dresden

Kubicek C (2001) Chapter 4.2, Citric acid. In: Ratledge C, Kristiansen B (eds) Basic biotechnology. Cambridge University Press, Cambridge, pp 305–315

Lozinov AB, Finogenova TV, Glazunova LM, Illarionova VI (1974) Growth limitation in *Candida lipolytica* cultures and supersynthesis of metabolites. Mikrobiologiya 43:786–790

Maldonado P, Desmarquest JP, Gaillardin C, Binet D (1976) Yeasts. Institut Français du Petrole. Patent US3930946

Mattey M (1992) The production of organic acids. Crit Rev Biotechnol 12:87–132

Mauersberger S, Kruse K, Barth G (2003) Induction of citric acid/isocitric acid and α-ketoglutaric acid production in the yeast Yarrowia lipolytica. In: Wolf KH, Breunig K, Barth G (eds) Nonconventional yeasts in genetics, biochemistry and biotechnology. Practical protocols. Springer, Berlin, pp 393–400

Moser PM, Greilberger J, Maier A, Juan H, Bücherl-Harrer C, Kager E (2007) Verwendung von alpha-Ketoglutarsäure und 5-Hydroxy-methylfurfural zur Reduktion von oxidativem Stress. C. Y. L. P. GmbH. Patent EP1842536 A1

Otto C, Yovkova V, Barth G (2011) Overproduction and secretion of α-ketoglutaric acid by microorganisms. Mini-review. Appl Microbiol Biotechnol 92:689–695

Otto C, Yovkova V, Aurich A, Mauersberger S, Barth G (2012) Variation of the by-product spectrum during α-ketoglutaric acid production from raw glycerol by overexpression of fumarase and pyruvate carboxylase genes in *Yarrowia lipolytica*. Appl Microbiol Biotechnol 95:905–917

Röhr M, Kubicek CP, Kominek J (1996) Citric acid. In: Rehm HJ, Reed G, Delweg D (eds) Biotechnology, vol 6, 2nd edn. VCH, Weinheim, pp 308–345

Sakaguchi K, Abe S, Beppu T (1960) Patent US2949404

Stottmeister U, Weissbrodt E (1991) Product formation by *Yarrowia lipolytica* – some generalizing aspects. In: Finogenova TV, Sharyshev AA (eds) Alkane metabolism and oversynthesis of metabolites by microorganisms. Center for Biological Research, USSR Academy of Science, Pushchino, Russia, pp 147–158

Stottmeister U, Behrens U, Weissbrodt E, Barth G, Franke Rinker D, Schulze E (1982) Nutzung von Paraffinen und anderen Nichtkohlenhydrat-Kohlenstoffquellen zur mikrobiellen Citronensäuresynthese. Z Allg Mikrobiol 22:399–424

Stottmeister U, Aurich A, Wilde H, Andersch J, Schmidt S, Sicker D (2005) White biotechnology for green chemistry: fermentative 2-oxocarboxylic acids as novel building blocks for subsequent chemical syntheses. J Ind Microbiol Biotechnol 32:651–664

Treton B, Le Dall MT, Heslot H (1978) Excretion of citric and isocitric acid by the yeast *Saccharomycopsis lipolytica*. Eur J Appl Microbiol Biotechnol 6:67–77

Tsugawa R, Okumura S (1969) Fermentation of n-paraffins by Yeast. Part II. a-ketoglutarate productivity of *Candida lipolytica* in various culture media. Agric Biol Chem 33:676–682

Tsugawa R, Nakase T, Kobayashi T, Yamashita K, Okumura S (1969a) Fermentation of n-paraffins by yeast. Part I. Fermentative production of a-ketoglutaric acid by *Candida lipolytica*. Agric Biol Chem 33:158–167

Tsugawa R, Nakase T, Kobayash T, Yamashita K, Okumura S (1969b) Fermentation of n-paraffins by yeast. III. a-ketoglutarate productivity of various yeast. Agric Biol Chem 33:929–941

Venter T, Kock JL, Botes PJ, Smit MS, Hugo A, Joseph M (2004) Acetate enhances citric acid production by *Yarrowia lipolytica* when grown on sunflower oil. Syst Appl Microbiol 27:135–138

Verseck S, Karau A, Weber M (2007) Fermentative Herstellung von alpha-Ketoglutarsäure. Evonik Degussa GmbH. Patent DE 10 2007 051 451.6

Verseck S, Karau A, Weber M (2009) Fermentative production of alpha-ketoglutaric acid. Evonik Degussa GmbH. Patent WO2009053489

Weissbrodt E, Gey M, Barth G, Weber H, Stottmeister U, Düresch R, Richter HP (1989) Verfahren zur Herstellung von 2-Oxoglutarsäure durch Hefen. Patent DD 267999

Yin X, Madzak C, Du G, Zhou J, Chen J (2012) Enhanced alpha-ketoglutaric acid production in Yarrowia lipolytica WSH-Z06 by regulation of the pyruvate carboxylation pathway. Appl Microbiol Biotechnol 96:1527–1537

Yu Z, Du G, Zhou J, Chen J (2012) Enhanced α-ketoglutaric acid production in *Yarrowia lipolytica* WSH-Z06 by an improved integrated fed-batch strategy. Bioresour Technol 114:597–602

Yuzbashev TV, Yuzbasheva EY, Sobolevskaya TI, Laptev IA, Vybornaya TV, Larina AS, Matsui K, Fukui K, Sineoky SP (2010) Production of succinic acid at low pH by a recombinant strain of the aerobic yeast *Yarrowia lipolytica*. Biotechnol Bioeng 107:673–682

Zeikus JG, Jain MK, Elankovan P (1999) Biotechnology of succinic acid production and markets for derived industrial products. Appl Microbiol Biotechnol 51:545–552

Zhou J, Zhou H, Du G, Liu L, Chen J (2010) Screening of a thiamine-auxotrophic yeast for alpha-ketoglutaric acid overproduction. Lett Appl Microbiol 51:264–271

Zhou J, Yin X, Madzak C, Du G, Chen J (2012) Enhanced α-ketoglutarate production in Yarrowia lipolytica WSH-Z06 by alteration of the acetyl-CoA metabolism. J Biotechnol 161:257–264

Production of Dicarboxylic Acids and Flagrances by *Yarrowia lipolytica*

Yves Waché

Abstract Yeasts are excellent biocatalysts in the field of alkane and fatty acids transformation into dicarboxylic acids and lactones. Their ability to produce some diacids through simple, less expensive and more environment friendly routes than chemical pathways and to produce particular diacids (e.g. unsaturated ones) but also to transform natural substrates into lactones with a natural label has made them the subject of many researches. Although *Candida* species were often first studied, the development of genetic tools, the knowledge of the genome and some genomic and biotechnological particularities of *Yarrowia lipolytica* resulted to interesting developments with this species. This chapter aims at presenting the family of compounds of interest of this field, the biotechnological strategies usually carried out with yeast biocatalysts and the developments obtained with *Y. lipolytica*. For dicarboxylic acids, the first complete strategies carried out with *Candida* sp. will be presented, followed by the works on *Yarrowia*. For lactones, the work on *Yarrowia*'s β-oxidation and on the production of γ-decalactone will be particularly detailed. The interactions of amphiphilic fatty acids and lactones with yeast membranes will be also presented as they are of importance for both processes and have not been much investigated in other yeast species.

Contents

1 Introduction 152
2 Compounds of Interest Resulting from β- and ω-Oxidation 153
  2.1 ω-Hydroxy Acids (1) and α,ω-Dicarboxylic Compounds (2) 153
  2.2 Volatile Aroma Lactones (1, 2, 3) 155

Y. Waché (✉)
Laboratoire PAM, Université de Bourgogne/AgroSup Dijon, 1, esplanade Erasme, 21000 Dijon, France

Laboratory of Food and Microbial Process (PAM), UMR AgroSup Dijon, University of Burgundy, Dijon, France
e-mail: ywache@u-bourgogne.fr

G. Barth (ed.), *Yarrowia lipolytica*, Microbiology Monographs 25,
DOI 10.1007/978-3-642-38583-4_6, © Springer-Verlag Berlin Heidelberg 2013

3 Production of Dicarboxylic Acids 158
3.1 Yeast Route to α,ω-Dicarboxylic Acids 158
3.2 Developments in *Yarrowia lipolytica* 160
4 Production of Lactones 161
4.1 Yeast for Production of Natural Lactones 161
4.2 Developments in *Yarrowia lipolytica* 162
5 Sensitivity of *Y. lipolytica* to Amphiphilic Compounds and Membrane Adaptation 165
6 Conclusion 167
References 167

1 Introduction

In most industrial fields, the needs for substrates and ingredients are becoming more and more specific with a requirement for a growing number of building blocks and specific compounds. In the petroleum and chemical era, the production of many compounds has been developed. However, these compounds do not fulfil all the needs in the today's world in regard to economical criteria, and moreover, they do usually not correspond to the environmental and health requirements of the evolving society. Although production of fine chemicals and ingredients is usually far more expensive through biotechnology than through chemistry, some production can become economically realistic when biological routes are more simple than chemical ones (e.g. for localised oxidation of hydrocarbons and for the synthesis of optically active compounds). When regarding environmental and health prospects, new advantages arise such as a lower energetic demand, lower environmental factor (the E-factor is defined as the mass ratio of waste to product in a given synthetic scheme), lower pollution with toxic metals or solvents, favourable product's life cycle analysis (LCA) and less hazardous processes (see Woodley 2008). Indeed, regulatory pressure is working for greener synthesis routes, but the consumer demand is maybe even more effective. Besides but related to the environment question, in some industrial fields the question of the natural origin of the compounds is becoming central. Some years ago, only some German and Swiss consumers were concerned by the natural origin of what they ate, but, although it took some years to extend this concern to the European continent, this idea has gained credit in North America and thereby became central for all food, health and cosmetic products all around the world. The word "natural" is legally defined in the frame of the American and European directives (e.g. for aromas: CFR 1990 and the European directive 88/388/CEE), and a substance can be considered as natural when it comes from a plant, animal or microbial origin with a physical, microbial or enzymatic process. Therefore, biotechnological routes may be, if they exclude any chemical steps, a way to get natural products.

In regard to these criteria (simpler and environmental friendly routes providing natural compounds), biocatalysis develops rapidly. In the case of alkane and lipid biotechnology, yeast are interesting tools because many yeast species are able to transform and utilise these substrates for growth (see Fickers et al. 2005). However, hydrophobic substrates are quite specific and the best known and thoroughly utilised yeast species *Saccharomyces cerevisiae* is not the most efficient one. For the biotechnology of hydrophobic compounds, strains belonging to other genera are employed such as *Candida*, *Rhodotorula*, *Pichia*, *Sporobolomyces* and of course *Yarrowia*.

The aim of this chapter is to present the production by *Yarrowia* of compounds of interest resulting from the oxidation of fatty acids or alkanes. The products have diverse applications interesting the polymer, cosmetics and food industries but share common metabolic pathways around β- and ω-oxidation. In most cases, pioneer developments were carried out with other species than *Y. lipolytica,* but the development of this species as a model organism for hydrophobic compounds metabolism enabled researchers to carry out interesting and original works on these subjects. This text will present the different compounds dealt with by this chapter, the corresponding metabolic pathways, the utilisation of yeast for biotechnological production and the developments obtained with *Y. lipolytica*.

2 Compounds of Interest Resulting from β- and ω-Oxidation

Several compounds of interest can be produced through β- or ω-oxidation; however, this chapter will focus on the production of acid alcohol and dioic acids which are direct precursors of lactones, macrocyclic compounds or building blocks of polymers (see Fig. 1). These compounds may be obtained from petroleum products and natural seed oils. As these substrates are usually carbon-chains lacking functionalities (e.g. hydroxyl or epoxy groups) which could give them wider industrial applications, processes have to oxidise substrates providing these functionalities.

2.1 ω-Hydroxy Acids (1) and α,ω-Dicarboxylic Compounds (2)

The first family of compounds is composed of α,ω-dicarboxylic compounds (2) and ω-hydroxy acids (1). These compounds have interesting properties due to their capacity to polycondensate and form polymeric structures. Condensation is based on the esterification reaction of an alcohol and an acid, resulting in an ester and a molecule of water. This reaction can concern an isolated alcohol functional group or the hydroxy part of a carboxylic group. Many diacids are used in complement of diamines to give rise to polyamides such as nylon (P) (3) or other polymers such as polyurethane. Sometimes intramolecular esterification is preferred yielding macrocycles resulting from the cyclisation of diacids (e.g. muscone (9), exaltone (8)) or macrocyclic lactones (e.g. Exaltolide (10)) (Fig. 1). These compounds are key musk odorants which are of central importance to the fragrance industry as they form the bottom note of perfume compositions (Frater et al. 1998) with warm, sweet, powdery or animal notes that are long-lasting, tenacious and substantive. The use of musks goes back to the late antiquity and is still very popular making

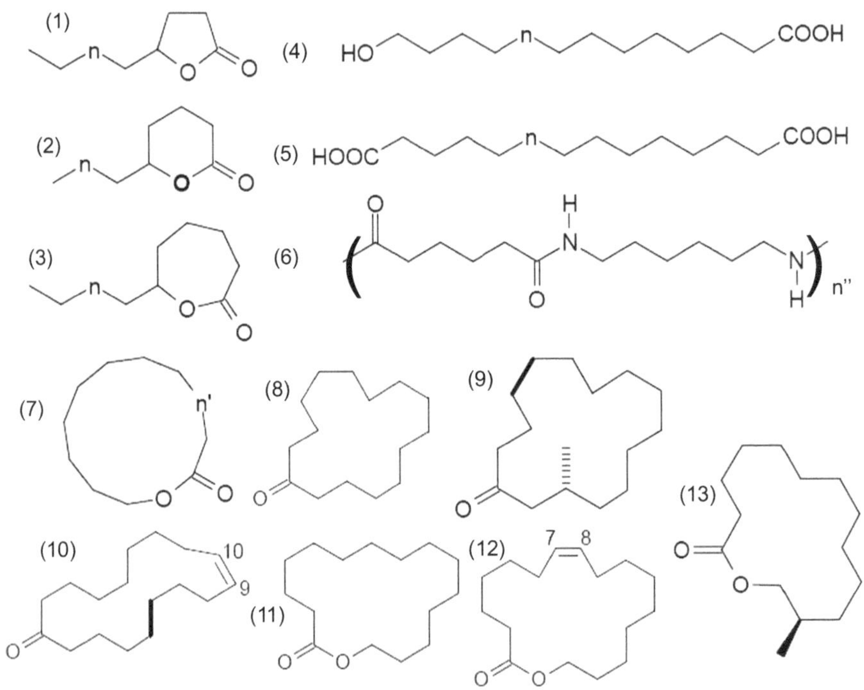

Fig. 1 General structures of some compounds presented in this study: (1) γ-lactones, (2) δ-lactones, (3) ε-lactones, (4) ω-hydroxy acids, (5) α,ω-dicarboxylic acids, (6) one unit of nylon polyamide, (7) macrocyclic lactones and some musk macrocyclic ketones or lactone: (8) exaltone, (9) muscone (principal odorous component of the secretion of the male musk deer *Moschus moschiferus*), (10) civetone, (11) thibetolide (Exaltolide), (12) ambrettolide, (13) *R*-12-methyl-13-tridecanolide (found in the Angelica root oil (*Archangelica officinalis* Hoffm. syn. *Angelica archangelica* L.))

natural extraction unable to respond to the demand. Indeed, musk fragrances are components that can be extracted from glands of some animals such as the Asian musk deer belonging to the family of *Moschidae* or other animals. Besides macrocyclic ketones that can be extracted from animal glands, compounds exhibiting similar sensorial notes but usually possessing an oxygen in the cycle (macrocyclic lactones) can be extracted from some plants. Apart from these natural perfume compounds, some unrelated artificial compounds (for instance the polycyclic galaxolide) exhibit strong musk odour. Although the synthesis of macrocyclic ketones has been realised for about one century, due to a high cost and to the existence of other musk-like compounds, it has not been industrially developed before the 1990s, corresponding to a debate over toxicologic effects of artificial musk compounds and to the first yeast-based processes of production. Now, macrocyclic musks are still expensive compared to polycyclic ones, but their production is nevertheless increasing.

Usually, dicarboxylic acids can be produced by chemical routes, often by cleavage of fatty acids (ozonolysis of the monounsaturated fatty acid, oleic acid or alkaline cleavage of the 12-hydroxylated ricinoleic acid) or through olefin metathesis of unsaturated oils (e.g. sunflower oil with Grubbs- or Hoveyda-type catalysts) (Marvey 2008; Schörken and Kempers 2009). However, these routes are not satisfying to produce unsaturated dicarboxylic acids. Yeast may be preferred to overcome this problem as well as for other specific compounds not easy to prepare from existing fatty acids.

It can be noted that dicarboxylic acids are precursors of polymers and macrocyclic musks, and an inter- or intramolecular condensation is still required. Although polycondensation is relatively easy to carry out, cyclisation is more difficult as it requires to work in diluted media. However, some ways are possible as described in Frater et al. (1998).

2.2 *Volatile Aroma Lactones (1, 2, 3)*

The second family corresponds to lactones of less than 12 carbons that possess strong sensorial properties usually with fruity and fatty notes and that are encountered in many fruits and fermented food. As it will be described later, the production of lactones from hydroxy fatty acids is relatively easy using yeast cells. However, the limiting step is the production of the related hydroxy fatty acids (Fig. 2). For instance, γ-decalactone can be produced by yeast from the unexpensive 12-hydroxylated ricinoleic acid (80 % of the acyl moieties of castor oil). It is even the highest biotechnological production of aroma compounds (Gatfield 1999). For other lactones, production depends on the availability of the substrate, but in some cases, biocatalysts can realise both hydroxylation and β-oxidation. This is particularly the case for the *Trichoderma* production of 6-pentyl-α-pyrone (an unsaturated δ-decalactone) from linoleic acid (Serrano-Carreon et al. 1993). In this case, lipoxygenase carries out the hydroperoxidation of the polyunsaturated fatty acid which is then reduced into a 13-hydroxy fatty acid, which yields lactone after four β-oxidation cycles. Thus, the production of natural lactone (usually from yeast production as the extraction from low-concentrated fruit is not economically realistic) depends on the existence of a hydroxylated substrate or of a possible hydroxylation step. Table 1 (from the site of The Good Scents Company) shows the number of aroma producers providing lactones with or without the natural label showing the difficulties to produce the compound from biocatalysis: for γ-deca- and octa-lactones, half of the aroma producers propose natural lactones (which may come from less producers), whereas nonalactones are more difficult to obtain in biotechnology with only one product proposed. The various routes to obtain lactones are given in Romero-Guido et al. (2011).

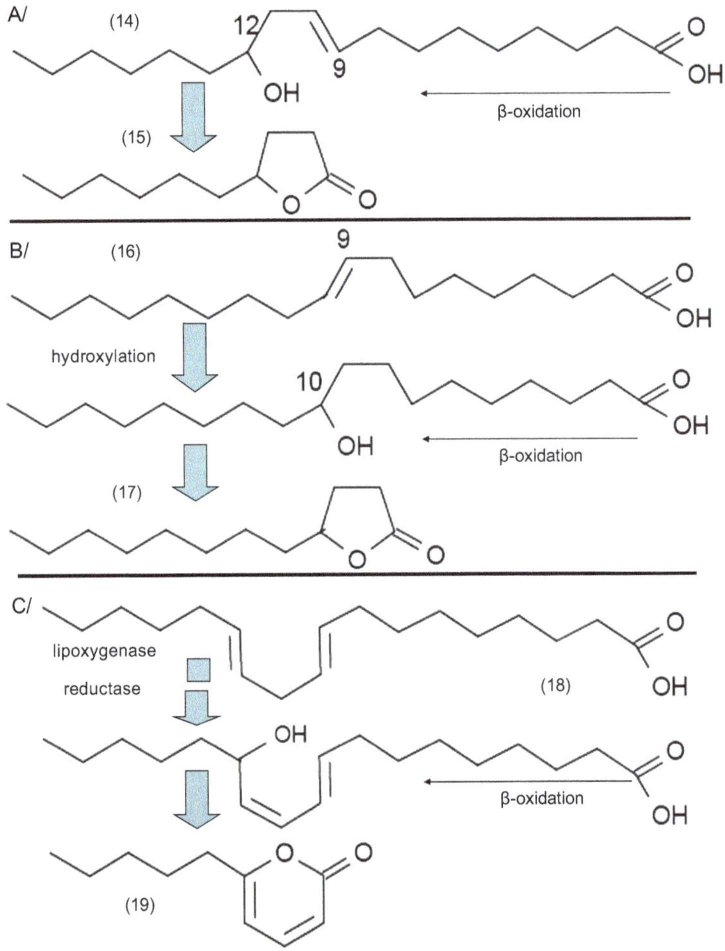

Fig. 2 Some lactones and their precursor. (**a**) Ricinoleic acid (14) and γ-decalactone (15), (**b**) oleic acid (16) and γ-dodecalactone (17), hydroxylation may be carried out by *Lactobacillus* strains, (**c**) linoleic acid (18) and 6-pentyl-α-pyrone (19)

Table 1 Aroma compounds from the family of lactones, number of providing firms, types of aromas and utilisation

	Number of providing firms			
Lactone	Total	Natural lactone	Type of odour	Utilisation
δ-C12	12	4	Tropical, peach, coconut, creamy	Apricot, butter, coconut, cream, cypress dairy fat, jasmin, lilac, lilas, syringa, milk, nut, peach, pear, plum, rose tropical, vanilla
γ-C12	15	3	Fruity, fatty peach	Apricot, butter, coconut, mango, papaya, peach, strawberry
δ-C10	14	6	Coconut	Apricot, butter, coconut, cream, dairy, mango, nectarine, nut, nut pecan, peach, pina colada, raspberry, strawberry tropical
γ-C10	24	12	Fruity, creamy, peach	Over 100 formulations among which are butter, cappuccino, caramel, cherry, chocolate, coconut, cranberry, cream, cucumber, durian, fig, frangipani, gardenia, gingerbread, grapefruit, guava, hibiscus, jacinthe, iris, kiwi, lily, linden, magnolia, mango, melon, mimosa, muguet, myrtle, neroli, orchid, passion fruit, peach, pina colada, mirabelle, plum, pumpkin, raspberry, wild rose, strawberry, tea, toffee, violet, wasabi
δ-C9	7	0	Coconut	Apricot, bread, butter, caramel, chocolate, cocoa, bakeapple, coconut, cream, curry, dairy, grape, hay, herbal, maple, nut, almond, passion fruit, peach, peppermint, pineapple, plum, raspberry, spearmint strawberry, tea, green, tea, tobacco, tabac tabaco, tomato, tropical vanilla, woody
γ-C9	19	1	Coconut, creamy	Amber, apricot, balsam, banana, caramel, cedarwood, chocolate, cocoa, coconut, fougere, fig, floral, fracas, fruit, tropical fruit, gardenia, grass sweet grass, honeysuckle, chevrefeuille, lavender, lilac, lilas, syringa, lychee, magnolia, maple, orange, orchid, oriental peach, pina colada, pine, plum, raspberry, sandalwood, strawberry, tonka, vanilla
δ-C8	4	0	Coconut	Apricot, blackberry, butter, butterscotch, caraway, chocolate, cocoa, christmas blends, coconut, cookie, cream, fat, honey, miel, mango, milk, oakwood, peach, raspberry, rootbeer, strawberry, tropical vanilla
γ-C8	21	10	Coconut	Apricot, blackberry, butter, butterscotch, caraway, chocolate, cocoa, christmas blends, coconut, cookie, cream, fat, honey, miel, mango, milk, oakwood, peach, raspberry, rootbeer, strawberry, tropical, vanilla

Data from Anonym (2010)

3 Production of Dicarboxylic Acids

This part presents first the various strategies carried out to increase the production of dicarboxylic compounds in *Candida* sp. and then the particular strategies attempted on *Yarrowia lipolytica*.

3.1 Yeast Route to α,ω-Dicarboxylic Acids

As mentioned above, these kinds of oxygenated compounds are obtainable from alkanes or lipids through ω-oxidation, and the alkane-assimilating yeast species are usually convenient catalysts. This possibility began to be explored in the 1970s (Kaneyuki and Ogata 1975; Kise and Furukawa 1983; Taoka and Uchida 1983). The genes involved in terminal oxidation and their induction were studied, especially in *Candida maltosa* and in other alkane utilising yeast, and their possible application for ω-oxidation of alkanes or fatty compounds (Mauersberger and Matiashova 1980; Mauersberger et al. 1981; Wiedmann et al. 1988; Schunck et al. 1989; Schunck et al. 1991; Ohkuma et al. 1995a, b; Ohtomo et al. 1996; Zimmer et al. 1996a, b, 2000). The first complete genetic engineering work was carried out later by Picataggio et al. (1992) who developed an approach with *Candida tropicalis*. In their strategy (summarised in Fig. 3), they first wanted to increase yields by inactivating the β-oxidation pathway. This was done by deleting the genes coding for acyl-CoA oxidase, the enzyme catalysing the first step of the pathway (Picataggio et al. 1993). The two copies of the genes *pox4* and *pox5* of the diploid yeast were deleted giving a strain able to transform 80–100 % of the alkane (C12 or C14) to the corresponding dicarboxylic acid. Once the substrate was completely redirected to the ω-oxidation pathway, their goal was to increase the specific production by amplifying the genes encoding the rate-limiting step of the pathway, the alkane monooxygenase system (AMOS) (Picataggio et al. 1991). Thus the P450*ALK1* and *CPR* genes coding for the cytochrome P450 and the NADPH-cytochrome P450-reductase enzymes, respectively, were introduced in several copies in the genome of the Δ*pox4*Δ*pox5* strain. The strain in which both genes had been coordinately amplified exhibited for the biotransformation of methyl myristate a 30 % increase in the productivity which yielded up to 2 g l^{-1} h^{-1} and 150 g l^{-1} in 92 h. However, further investigation revealed that only the reductase enzyme has been amplified and even by changing the promoter of *alk1*, the hydroxylase activity could not be increased. After that, new works in the field implied to take into account this patented process that is to say use novel routes, improve this process (for instance with more stable mutant strains) or explore new applications. Several improvements have been patented, and a completely different strategy was followed in the group of Ohta and Takagi with a *Candida maltosa* strain. Through repeated mutagenesis and screening for higher DCA production, they selected strains overproducing dicarboxylic acids (Fig. 4). From the original strain producing less than 5 g l^{-1} brassylic acid from *n*-tridecane, they obtained after several steps strains producing up to 165 g l^{-1}. These mutations

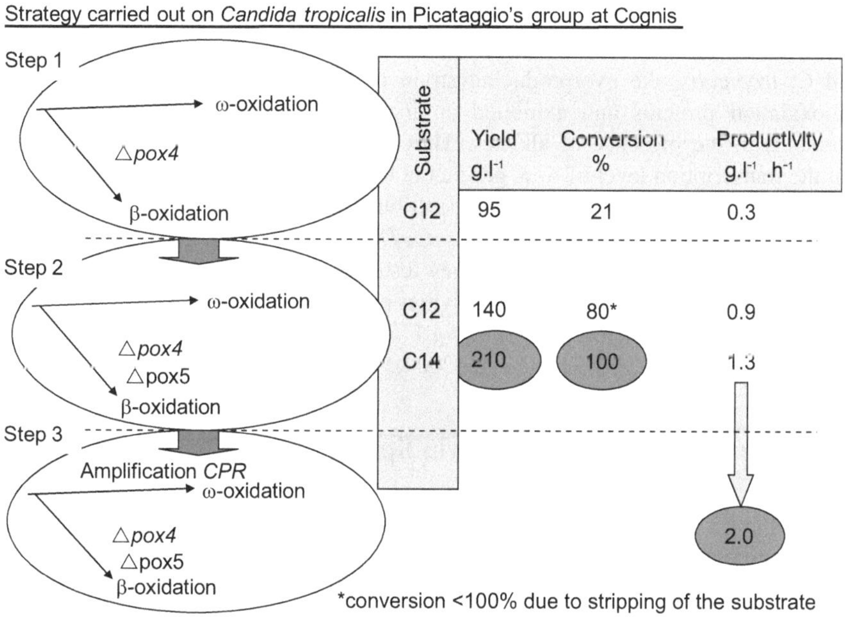

Fig. 3 Strategy carried out on *Candida tropicalis* in Picataggio's group at Cognis. From a *Candida tropicalis* strain, genes were deleted or amplified (steps 1–3 on the *left*), and the results in the biotransformation of C12 or C14 alkanes are given on the *right side*

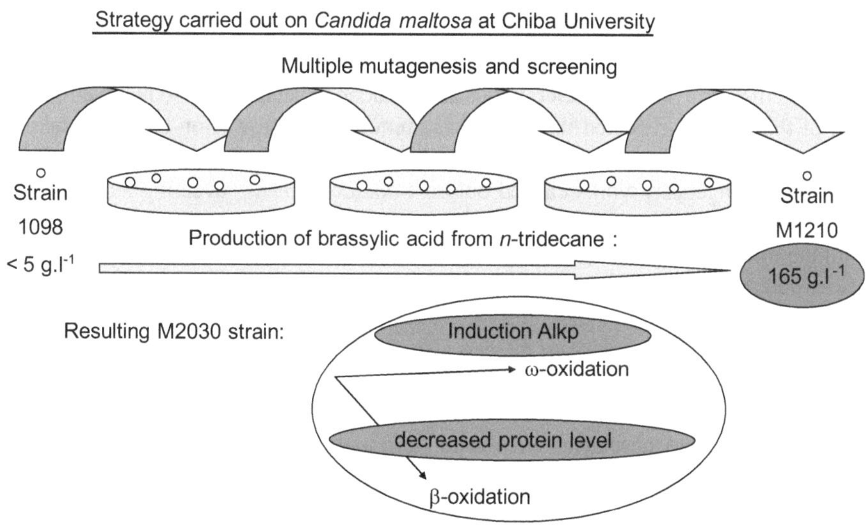

Fig. 4 Strategy carried out on *Candida maltosa* at Chiba University. From the strain 1098, multiple mutagenesis and selection steps for the production of brassylic acid from *n*-tridecane were carried out and the strain 1210 was obtained. The characterisation of the strain is shown at the *bottom* of the figure

were carried out without determining mutation points, but further analysis of the latter strain (Kogure et al. 2007) revealed that, similarly to the overproducing strain of *C. tropicalis*, the overproducing strain of *C. maltosa* had decreased level of β-oxidation proteins and exhibited an increased induction of synthesis of Alk proteins in the presence of alkanes. They identified the implication of regulation at the transcription level of *ALK* genes, but the regulation system was still complex.

Besides these two examples, many contributions have been added to the production of DCA with several species of *Candida*, *Debaryomyces*, *Saccharomyces*, *Schizosaccharomyces* and *Pichia* genera. They resulted in conversion of over 90 % of the substrate to diacids at specific productivities over 1 g l^{-1} h^{-1} and final concentrations of about 30–40 g l^{-1} reaching 75–80 g l^{-1} in certain cases (Zimmer et al. 1996a; Mobley and Shank 2000; Zhang et al. 2004).

3.2 Developments in Yarrowia lipolytica

Studies in *Candida* species resulted to performing processes, especially due to high potentialities of wild-type strains and, therefore, to the advanced knowledge of their oxidation system. However, interest arose also to *Yarrowia lipolytica*, as a species subject of many research attentions and moreover able to grow on alkanes. Indeed, the metabolic particularities of *Y. lipolytica* strains as well as the developments of molecular tools and, later, the genome knowledge motivated researchers to investigate its capacities in this field.

As detailed in Fukuda and Ohta (2013), *Y. lipolytica* strains possess also an important family of CytP450 isoforms of the CYP52 gene family (*ALK1 to ALK8* in strain CX161-1B and up to 12 genes in strain E150) (Fickers et al. 2005). The different isoforms exhibited different substrate selectivities and induction specificities. Sequential disruption of these genes revealed that *ALK1* was required for growth on decane, whereas *ALK1* and *ALK2* were necessary for growth on hexadecane (Iida et al. 1998, 2000). After the mutagenesis works carried out on *Candida* maltosa revealing an important role of AMOS regulation in the overproduction of dicarboxylic acids, works dealing with the induction of these genes by the different carbon sources in *Y. lipolytica* were also of interest. Sumita et al. (2002a, b) identified thus two alkane-responsive elements (ARE1 and 2) forming an alkane-responsive region (ARR1). The role of *PEX10*, *PEX5* and *PEX6* in the repression of *ALK1* was also observed. Of particular interest was also the identification of the yeast alkane-signalling gene *YAS1* encoding a transcription factor for alkane signalling which is essential for the induction of *ALK1* (Yamagami et al. 2004). Despite these genes involved in terminal oxidation, *Y. lipolytica* possesses also an important family of isoforms of the gene coding for the enzyme catalysing the first step of β-oxidation (*POX1* to *6* coding for Aox1 to 6) (Wang et al. 1999a, b) and of genes coding for lipases (see Fickers et al. 2013).

In total, *Y. lipolytica* possesses positive and negative points for its use in the production of diacids. Advantages come from the ability to hydrolyse triglycerides with its extracellular lipases without the use of chemical hydrolysis, from the knowledge of the AMOS system and its possibly simpler regulation than for the

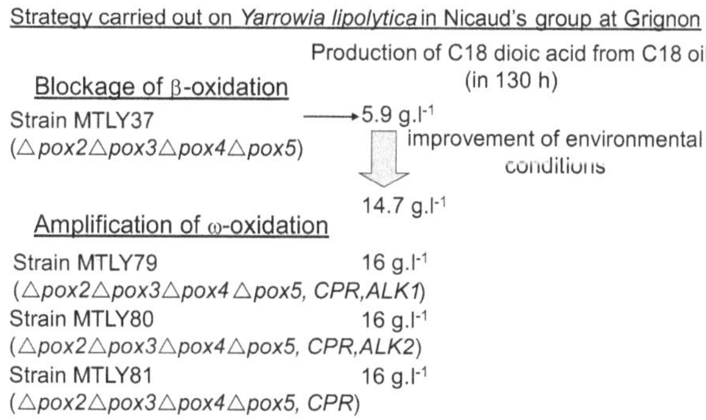

Fig. 5 Strategy carried out on *Yarrowia lipolytica* in Nicaud's group at Grignon for the production of C18 dioic acid from C18 fatty acid. Mutants are presented on the *left side* and results on the *right side*

one of *Candida* sp. Moreover, the possibility to work with a haploid strain for which practical genetic tools have been developed is also an advantage. However, the important family of *POX* genes coding for the undesired substrate degrading acyl-CoA oxidase plays a negative role. Eventually, a process has been disclosed recently for the production of long-chain dicarboxylic acids from fatty acids (or corresponding oils) (Nicaud et al. 2006) (Fig. 5). This process was based on the construction of an efficient strain with deletion of all active *pox* genes (at least *pox2*, *pox3*, *pox4* and *pox5*) and the overexpression of *CPR*, *ALK1* and/or *ALK2* behind the promoter of p*POX2*.

4 Production of Lactones

The main lactone investigated is γ-decalactone. Here a brief history of its production with other species will be presented, and the important developments carried out with *Yarrowia lipolytica* will be described.

4.1 Yeast for Production of Natural Lactones

The interest of yeast biotechnologists for lactone arose in the 1960s after results obtained by a group of nutritionists in Japan who studied the catabolism of hydroxylated fatty acids in different organisms (Okui et al. 1963a–c). From ricinoleic acid (14), for most cells, oxidation was limited to 2–3 cycles resulting in fatty acids of C12 to C16, but for a yeast strain, metabolism went further and γ-decalactone (15) accumulated. From the compounds accumulating, fatty acids possessing 2, 4, 6, 8 or 10 carbons less than the substrate with the hydroxyl group nearer the carboxylic group, it became evident that the pathway of transformation went through β-oxidation.

From that result, several works were carried out on the production of γ-decalactone from yeast which were often limited to the screening of yeast strains and media (reviewed in Endrizzi et al. 1996). Various genera were thus selected for their potentialities including *Sporobolomyces*, *Pichia*, *Candida* and *Rhodotorula*.

For this production, strains of *Y. lipolytica* were often encountered in the top producing strains. Most of these overproducing strains were selected for production processes and metabolic studies have been carried out only on less performing strains. From a study on four different *Sporidiobolus* strains belonging to different species, Blin-Perrin et al. (2000) suggested that the diversity of lactone accumulation was related to the diversity of metabolic organisation. Indeed, from the intermediates identified in the medium (with whole cell or cell-free enzymatic transformation), these authors deduced that for cells having a free intermediate model, accumulation was possible whereas it was low for cells having a leaking pipe model and no lactone was detected for cells possessing a channelled β-oxidation.

4.2 Developments in Yarrowia lipolytica

Contrary to what has been done for dicarboxylic acid production, *Y. lipolytica* strains have been selected for the production of γ-decalactone first because of their good results during screening. Then, the fact that this yeast is becoming a model organism for the study of lipid metabolism has increased the interest for this species. According to Gatfield (1999), this yeast is the one accumulating the highest amounts of γ-decalactone; however, some metabolic particularities have to be taken into account for its use in biotransformation.

4.2.1 Chain-Length Lowering

When analysing the accumulation of intermediates of the β-oxidation of ricinoleic acid, the accumulation of γ-decalactone is usually observed, but this step is usually followed by a degradation of this compound by the cells themselves. Indeed, β-oxidation is normally a pathway degrading fatty acyl-CoA from 18 (or more) carbons to two- or three-carbon units. In fact, this ideal pathway may be that efficient for the mitochondrial β-oxidation in animal cells, but in the peroxisomal pathway occurring in yeast β-oxidation, the number of oxidation cycles depends on the environmental conditions and on many regulating factors (chain length of substrates, concentrations of acyl-CoA, CoA, cofactor pools and regeneration, etc. (Bartlett et al. 1990)). In regard to the production of decalactone, conditions have to be suitable to enable the occurring of four cycles. The fact that a γ-hydroxy acid tends to lactonise readily is important to explain the production, but it does not prevent the cyclised lactone from re-entering β-oxidation. This degradation possibility is a major problem encountered by biotechnologists, and several works have been carried out to extract γ-decalactone during production (Druaux et al. 1995; Dufossé et al. 1997, 1999; Souchon et al. 1998). Indeed, the strategy consisting in modifying the strain to avoid degradation is usually impossible as β-oxidation is a pathway in which the same enzymes are active from C18 to C4 and deleting the gene coding for one β-oxidation enzyme would simply stop the pathway at the C18 level (as done for the production of dicarboxylic acids). The case of *Y. lipolytica* is somewhat particular

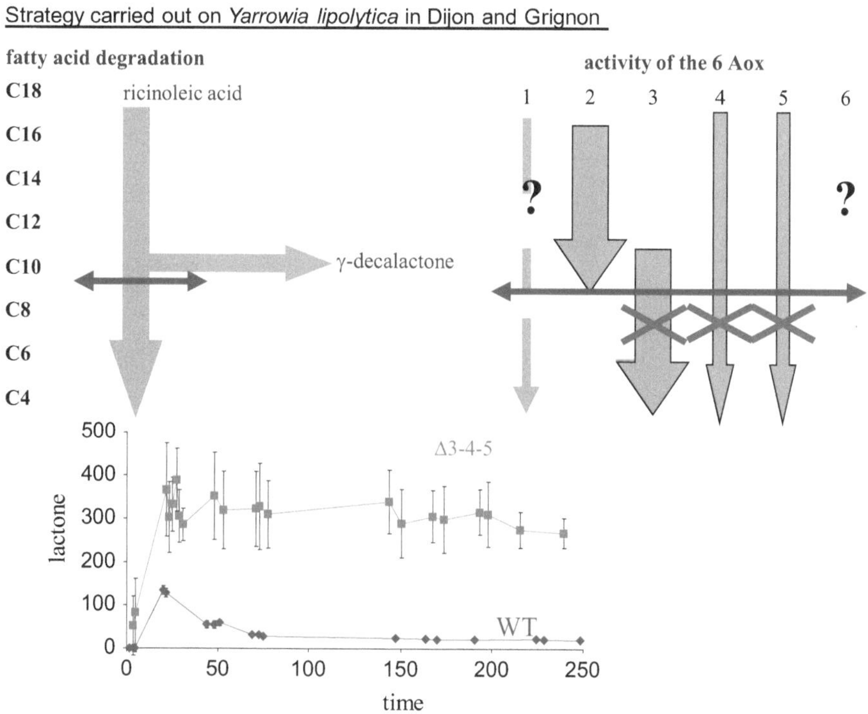

Fig. 6 Strategy carried out on *Yarrowia lipolytica* to favour the exit of metabolites at the C10 level corresponding to the production of γ-decalactone. C18 to C4: fatty acids from 18 to 4 carbons, on the *left*, activity of the 6 acyl-CoA oxidases (Aox) noted 1–6. *Down*, accumulation of lactone by the wild-type strain (WT) and by the mutant deleted for *pox3 pox4* and *pox5*

due to the presence of six isoforms of acyl-CoA oxidases with different substrate selectivity (Wang et al. 1999a, b). A strategy was thus carried out by sequential disruptions to study the role of each acyl-CoA oxidase in the catabolism of ricinoleic acid and the production of γ-decalactone (Pagot et al. 1998; Waché et al. 2000, 2001, 2002; Groguenin et al. 2004). By deleting all the genes coding for acyl-CoA oxidases active on short-chain acyl-CoA, it was possible to increase notably the production of γ-decalactone and to avoid further degradation (Fig. 6).

4.2.2 Exit Between Two β-Oxidation Cycles

One particularity of *Y. lipolytica* is its capacity to accumulate other lactones during the degradation of ricinoleic acid (Fig. 7) (Gatfield et al. 1993). The main lactone produced, 3-hydroxy-γ-decalactone, has no sensorial properties, while the other two decenolides exhibit some interesting and strong fruity notes for one and fungi notes for the other. Although the amount of these particular lactones can be low during screening, it tends to increase during scaling-up showing the impact of environmental

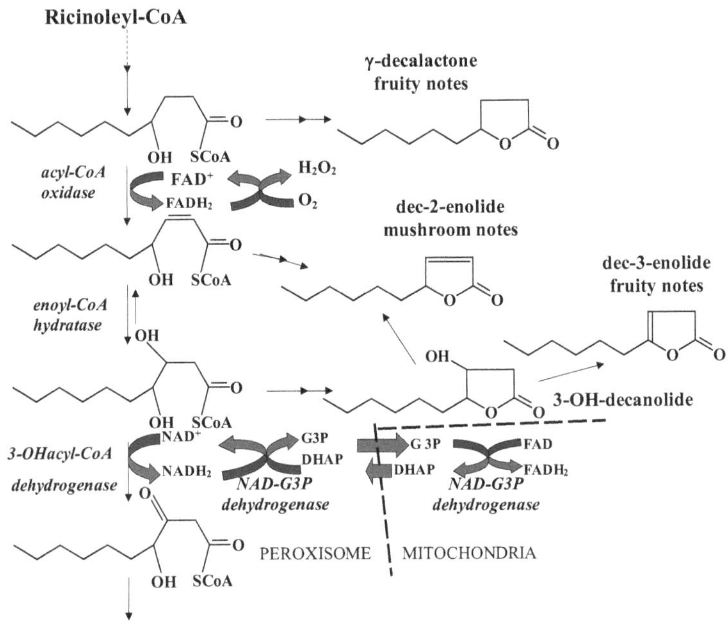

Fig. 7 β-Oxidation loop at the C10 level of the degradation of ricinoleyl-CoA, enzymatic activities, oxygen and cofactor requirements and reaction products

conditions on their accumulation. These lactones result from exit of metabolites at different steps of the β-oxidation cycle (Waché et al. 2001, 2003; Groguenin et al. 2004; Escamilla García et al. 2007a, b, 2009) (Fig. 7) with only the 3-keto-lactone lacking. However, the presence of dec-3-enolide in the mean time than dec-2-enolide suggests that these two compounds could derive from the hydroxy lactone and not from the β-oxidation cycle. From the presence of these intermediates, it can be suggested that, contrary to what occurs in other yeast, the β-oxidation cycle can be divided in two parts (either due to enzyme location/organisation or through enzyme activity (β-oxidation or coenzyme A ester hydrolysis)) and intermediates can exit at the two oxidation levels (oxidase and dehydrogenase). Indeed modifying environmental or genetic conditions for oxidation exhibits a high impact on the accumulation of these two lactones (Fig. 8), while the accumulation of the two decenolides is more or less maintained showing that dehydration of the hydroxy lactone depends on another rate-limiting step. When modifying oxygenation (Escamilla García et al. 2007a, 2009), several levels of oxygenation are observable: at medium oxygenation, β-oxidation functions well while decreasing oxygen increases the accumulation of the hydroxylated lactone suggesting that oxygenation is too low to permit the re-oxidation of NADH. When still decreasing oxygenation to very low levels, oxygen becomes insufficient to enable oxidation by the acyl-CoA oxidase (Aox), and γ-decalactone accumulates while the growth rate decreases. When increasing oxygenation to very high level, accumulation of lactones increases slightly showing

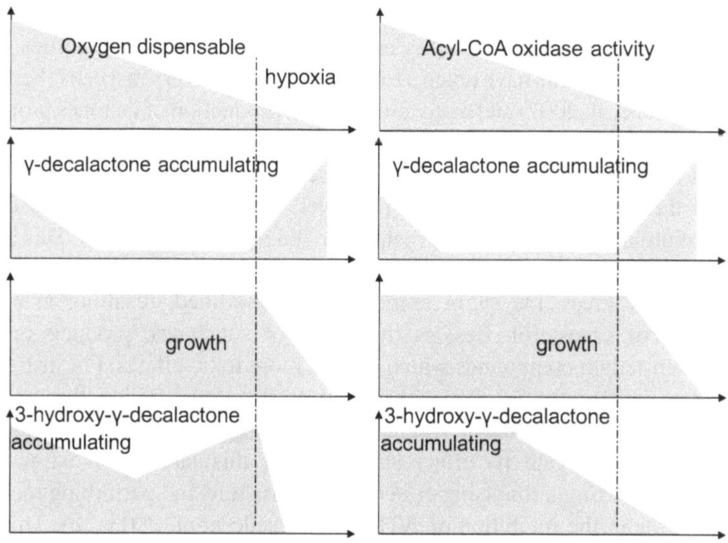

Fig. 8 Effect of oxygenation (*left*) and acyl-CoA oxidase activity (*right*) on the accumulation of γ-decalactone and cell growth and accumulation of 3-hydroxy-γ-decalactone

that conditions are not optimal. Results obtained with mutants with various Aox activities on C18-C10 acyl-CoAs are consistent with oxygenation results: when decreasing Aox activity, the amount of 3-hydroxy lactone accumulating decreases, and when the activity is low, the growth rate and γ-decalactone accumulation decrease (Escamilla García et al. 2007b).

In conclusion, *Y. lipolytica*'s β-oxidation is normally highly sensitive to NAD regeneration (exit of intermediates at the 3-hydroxy-acyl-CoA dehydrogenase step), but in cases of hypoxia or very low acyl-CoA oxidase activity, oxidation becomes the rate-limiting step of the pathway and γ-decalactone accumulates.

5 Sensitivity of *Y. lipolytica* to Amphiphilic Compounds and Membrane Adaptation

In all processes dealing with the biotransformation of hydrophobic and amphiphilic substrates into amphiphilic products, the potential toxic effect of these compounds has to be taken into account as well as their efflux into the medium. Alkanes and fatty acids tend to be extracted into the hydrophobic and lipid-rich parts of the cells. They can thus perturb the function of membranes and especially of the plasma membrane. Different effects can be exhibited. Acids and alcohols can transport protons through the membrane depending on the p*K*a of the acid, decreasing thereby the proton motive force which results in an inhibiting effect on secondary transporters and in changing cell homeostasis. Moreover, the extraction of these compounds into the membrane can modify the physicochemical properties of this cell component, modifying its fluidity and perturbing its barrier properties.

Although some authors have noted a toxicity of oleic acid on yeast cells (Feron et al. 1997; Lockshon et al. 2007) and many processes of production of lactones propose the utilisation of esterified forms of the substrates (methyl ricinoleate or castor oil), it seems that these sources are rapidly hydrolysed by *Y. lipolytica*'s lipases, but the toxicity of these acids is not obvious. However, methyl oleate when added to the culture medium increases rapidly the membrane fluidity (Ta et al. 2012). This increase is due to the immediate incorporation of oleate moieties into the membrane, but, as a response to this stress, the sterol homeostasis is modified, resulting in a higher concentration of ergosterol. Besides the effect of substrates, products are often medium-chain-length compounds which exhibit more toxic effects. For instance, the toxicity of γ-decalactone is often considered as the step limiting the process of production (Endrizzi et al. 1996; Waché et al. 2003, 2006). Some works have been carried out to investigate its effect showing that this lactone is extracted into membranes, provoking a fluidising effect on this structure and perturbing membrane energetics through the inhibition of ATPases (Aguedo et al. 2003a, b). This effect depends on the length of the apolar part of the lactone with dodecalactone being more toxic than shorter chain-length lactones (Aguedo et al. 2002). Moreover, it was shown recently that γ-dodecalactone provoked the depletion of ergosterol in a way similar to methyl-β-cyclodextrin and that depleted cells were more sensitive to the lactone shock (Ta et al. 2010).

However, due to practical aspects, these studies have been carried out on cells grown on glucose media. In a recent study, we have investigated the effect of an amphiphilic pre-shock (methyl oleate) on cell sensitivity to amphiphilic medium-chain-length lactones (Ta et al. 2010). An increase in resistance was observed with *Y. lipolytica* and other lipophilic species, while other yeast species were less resistant after the pre-shock. With *Y. lipolytica*, it was shown that although the incorporation of oleyl moieties into the membrane made it more fluid, it also gave it a higher buffer capacity toward the incorporation of γ-dodecalactone. This was imputable to the location of the long-chain oleyl moieties into the membrane, fluidising both the polar edge and the apolar core of the membrane and modifying the polarity of the edge (Ta et al. 2010). The presence of oleyl could thus reinforce the membrane against a toxic interdigitated phase transition which could be induced by the presence of short-chain compounds (Weber and De Bont 1996). Moreover, a preculture on methyl oleate induced an increase in the concentration of ergosterol and protection against the dodecalactone-induced depletion of ergosterol, suggesting a different structure of the membranes of oleate-grown cells with the possible organisation of ergosterol in microdomains protected from lactones (Ta et al. 2010).

The extraction of medium- or long-chain α,ω-dicarboxylic acids in cell membranes has not been investigated, but the case might be more complex than the case of monopolar fatty acids or lactones due to the presence of two apolar extremities. In comparing with some carotenoids oxygenated at both extremities, the localisation of such compounds into membranes depends on the thickness of the hydrophobic core of the membrane. If this thickness corresponds to the length between the two polar extremities of the molecule, compounds are soluble into membranes; otherwise, the solubility decreases (Gruszecki and Sielewiesiuk 1990; Wisniewska et al. 2006). However, it is likely that C10 to C18 dicarboxylic acids behave differently than longer carotenoids. Nevertheless, in regard to the concentrations reached in the biotransformation medium, the toxicity of dicarboxylic acids is probably negligible although their transport across membranes could be improved into biotechnological processes.

6 Conclusion

The possibility of yeast to oxygenate saturated carbon chains presents some advantages over traditional chemical routes. This led to many biotechnological developments for the production of some hydroxylated or dicarboxylic acids and of lactones. Due to the properties of these compounds as building blocks in synthesis, polymer units, flavour and fragrances, antimicrobial, etc., the demand of such compounds is increasing and, despite the relatively high price of cultivating microorganisms, yeast biotransformation can increase the diversity of the producible compounds. The interest of *Y. lipolytica* compared to *Candida* species depends on the processes with higher production of lactones and lower ones of dicarboxylic acids. However, when considering the demand for new compounds in these families and the developments of tools for metabolic engineering, *Y. lipolytica* is a promising species in the field. However, to improve the engineering of this species, data are still required in metabolic biochemistry (e.g. regulation of the β-oxidation pathway) and cell biophysics (especially at the membrane and cell wall levels).

References

Aguedo M, Beney L, Waché Y, Belin J-M, Gervais P (2002) Interaction of odorous lactones with phospholipids: implications in toxicity towards producing yeast cells. Biotechnol Lett 24: 1975–1979

Aguedo M, Beney L, Waché Y, Belin JM (2003a) Interaction of an odorant lactone with model phospholipid bilayers and its strong fluidizing action in yeast membrane. Int J Food Microbiol 80:211–215

Aguedo M, Beney L, Waché Y, Belin JM (2003b) Mechanisms underlying the toxicity of lactone aroma compounds towards the producing yeast cells. J Appl Microbiol 94:258–265

Anonym (2010) The Good Scent Company. http://www.thegoodscentscompany.com/data/rw1012281.html#tosupp. Accessed 18 Jan 2010

Bartlett K, Hovik R, Eaton S, Watmough NJ, Osmundsen H (1990) Intermediates of peroxisomal beta-oxidation. Biochem J 270:175–180

Blin-Perrin C et al (2000) Metabolism of ricinoleic acid into gamma-decalactone: beta-oxidation and long chain acyl intermediates of ricinoleic acid in the genus *Sporidiobolus* sp. FEMS Microbiol Lett 188:69–74

Druaux C, Lubbers S, Charpentier C, Voilley A (1995) Effects of physico-chemical parameters of a model wine on the binding of gamma-decalactone on bovine serum albumin. Food Chem 53:203–207

Dufossé LH, Souchon I, Feron G, Latrasse A, Spinnler HE (1997) Strategies to overcome toxicity during flavour production by micro-organisms: the case of g-decalactone from *Sporidiobolus salmonicolor*. Riv. Ital. EPPOS Spec. Num.:284–298
Dufossé L, Souchon II, Feron G, Latrasse A, Spinnler HE (1999) In situ detoxification of the fermentation medium during gamma-decalactone production with the yeast *Sporidiobolus salmonicolor*. Biotechnol Prog 15:135–139
Endrizzi A, Pagot Y, Le Clainche A, Nicaud J-M, Belin J-M (1996) Production of lactones and peroxisomal beta-oxidation in yeasts. Crit Rev Biotechnol 16:301–329
Escamilla García E, Belin J-M, Waché Y (2007a) Use of a Doehlert factorial design to investigate the effects of pH and aeration on the accumulation of lactones by Yarrowia lipolytica. J Appl Microbiol 103:1508–1515
Escamilla García E, Nicaud J-M, Belin J-M, Waché Y (2007b) Effect of acyl-CoA oxidase activity on the accumulation of gamma-decalactone by the yeast *Yarrowia lipolytica*: a factorial approach. Biotechnol J 2:1280–1285
Escamilla García E et al (2009) Production of 3-hydroxy-gamma-decalactone, the precursor of two decenolides with flavouring properties, by the yeast *Yarrowia lipolytica*. J Mol Catal B 57: 22–26
Feron G, Dufossé L, Mauvais G, Bonnarme P, Spinnler HE (1997) Fatty acid accumulation in the yeast *Sporidiobolus salmonicolor* during batch production of gamma-decalactone. FEMS Microbiol Lett 149:17–24
Fickers P et al (2005) Hydrophobic substrate utilization by the yeast *Yarrowia lipolytica*, and its potential applications. FEMS Yeast Res 5:527–543
Fickers P, Sauveplane V, Nicaud JM (2013) The lipases from *Y. lipolytica*: genetics, production, regulation and biochemical characterisation. In: Barth G (ed) Yarrowia lipolytica. Springer, Heidelberg
Fráter G, Bajgrowicz JA, Kraft P (1998) Fragrance chemistry (Review Article). Tetrahedron 54 (27):7633–7703
Fukuda R, Ohta A (2013) Utilization of hydrophobic substrates by Y*arrowia lipolytica*. In: Barth G (ed) Yarrowia lipolytica. Springer, Heidelberg
Gatfield IL (1999) Biotechnological production of natural flavor materials. In: Teranishi R, Wick EL, Hornstein I (eds) Flavor chemistry, thirty years of progress. Kluwer, Plenum, New York, pp 211–227
Gatfield IL, Güntert M, Sommer H, Werkhoff P (1993) Some aspects of the microbiological production of flavor-active lactones with particular reference to gamma-decalactone. Chem Mikrobiol Technol Lebensm 15:165–170
Groguenin A et al (2004) Genetic engineering of the beta-oxidation pathway in the yeast *Yarrowia lipolytica* to increase the production of aroma compounds. J Mol Catal B 28:75–79
Gruszecki WI, Sielewiesiuk J (1990) Orientation of xanthophylls in phosphatidylcholine multibilayers. Biochim Biophys Acta 1023:405–412
Iida T, Ohta A, Takagi M (1998) Cloning and characterization of an *n*-alkane-inducible cytochrome P450 gene essential for *n*-decane assimilation by *Yarrowia lipolytica*. Yeast 14:1387–1397
Iida T, Sumita T, Ohta A, Takagi M (2000) The cytochrome P450ALK multigene family of an n-alkane-assimilating yeast, Yarrowia lipolytica: cloning and characterization of genes coding for new CYP52 family members. Yeast 16:1077–1087
Kaneyuki H, Ogata K (1975) Method for producing dicarboxylic acids by microorganisms. Patent US3912586
Kise S, Furukawa T (1983) Production of dicarboxylic acid. Patent JP58121797
Kogure T, Horiuchi H, Matsuda H, Arie M, Takagi M, Ohta A (2007) Enhanced induction of cytochromes P450alk that oxidize methyl-ends of n-alkanes and fatty acids in the long-chain dicarboxylic acid-hyperproducing mutant of Candida maltosa. FEMS Microbiol Lett 271: 106–111
Lockshon D, Surface LE, Kerr EO, Kaeberlein M, Kennedy BK (2007) The sensitivity of yeast mutants to oleic acid implicates the peroxisome and other processes in membrane function. Genetics 175:77–91

Marvey BB (2008) Sunflower-based Feedstocks in Nonfood Applications: Perspectives from Olefin Metathesis. Int J Mol Sci 9:1393–1406

Mauersberger S, Matiashova RN (1980) Cytochrome P-450 content in yeast cells during growth on hexadecane. Mikrobiologiia 49:571–577

Mauersberger S, Schunck WH, Muller HH (1981) The induction of cytochrome P-450 in Lodderomyces elongisporus. Z Allg Mikrobiol 21:313–321

Mobley DP, Shank GK (2000) Method for high specific bioproductivity of alpha, omega-alkane dicarboxylic acids using *Candida tropicalis*. Patent WO0017380

Nicaud J, Theveniau F, Le Dall M, Marchal R (2006). Production d'acides dicarboxyliques par des souches mutantes améliorées de *Yarrowia lipolytica*. Patent FR2879215

Ohkuma M, Masuda Y, Park SM, Ohtomo R, Ohta A, Takagi M (1995a) Evidence that the expression of the gene for NADPH-cytochrome P-450 reductase is n-alkane-inducible in Candida maltosa. Biosci Biotechnol Biochem 59:1328–1330

Ohkuma M, Muraoka S, Tanimoto T, Fujii M, Ohta A, Takagi M (1995b) CYP52 (cytochrome P450alk) multigene family in Candida maltosa: identification and characterization of eight members. DNA Cell Biol 14:163–173

Ohtomo R, Kobayashi K, Muraoka S, Ohkuma M, Ohta A, Takagi M (1996) Peroxisome proliferators activate cytochrome P450 genes in an alkane-assimilating yeast, Candida maltosa. Biochem Biophys Res Commun 222:790–793

Okui S, Uchiyama M, Mizugaki M (1963a) Metabolism of hydroxy fatty acids: 1. Metabolic conversion of ricinoleic acid by a certain microorganism to 8-D-(+)-hydroxy tetradec-cis-5-enoic acid. J Biochem 53:265–270

Okui S, Uchiyama M, Mizugaki M (1963b) Metabolism of hydroxy fatty acids: 2. Intermediates of the oxidative breakdown of ricinoleic acid by Genus Candida. J Biochem 54:536–540

Okui S, Uchiyama M, Mizugaki M, Sugawara A (1963c) Characterization of hydroxy acids in depot fat after feeding of ricinoleic acid. Biochim Biophys Acta 70:344–346

Pagot Y, Le Clainche A, Nicaud J-M, Waché Y, Belin J-M (1998) Peroxisomal Beta-oxidation activities and gamma-decalactone production by the yeast *Yarrowia lipolytica*. Appl Microbiol Biotechnol 49:295–300

Picataggio S, Rohrer T, Eirich LD (1991) Method for increasing the omega-hydroxylase activity in *Candida tropicalis*. Patent WO9114781

Picataggio S et al (1992) Metabolic engineering of *Candida tropicalis* for the production of long chain dicarboxylic acids. Biotechnology 10:894–898

Picataggio S, Deanda K, Dudley E (1993) Site-specific modification of the candida tropicals genome. Patent US5254466 (A)

Romero-Guido C, Belo I, Ta TMN, Cao-Hoang L, Alchihab M, Gomes N, Gómez-Díaz D, Thonart P, Teixeira JA, Destain J, Waché Y (2011) Biochemistry of lactone formation in yeast and fungi and its utilization for the production of flavour and fragrance compounds. Appl Microbiol Biotechnol 89:535–547

Schörken U, Kempers P (2009) Lipid biotechnology: industrially relevant production processes. Eur J Lipid Sci Technol 111:627–645

Schunck WH et al (1989) Molecular cloning and characterization of the primary structure of the alkane hydroxylating cytochrome P-450 from the yeast Candida maltosa. Biochem Biophys Res Commun 161:843–850

Schunck WH et al (1991) Comparison of two cytochromes P-450 from *Candida maltosa*: primary structures, substrate specificities and effects of their expression in *Saccharomyces cerevisiae* on the proliferation of the endoplasmic reticulum. Eur J Cell Biol 55:336–345

Serrano-Carreon L, Hathout Y, Bensoussan M, Belin JM (1993) Metabolism of linoleic acid or mevalonate and 6-pentyl-alpha-pyrone biosynthesis by Trichoderma species. Appl Environ Microbiol 59:2945–2950

Souchon I, Spinnler HE, Dufossé L, Voilley A (1998) Trapping of gamma-decalactone by adsorption on hydrophobic sorbents : application to the bioconversion of methyl ricinoleate by the yeast *Sporidiobolus salmonicolor*. Biotechnol Tech 12:109–113

Sumita T, Iida T, Hirata A, Horiuchi H, Takagi M, Ohta A (2002a) Peroxisome deficiency represses the expression of n-alkane-inducible YlALK1 encoding cytochrome P450ALK1 in *Yarrowia lipolytica*. FEMS Microbiol Lett 214:31–38

Sumita T, Iida T, Yamagami S, Horiuchi H, Takagi M, Ohta A (2002b) YlALK1 encoding the cytochrome P450ALK1 in *Yarrowia lipolytica* is transcriptionally induced by n-alkane through two distinct cis-elements on its promoter. Biochem Biophys Res Commun 294:1071–1078

Ta TMN et al (2010) New insights into the effect of medium chain length lactones on yeast membranes. Importance of the culture medium. Appl Microbiol Biotechnol 87:1089–1099

Ta TMN et al (2012) A shift to 50°C provokes death in distinct ways for glucose- and oleate-grown cells of *Yarrowia lipolytica*. Appl Microbiol Biotechnol. doi:10.1007/s00253-011-3537-3

Taoka E, Uchida S (1983) Production of long-chain dicarboxylic acid from fat or oil. Patent JP58165795

Waché Y et al (2000) Involvement of acyl-CoA oxidase isozymes in biotransformation of methyl ricinoleate into γ-decalactone by *Yarrowia lipolytica*. Appl Environ Microbiol 66:1233–1236

Waché Y, Aguedo M, Choquet A, Gatfield I, Nicaud J-M, Belin J-M (2001) Role of β-oxidation enzymes in the production of γ-decalactones from methyl ricinoleate. Appl Environ Microbiol 67:5700–5704

Waché Y, Aguedo M, LeDall M-T, Nicaud J-M, Belin J-M (2002) Optimization of *Yarrowia lipolytica*'s β-oxidation pathway for lactones production. J Mol Catal B Enzym 19–20: 347–351

Waché Y, Aguedo M, Nicaud J-M, Belin J-M (2003) Catabolism of hydroxyacids and production of lactones by the yeast *Yarrowia lipolytica*. Appl Microbiol Biotechnol 61:393–404

Waché Y, Husson F, Feron G, Belin J-M (2006) Yeast as an efficient catalyst for the production of flavour and fragrances from lipids. Antonie Van Leeuwenhoek 89:405–416

Wang H et al (1999a) Evaluation of acyl CoA oxidase (Aox) isozyme function in the n-alkane-assimilating yeast *Yarrowia lipolytica*. J Bacteriol 181:5140–5148

Wang H, Le Dall M-T, Waché Y, Laroche C, Belin J-M, Nicaud J-M (1999b) Cloning, sequencing and characterization of five genes coding for Acyl-CoA oxidase isozymes in the yeast Yarrowia lipolytica. Cell Biochem Biophys 31:165–174

Weber FJ, De Bont JAM (1996) Adaptation mechanisms of microorganisms to the toxic effects of organic solvents on membranes. Biochim Biophys Acta 1286:225–245

Wiedmann B, Wiedmann M, Mauersberger S, Schunck WH, Muller HG (1988) Oxygen limitation induces indirectly the synthesis of cytochrome P-450 mRNA in alkane-growing Candida maltosa. Biochem Biophys Res Commun 150:859–865

Wisniewska A, Widomska J, Subczynski WK (2006) Carotenoid-membrane interactions in liposomes: effect of dipolar, monopolar, and nonpolar carotenoids. Acta Biochim Pol 53:475–484

Woodley JM (2008) New opportunities for biocatalysis: making pharmaceutical processes greener. Trends Biotechnol 26:321–327

Yamagami S, Morioka D, Fukuda R, Ohta A (2004) A basic helix-loop-helix transcription factor essential for cytochrome p450 induction in response to alkanes in yeast Yarrowia lipolytica. J Biol Chem 279:22183–22189

Zhang Y, Wilson RC, Craft DL, Eirich DL, Frayer RW (2004) Use of *POX4* promoter to increase gene expression in *Candida tropicalis*. Patent WO2004013336

Zimmer T, Kaminski K, Schunck W-H, Kaergel E, Scheller U, S M (1996a) Process for hydroxylating long-chain alkanes, fatty acids and other alkyl compounds. Patent WO9627678

Zimmer T, Ohkuma M, Ohta A, Takagi M, Schunck WH (1996b) The CYP52 multigene family of Candida maltosa encodes functionally diverse n-alkane-inducible cytochromes P450. Biochem Biophys Res Commun 224:784–789

Zimmer T, Ogura A, Takewaka T, Zimmer RM, Ohta A, Takagi M (2000) Gene regulation in response to overexpression of cytochrome P450 and proliferation of the endoplasmic reticulum in Saccharomyces cerevisiae. Biosci Biotechnol Biochem 64:1930–1936

Cytochrome P450 Expression in *Yarrowia lipolytica* and Its Use in Steroid Biotransformation

Stephan Mauersberger, Ludmila A. Novikova, and Vladimir M. Shkumatov

Abstract This review is a first attempt to systematize data on the potential of transgenic *Y. lipolytica* to catalyze diverse reactions of steroid transformation. The yeast *Y. lipolytica* was tested as host for P450-catalyzed biotransformation of steroids, including the mammalian P450scc system (three components) and/or the two-component P450c17 system, being functionally active with yeast NADPH-P450 reductase (CPR). New strategies for the construction of recombinant *Y. lipolytica* strains containing several expression cassettes containing heterologous cDNA (up to 4, in five vectors) under control of the isocitrate lyase (*ICL1*) promoter have been developed. Characteristics of recombinant *Y. lipolytica* strains functionally expressing the P450scc system and/or P450c17, being functionally active with yeast NADPH-P450 reductase (*YlCPR*), are presented.

Functional expression of P450 systems in yeasts was proved by biotransformation of cholesterol (Cho) or by 17α-hydroxylation of progesterone (Pro) or pregnenolone (Pre). Strains coexpressing the P450scc system and P450c17 exhibited a high biotransformation capacity of Pro into 17α-hydroxyprogesterone (17HPro); the conversion of Cho to Pre and 17α-hydroxypregnenolone occurred rather slowly. For selected P450c17 expressing *Y. lipolytica* strains, the cultivation conditions

S. Mauersberger
Institute of Microbiology, Technische Universität Dresden, 01062 Dresden, Germany
e-mail: stephan.mauersberger@tu-dresden.de

L.A. Novikova
Belozersky Institute of Physico-Chemical Biology, Lomonosov Moscow State University, 119991 Moscow, Russia
e-mail: novik@genebee.msu.su

V.M. Shkumatov (✉)
Institute of Physico-Chemical Problems, Belarusian State University, 220030 Minsk, Belarus

Department of Chemistry, Laboratory of Biochemistry of Medical Drugs, 220030 Minsk, Belarus
e-mail: vlad.shkumatov@tut.by

G. Barth (ed.), *Yarrowia lipolytica*, Microbiology Monographs 25,
DOI 10.1007/978-3-642-38583-4_7, © Springer-Verlag Berlin Heidelberg 2013

(induction, bioconversion) were optimized for high product yield (up to 95 % 17HPro) and reduction of the diol side-product formation from 19–22 % to 1–2 % without gene destruction.

The results obtained could be used for elaboration of new biotechnological approaches with using recombinant yeast strains for synthesis of pharmaceutically active steroids and for screening of compounds which inhibit the P450c17 enzyme activity, playing important roles in the development of hormonal carcinogenesis.

Contents

1 Introduction ... 173
2 Heterologous Expression of Cytochromes P450 in Yeasts ... 175
  2.1 Overview on Heterologous Cytochrome P450 Expression in Yeasts and Filamentous Fungi ... 175
  2.2 Reconstruction of Mammalian Steroid Synthesis in *Saccharomyces cerevisiae* ... 183
3 Transgenic *Yarrowia lipolytica* Strains in Steroid 17α-Hydroxylation ... 185
  3.1 Endogenous Yeast Enzymes Involved in Steroid Biotransformations ... 185
  3.2 Recombinant *Yarrowia lipolytica* Strains Expressing Cytochrome P450c17 and CPR ... 187
  3.3 Overexpression of Cytochrome P450c17 and YlCPR Induces ER Proliferation in *Yarrowia lipolytica* ... 195
  3.4 Functional Activity of Genetically Modified *Yarrowia lipolytica* in Steroid 17α-Hydroxylation ... 197
  3.5 Optimisation of Progesterone to 17α-Hydroxyprogesterone Bioconversion ... 205
4 Transgenic *Yarrowia lipolytica* Strains for Sterol Side-Chain Cleavage and 17α-Hydroxylation ... 207
  4.1 Substrate Specificity of Cytochrome P450scc ... 208
  4.2 Strain Construction for Coexpression of the Cytochrome P450scc System and Cytochrome P450c17 ... 210
  4.3 Sterol and Steroid Bioconversion with Recombinant *Yarrowia lipolytica* Coexpressing AdR, Adx, and Cytochromes P450scc and P450c17 ... 214
5 Conclusions ... 217
References ... 219

Abbreviations

3β-HSD	3β-hydroxysteroid dehydrogenase/Δ5,4 isomerase (3β-hydroxy-5-ene steroid dehydrogenase, EC 1.1.1.145)
AdR	FAD-containing NADPH-adrenodoxin reductase (EC 1.18.1.2)
Adx or shortly Ax	Adrenodoxin, [2Fe-2S]-ferredoxin from adrenal cortex
AxP	Fusion of human Adx and P450scc
C.	*Candida*
CPR	NADPH-cytochrome P450-reductase
DHEA	Dehydro*epi*androsterone
E.	*Escherichia*

HSD	Hydroxysteroid dehydrogenase
P450 or CYP	Cytochrome P450
P450c11 (CYP11B1, P45011β)	Cytochrome P450 11β-hydroxylase
P450c17 (CYP17A1)	Shortly P17 (CYP17, P45017α), cytochrome P450 17α-hydroxylase/17,20-lyase (EC 1.14.99.9)
P450c18 (CYP11B2)	Cytochrome P450 aldosterone synthase
P450c19 (CYP19, P450arom)	Cytochrome P450 aromatase
P450c21 (CYP21)	Cytochrome P450 21-hydroxylase
P450scc (CYP11A1)	Shortly Pb (bovine) or Ph (human), cytochrome P450 cholesterol hydroxylase/20,22-lyase (EC 1.14.15.6)
S.	*Saccharomyces*
Y.	*Yarrowia*

1 Introduction

Steroid substances, which include steroid alkaloids, glycosides and saponins of plants, steroid hormones, vitamin D, bile acids of animals, as well as various insecticides, fungicides, and plant growth regulators, find increasing use in medicine and agriculture. Steroid compounds are conventionally obtained by extraction from plants and animal tissues, complete chemical synthesis, and combined chemical and enzymatic synthesis. The above approaches have made available a broad spectrum of steroid drugs and hormones. Several microbial bioconversions of steroids and sterols have been reported ever since focusing mainly on steroid hydroxylations, Δ^1-dehydrogenation, and sterol side-chain cleavage (Sedlaczek 1988; Ahmad et al. 1992; Fernandes et al. 2003). These biotransformations, mostly associated with chemical synthesis steps, have provided adequate tools for the large-scale production of natural or modified steroid analogues. Fungal biotransformation of steroids is among the earliest examples of biocatalysis for producing stereo- and site-specific products, including commercially important cytochrome P450-mediated steroid hydroxylations. Such biotransformations of steroids are of applied interest due to the economic importance of stereo- and regiospecific reactions to produce steroidal products and vitamin D (Sedlaczek 1988; Pajic et al. 1999; Fernandes et al. 2003).

Cytochrome P450 (P450 or CYP) enzymes constitute a large, ubiquitous family of heme-thiolate monooxygenases (CYP gene superfamily) that are involved in the oxidative metabolism of a wide variety of endo- and xenobiotic chemicals (Ortiz de Montellano 2005; Bernhardt 2006). Most P450 systems are composed of a P450 monooxygenase and one (class II) or two (class I) additional proteins, constituting an electron transfer chain. These P450 system components are either expressed as individual genes or linked resulting in a single peptide as self-sufficient P450s (class III).

Eleven proteins are directly involved in the steroidogenic pathway from cholesterol to steroid hormones in mammal, among which are six P450s, 17β-hydroxysteroid dehydrogenases (17β-HSD), 3β-hydroxysteroid dehydrogenases/Δ5,4 isomerase (3β-HSD), and three electron transfer proteins (Bernhardt 2006). The P450s are membrane-bound proteins associated with either the mitochondrial inner membranes (P450scc, P450c11, P450c18—type I P450 enzymes, which receive reducing equivalents via electron transfer chains consisting of adrenodoxin, a [2Fe-2S] ferredoxin, Adx, and NADPH-adrenodoxin reductase, a FAD-flavoprotein, AdR) or the endoplasmic reticulum membranes (P450c17, P450c21, P450c19—type II P450 enzymes, receiving reducing equivalents from a single FAD/FMN-flavoprotein, NADPH-cytochrome P450 reductase (CPR)).

By heterologous expression of selected P450s, it is possible to combine the strict regio- and stereoselectivity of steroidogenic enzymes and biotechnological advantages of microorganisms. The use of recombinant microorganisms, synthesising a few enzymes involved in steroidogenesis, can afford an opportunity to realise several consecutive reactions to a single stage that will simplify so essentially the technology of steroid drug synthesis. There are two limitations for widespread using of biotechnological approaches for synthesis of steroids. The *first* is the fact that natural microorganisms, which are used for synthesis of target steroids, usually have their own enzymatic systems producing unwanted by-products. This basic disadvantage can be overcome taking advantage of heterologous expression of substrate-specific steroidogenic enzymes (including mammalian P450 systems) in selected microbial hosts. The *second* restriction consists in low solubility of steroids in aqueous environments. Different means are for overcoming of this limitation (liquid two-phase systems, organic additives), which may complicate technological processes. However, one perspective for improvement of mass-transfer parameters of hydrophobic steroids into the cells is to use alkane-utilizing yeasts, like *Yarrowia (Y.) lipolytica*, which is capable to utilise very efficiently hydrophobic substrates, like lipids, fatty acids or alkanes. This yeast could be used for bioconversion of such hydrophobic compounds into valuable products, like dicarboxylic acid (DCA), for flavour or aroma (lactones), for bioremediation purposes, as well as for degradation of triglycerides into organic acids (citric, isocitric, or 2-ketoglutaric acids; Barth and Gaillardin 1996; Fickers et al. 2005; Bankar et al. 2009; Beopoulos et al. 2009, 2011; Thevenieau et al. 2009, 2010; Coelho et al. 2010). Therefore, there probably exist fundamental advantages of this yeast for synthesis of steroids in comparison with *Escherichia (E. coli)* and the yeast *Saccharomyces (S. cerevisiae)*.

Alkane-utilising yeasts, like *Y. lipolytica* or *Candida* spp., exhibit a high catalytic activity of their alkane- or fatty acid-inducible P450 systems (*ALK* genes of CYP52 family, alkane or fatty acid ω-hydroxylase activities with turnover numbers of 1–2 μmol/nmol P450 × min (calculated in vivo P450 activity in alkane-utilising cells), catalysing terminal hydroxylations of *n*-alkanes or fatty acids). This high P450 activity is supported by an efficient subcellular organisation of the substrate and product transport processes (inducible uptake and excretion systems for hydrophobic substrates, ER–peroxisomes interaction for substrate and product transport processes, like fatty acids and DCA), a proliferation of the ER during growth on alkanes, and the efficiency of electron transfer systems (alkane- or fatty-acid-inducible higher content of microsomal electron transfer components, NAD(P)H-dependent P450 reductases, and cytochrome b_5 then in *S. cerevisiae*) to host-own P450s (Mauersberger et al. 1987, 1996; Schunck et al. 1987a, b; Fickers et al. 2005; cf. Mauersberger 2013), which might also support the function of heterologous P450 functionally expressed in these yeasts. Therefore, the putative advantages of a hydrocarbon-assimilating yeast cell as a host for heterologous P450, catalysing biotransformation reactions with hydrophobic substrates, were tested by functional expression of selected P450s in *Y. lipolytica* (e.g. P450c17, P4501A1, P450scc, P4502D6, and P4503A4; cf. Sect. 2.1, Table 1). Indeed, comparison of functional P450c17 expression in *Y. lipolytica* and in the commonly used yeast *S. cerevisiae* revealed significant advantages of alkane-assimilating yeast cells for P450-catalyzed biotransformations of hydrophobic substrates (Juretzek et al. 2000b). In this contribution, the present knowledge on heterologous expression of different P450 forms in the yeast *Y. lipolytica* will be given, ranging from the first functional expression of bovine P450c17 in middle of 1990 to recent data on evaluation and application of this promising yeast host for P450-catalyzed biotransformation of mainly hydrophobic substrates, including steroids and sterols by different P450 forms.

2 Heterologous Expression of Cytochromes P450 in Yeasts

To make use of the high regio- and stereoselectivity of hydroxylations by P450 systems for biotransformation of hydrophobic substrates, the P450 enzymes need to be functionally expressed in an appropriate host cells. Several recombinant P450 expression systems have been investigated in the past 25 years, including mammalian and *Baculovirus*-infected insect cell systems and bacterial (mainly *Escherichia coli*), fungal (mainly *Saccharomyces cerevisiae*), and plant expression systems (cf. Table 1 for earlier review ref; Dumas et al. 2006; Novikova et al. 2009; Cornelissen et al. 2012). Preferred hosts are *E. coli* and *S. cerevisiae*. Other suitable host microorganisms are found also among bacteria (*Bacillus*, *Pseudomonas*, *Streptomyces* species), yeasts (*Pichia pastoris*, *Schizosaccharomyces pombe*, *Kluyveromyces lactis*, *Candida* spp., *Y. lipolytica*), and filamentous fungi (*Fusarium verticillioides*; cf. Sect. 2.1, Table 1).

2.1 *Overview on Heterologous Cytochrome P450 Expression in Yeasts and Filamentous Fungi*

A short overview on established P450 expression systems in yeasts and filamentous fungi is presented in Table 1. A brief description of P450 expression data in *Y. lipolytica* will be given at the end of this section. The first functional expression

Table 1 Host–vector systems for expression of cytochromes P450 in yeasts and filamentous fungi

Yeast species Vectors	Promoters/ terminators	Expressed genes or cDNAs	References and remarks
Saccharomyces cerevisiae			
Mostly used host-vector system with high copy vectors (e.g. YEp51, with 2μ parts)	*ADH1* *CUP1* *PHO5* *GAL7* *GAL1-GAL10* *GAL-CYC* *PGK*	Different P450 systems from fungi, plants, insects, vertebrates	First report: Oeda et al. (1985) Reviewed by: Urban et al. (1994) Renaud et al. (1993) Gonzalez and Korzekwa (1995) Pompon et al. (1996, 1997) Szczebara et al. (2003), Dumas et al. (2006), Novikova et al. (2009)
Schizosaccharomyces pombe			
ARS-expression vectors (e.g. pREP1, pTL2M1, pNMT1), integrative vectors (pCAD1)	*nmt1* (*thi3*) *hcmv* *nmt1* *nmt41*	h*CYP2C11* (P450M1) h*CYP2C9*, h*CYP2C19* h*CYP11B2* (P450aldo) h*CYP11B1* (P45011β) h*CYP21*, h*CYP17* h*CYP2D6*, h*CPR* h*CYP4Z1*, h*CPR* h*CYP2B6*, h*CYP2C9* h*CYP2C19*, h*CYP3A4*	Yamazaki et al. (1993) Yasumori (1997), Yasumori et al. (1999) Bureik et al. (2002), Ehmer et al. (2002) Bureik et al. (2004), Hakki et al. (2008) Drăgan et al. (2005, 2006a, b) Zöllner et al. (2010) Peters et al. (2007, 2009a, b) Zöllner et al. (2009) Bureik (2008, pers. communication)
Kluyveromyces lactis			
Integrative vector pGB: pScADH1-Tn5 (G418) pScADH1-hygB	*LAC4* (lactase)	b*CYP11A1* (P450scc) b*ADX*, b*ADR* b*CYP17*, b*CYP21* b*CPR*	Slijkhuis et al. (1989, 1990–2004), patent series; Menke et al. (1990) (functional, low-level expression)
Candida maltosa[a]			
High-copy *ARS* vectors (pNGH2-ALK)	*PGK1* *ALK* *GAL1-GAL10*	Homologous P450: Cm*ALK1*, Cm*ALK3* Cm*ALK1* to Cm*ALK8* (*CYP52A3*)	Masuda et al. (1994) Ohkuma et al. (1995a, b) Park et al. (1997) Fallon et al. (2004)
Candida tropicalis[b]			
Integrative vectors pCU3ALK1, pCU3RED	Homologous Promoters	Homologous P450: Ct*ALK1* and Ct*CPR* *CYP52A2A*	Picataggio et al. (1992) Fallon et al. (1999, 2004) Wilson et al. (2001)

(continued)

Table 1 (continued)

Yeast species Vectors	Promoters/ terminators	Expressed genes or cDNAs	References and remarks
Pichia pastoris			
Integrative expression vectors (e.g. pPICZA)	*AOX1*	*CYP17* shark, human	Trant (1996), Kolar et al. (2007)
		CYP2L1 lobster	Storbeck et al. (2008)
		CYP79D1/D2 cassava	Boyle et al. (1998) Andersen et al. (2000)
		CYP from plants	Andersen and Møller (2002)
		Cm*ALK1*-Cm*ALK8*	Fallon et al. (2004)
		h*CYP2D6,* h*CPR*	Dietrich et al. (2005), Geier, Braun & Glieder (TU Graz, pers. commun.)
		Pc*CYP1f* (*CYP53* like)	
		CYP85A2 *Arabidopsis*	Matsuzaki and Wariishi (2005)
		OsKO2 rice (*CYP701A*) and fungal Ph*CPR*	Katsumata et al. (2008) Ko et al. (2008)
Yarrowia lipolytica			
Low-copy (*ARS-CEN*) vectors, integrative multicopy (*ura3d4*) vectors (rDNA, LTR zeta)	*ICL1*	Cm*CYP52A3* (P450Cm1)	Prinz (1995), Mauersberger et al. (1995)
		b*CYP17,* Yl*CPR* (functional, lc: low level, mc: high expr. level)	Juretzek (1999), Juretzek et al. (1997, 1999, 2000b) Gerber (1999), Förster (2001)
			Shkumatov et al. (1998, 2003, 2006, 2007)
			Mauersberger et al. (2002, 2005)
			Novikova et al. (2009)
pYEG1 integrative low-copy (*LEU2, ura3d1*) and multicopy *(ura3d4)* vectors	*POX2*/ *XPR2t*	Plant HPO lyase (*CYP74*)	Bourel et al. (2004), Santiago-Gómez et al. (2007)
	POX2, ICL1	h*CYP1A1,* Yl*CPR*	
	ICL1	h*CYP11A1* (h/ bP450scc)	Nthangeni et al. (2004) Yovkova (2006)
		h*ADR,* h*ADX,* b*CYP17*	Novikova et al. (2008 – present., posters)
			Novikova et al. (2009)
	POX2, ICL1	Rm*CYP53B1,* Yl*CPR*	Shiningavamwe et al. (2006) Theron (2007)
	(*ICL1*) *ICL1*	Yl*1Alk2 to 11,* Yl*CPR*	Thevenieau (2006) Thevenieau et al. 2009, 2010
		h*CYP2D6,* h*CPR,* Yl*CPR*	Braun et al. (2012) Braun et al. (2012)
		h*CYP3A4*	

(continued)

Table 1 (continued)

Yeast species Vectors	Promoters/ terminators	Expressed genes or cDNAs	References and remarks
Fusarium verticillioides			
	FUM8	Fs*TRI1*, Fg*TRI1* Fg*TRI4*, Mr*TRI4*	McCormick et al. (2006)

The summary is with some emphasis on steroidogenic P450; the references are selected examples, without being comprehensive for all P450s and authors

Abbreviation: *PGK*, phosphoglycerate kinase gene; OsKO2, rice *ent*-kaurene oxidase (*CYP701A*); hcmv, human cytomegalovirus promoter; nmt1, strong promoter induced in absence of thiamine; *LAC4*, gene for lactase, the β-galactosidase of *K. lactis;* h, human, b, bovine, Cm, *C. maltosa*; Ct, *C. tropicalis*; Pc, *Phanerochaete chrysosporium*; Ph, *Phaeosphaeria* spec. L487; Rm, *Rhodotorula minuta*; Rr, *Rhodotorula retinophila*; Bm, *Bacillus megaterium*; Fs, *Fusarium sporotrichioides*; Fg, *Fusarium graminearum*; Yl, *Y. lipolytica*

[a]Not useful for heterologous P450 expression due to a deviation of the universal genetic code observed in *Candida* yeasts (Zimmer and Schunck 1995)

[b]Amplification (two or more copies) of the homologous P450 *ALK1* and *CPR* genes, resulting in 30 % increased DCA productivity

of mammalian P450s was demonstrated in 1980s in baker's yeast *S. cerevisiae* (Oeda et al. 1985; Sakaki et al. 1989, 1990, 1991; cf. Table 1). The mutant yeasts combine the ease of handling of single-cell microbial systems with the specific features of eukaryotic cells. Since then numerous mammalian and non-mammalian *CYP* genes have been expressed in the commonly used yeast *S. cerevisiae* and more recently also in the non-conventional yeasts *P. pastoris*, *S. pombe*, *K. lactis*, and *Y. lipolytica* (cf. Table 1 for ref.). In particular, numerous studies have been published on heterologous expression of individual or coexpression of several mammalian steroidogenic P450s, i.e. microsomal P450c17, P450c19, P450c21, mitochondrial P450scc and P450c11 (P450 expressed in their natural or artificial fused forms with respective electron transfer components, including also P450 coexpression with 3β-HSD, Adx, CPR, or cytochrome b_5), using besides *E. coli* mainly the yeast *S. cerevisiae* (for ref. of reviews cf. Table 1). The first successful expression of microsomal steroidogenic P450 enzymes was based on the fact that yeast CPR can support the activities not only of host-own but also of expressed heterologous P450s, in particular P450c17 and P450c21 (Sakaki et al. 1989, 1991). These recombinant yeast cells expressing heterologous P450s were used for the characterisation of individual CYP forms and the interaction with coexpressed redox partners and were applied for P450-catalyzed biotransformations of predominantly hydrophobic substrates (including steroids) with whole cells or cell fractions. These whole cells ("yeast cell factories") can be easily used for biotransformations, either in single or in multistep reactions, to deal with inherent stability problems of P450 enzymes and regeneration of NADPH.

After first successful reports in 1990s, application of *S. pombe* and *P. pastoris* for heterologous P450 expression developed rapidly in the last decade. In particular, *S. pombe* was successfully used in heterologous P450 expression studies and applied for P450-catalyzed biotransformations of steroids and other substrates.

Otherwise, K. lactis was tested for P450scc, P450c17, and P450c21 expression in first studies only. Heterologous P450 expression in the well-developed host–vector systems for *Hansenula polymorpha* (*Pichia angusta*) and *Arxula adeninivorans* was not yet reported. In addition to the widely used yeast host–vector systems, first reports on heterologous expression of P450 in filamentous fungi appeared. The P450 monooxygenase-encoding *TRI1* and *TRI4* genes of *Fusarium sporotrichioides* (*FsTRI1*), *Fusarium graminearum* (*FgTRI1*, *FgTRI4*), were heterologously expressed in the trichothecene-nonproducing species *Fusarium verticillioides* under promoter control of the fumonisin biosynthetic gene *FUM8* to study their function in the trichothecene mycotoxins (McCormick et al. 2006). The alkane-utilising *Candida* yeasts (*C. maltosa*, *C. tropicalis*) are not very useful for heterologous P450 expression due to a deviation of the universal genetic code observed in these yeasts (Zimmer and Schunck 1995). However, for the yeasts *C. maltosa*, *C. tropicalis*, and *Y. lipolytica*, a gene-dose-dependent overexpression of host-own P450 and NADPH-P450 reductases was performed, in particular to increase the alkane- or fatty acid-hydroxylating P450 activities (CYP52 family) involved in the formation of dicarboxylic acids (DCA) derived from alkanes or fatty acids (for ref. cf. Table 1).

Expression of Heterologous P450 in *Yarrowia lipolytica*. In contrast to the *Candida* yeasts, functional heterologous P450 expression was successfully demonstrated for the hydrocarbon-assimilating yeast *Y. lipolytica*. In this section, a brief description of all P450 expression studies in *Y. lipolytica* will be given. Details on the expression of P450c17 and the P450scc system will be presented in Sects. 3 and 4. At least eight heterologous P450 (CYP1A1, CYP2D6, CYP3A4, CYP11A1, CYP17A, CYP52A3, CYP53B1, CYP74) and three heterologous electron transfer components (all human NADPH-P450 reductase, hCPR; adrenodoxin, hADX; NADPH-adrenodoxin reductase, hADR; additionally, homologous YlCPR overexpression) were expressed in the yeast *Y. lipolytica*. In total 12 P450 and related electron transfer proteins were expressed under control of different promoters. Among them are five mammalian P450 proteins (CYP1A1, CYP2D6, CYP3A4, CYP11A1, human and bovine CYP17), one plant P450 (CYP74), two yeast proteins (CYP52A3 of *C. maltosa*, CYP53B1 of *Rhodotorula minuta*), and three human electron transfer proteins (hCPR, hADX, hADR), and additionally, the host-own P450 reductase (YlCPR) was overexpressed in several cases (cf. Table 1). The heterologous P450s expressed in *Y. lipolytica* summarized in Table 1 are the following:

1. The *CYP52A3* (*ALK1*) gene encoding the alkane hydroxylating P450Cm1 of *C. maltosa* (Schunck et al. 1991) was expressed under pICL1 control in *Y. lipolytica* using a low-copy *ARS-CEN* replicative vector, comparable to pIC17α shown in Fig. 1b (Prinz 1995; Mauersberger et al. 1995; unpublished meeting reports; Juretzek et al. 1997). The P450Cm1 protein was detected in Western blots; determination of the P450Cm1 activity was hindered due to the interference with host-own P450 ALK.

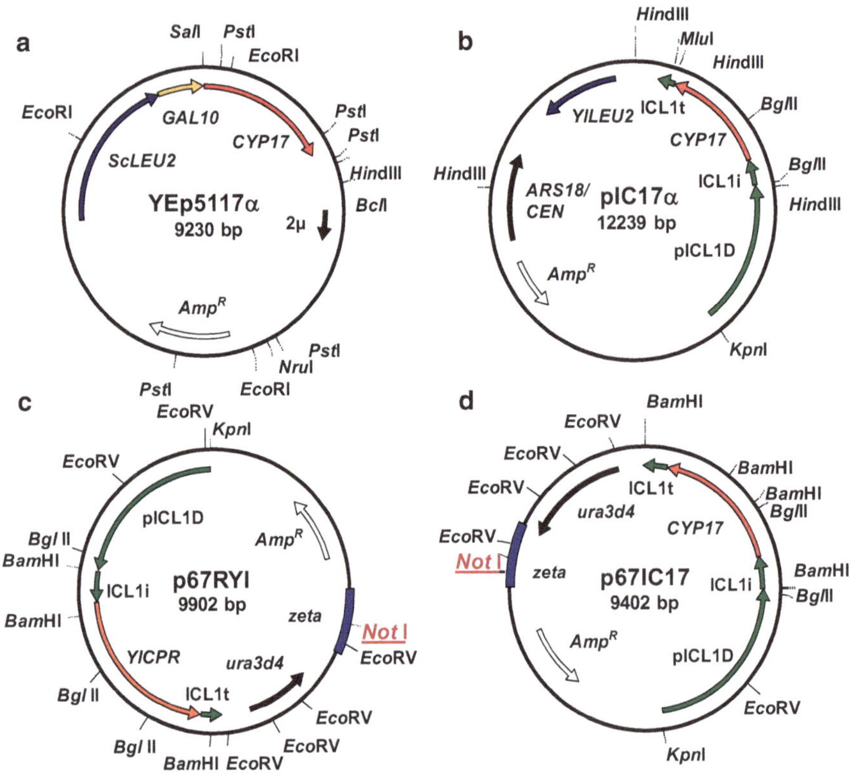

Fig. 1 Replicative and integrative vectors for the heterologous expression of bovine adrenal cytochrome P450c17 (*CYP17* cDNA) in the yeasts *Saccharomyces cerevisiae* (**a**) and *Yarrowia lipolytica* (**b, d**) and for the overexpression of the yeast NADPH-cytochrome P450 reductase (*YlCPR*) in *Yarrowia lipolytica* (**c**). (**a**) YEp5117α: high-copy, 2μ-based replicative expression vector for P450c17 in *S. cerevisiae*; *GAL10*, strong galactose-inducible promoter; *Amp*^{R}, ampicillin resistance genes for selection in *E. coli*; *CYP17*, cDNA for bovine P450c17; *ScLEU2*, yeast selection marker gene. (**b**) pIC17α: low-copy (1–2) *ARS18/CEN* expression vector; pICL1D, full-length, strong and regulated *ICL1* promoter D (induced by alkanes, fatty acids, ethanol or acetate, almost 90–95 % repressed by glucose); ICL1i, intron in the *ICL1* gene; ICL1t, *ICL1* terminator; *YlLEU2*, selection marker in *Y. lipolytica*. (**c**) p67RYl: multicopy (at least 8–10) LTR *zeta*-based integrative vector for pICL1-controlled high-level expression of the host-own ER-resident NADPH-P450 reductase (*YlCPR* gene) in *Y. lipolytica; ura3d4*, defective, promoter-truncated *URA3* gene as multicopy selection marker in *Y. lipolytica; zeta*, long-terminal repeat LTR *zeta* of the *Y. lipolytica* retrotransposon Ylt1 as vector integration targeting sequence after its linearisation by *Not*I prior transformation. (**d**) p67IC17: comparable multicopy LTR *zeta*-based integrative vector for high-level expression of the bovine P450c17 in *Y. lipolytica* under pICL1 control these and comparable expression vectors and their use for gene-dose dependent high-level heterologous protein expression in *S. cerevisiae* and *Y. lipolytica* were described in Juretzek (1999), Juretzek et al. (2000a, b, 2001), and Shkumatov et al. (1998, 2002)

2. The functional heterologous expression of bovine steroidogenic P450c17 (*CYP17A* cDNA) in the yeast *Y. lipolytica*, which is naturally well adapted to the utilization of hydrophobic substrates, was first studied to test the assumed advantages of an alkane-utilising yeast as host for heterologous P450, catalyzing biotransformation reactions with hydrophobic substrates (cf. Sects. 1 and 3.2, Table 1). For this purpose, the non-conventional oleaginous yeast *Y. lipolytica* was selected, which is non-pathogenic and phylogenetically very distant from the commonly studied *S. cerevisiae*, and for which the main genetic engineering tools were available. Bovine P450c17 was the first mammalian P450 expressed in *Y. lipolytica* (Table 1). In particular, functional expression of the ER-resident and steroid transforming P450c17 under control of the strong and regulated isocitrate lyase promoter pICL1 in *Y. lipolytica* was established and used for steroid biotransformation with recombinant yeast cells (Juretzek and Mauersberger et al. 1995, unpublished meeting reports; Shkumatov et al. 1998, 2003, 2006; Juretzek 1999; Juretzek et al. 1999, 2000b; Novikova et al. 2009; Table 1). These first studies of P450c17 expression in *Y. lipolytica* indicated several advantages (hydrophobic substrate uptake and transport, cellular properties supporting P450-catalyzed reactions) of the alkane-assimilating yeast cells for P450-catalyzed biotransformations of hydrophobic steroid substrates in comparison with *S. cerevisiae* (cf. Sects. 3 and 4 for further details).
3. Subsequently, functional expression of first human (second mammalian) P4501A1 (CYP1A1, involved in drug oxidation and catalysing 7-ethoxyresorufin O-deethylase activity, EROD) in *Y. lipolytica* under control of pPOX2 promoter, with or without overproduction of host-own NADPH-P450 reductase (*YlCPR*) expressed under pICL1 or pPOX2 promoters' control was demonstrated (Nthangeni et al. 2004). Significantly (up to 50-fold) increased P4501A1 activity in whole-cell biotransformation of 7-ethoxyresorufin to resorufin (7-hydroxy-3H-phenoxiazin-3-one) due to *CYP1A1* copy number increase and *YlCPR* coexpression under these promoters was observed.
4. The first heterologous expression of a plant P450 in *Y. lipolytica* was shown using the green bell pepper HPO lyase gene (CYP74B) under the control of the pPOX2 using non-homologous LTR *zeta*-based integration into the genome (Bourel et al. 2004; Santiago-Gómez et al. 2007). The expression of this unusual P450 (170 kDa) with HPO lyase activity (fatty acid hydroperoxide lyase, HPL) in *Y. lipolytica* resulted in the production of high yields of volatile lipid-derived C6-aldehydes (hexanal and *trans*-2-hexenal, components of green notes aroma), much higher than using the plant system. The derived volatile products are used industrially to reconstitute the "fresh green odour" of fruits and vegetables lost during processing (Fickers et al. 2005; Santiago-Gómez et al. 2007). Thus, it was demonstrated that *Y. lipolytica* could be a useful host for the expression of HPO lyase and a simple process that could yield high quantities of C6-aldehydes was established with the recombinant yeast.

5. CYP53B1 (P45053B1) from *Rhodotorula minuta* encoding a benzoate para-hydroxylase, completely absent in the host, was functionally expressed in *Y. lipolytica* strain E150 after multicopy integration of expression cassettes under control of pPOX2, with and without pICL1-controlled coexpression of the *YlCPR* (Shiningavamwe et al. 2006). Whole-cell biotransformation of benzoic acid to *para*-hydroxybenzoic acid (pHBA) was used to analyse the hydroxylase activity of the recombinant *Y. lipolytica* cells, which was one of the highest hydroxylation activities thus reported for whole-cell biotransformation studies carried out with yeasts expressing foreign CYP450s.
6. The functional expression of the cholesterol side-chain cleavage P450scc system, a three-component, class I P450 system, consisting of CYP11A1 (human or bovine), and the electron transfer proteins NADPH-adrenodoxin reductase (human AdR) and adrenodoxin (human Adx) with coexpression of P450c17 (CYP17A) in *Y. lipolytica* was established in frame of an INTAS project in our three laboratories (cf. Table 1 for ref.; Yovkova 2006; Novikova et al. 2008, 2009; details will be presented in Sect. 4).
7. More recently coexpression of human *CYP2D6* or *CYP3A4* genes together with human P450 reductase (hCPR) or the host-own P450 reductase (YlCPR) in *Y. lipolytica* H222-S4 was shown (Braun et al. 2012; cf. Table 1). With these recombinant whole-cell biocatalysts, the potential of the hydrocarbon-assimilating yeast *Y. lipolytica* for the bioconversion of poorly soluble hydrophobic steroids (testosterone, 17α-testosterone, progesterone) was tested. Additionally, two-liquid biphasic culture systems (aqueous and organic solvent phases) were evaluated to increase the substrate availability. Best bioconversion results were observed in a bioreactor employing a biphasic system with the organic solvent and *Y. lipolytica* carbon source ethyl oleate (compared to bis-ethylhexyl phthalate, BEHP, or dibutyl phthalate, DBP) for the whole-cell bioconversion of progesterone. Multicopy transformants showed a 50–70-fold increase of P450 activity as compared to single-copy strains, and coexpression of human CPR gene resulted in a 4–10-fold higher specific P450 activity compared to co-overexpression of the Yl*CPR* gene. These results demonstrated the high potential of P450 expressing *Y. lipolytica* cells for biotransformations of hydrophobic steroid substrates in two-liquid biphasic systems. Especially organic solvent phases which can be efficiently taken up and metabolised by the cell enable more efficient bioconversion as compared to aqueous systems and even enable high-yield long-time processes.

Thus, the stable high-level and functional expression of heterologous P450s together with its NADPH-P450 reductase opens new perspectives for further improvement of the efficiency of biotransformation reactions with recombinant *Y. lipolytica* cells, a system which seems to be useful especially for bioconversion of hydrophobic substrates.

2.2 *Reconstruction of Mammalian Steroid Synthesis in* Saccharomyces cerevisiae

In a long-term project (cf. reviews Dumas et al. 2006; Brocard-Masson and Dumas 2006 and ref. therein) first the two initial stages of mammalian steroidogenesis were reconstituted in *S. cerevisiae*, which were realised by the bovine P450scc system and human 3β-hydroxysteroid dehydrogenases/isomerase (3β-HSD) coexpressed with a plant sterol Δ7-reductase, leading to self-sufficient biosynthesis of pregnenolone and progesterone during growth of the engineered yeast cells on glucose or ethanol (Duport et al. 1998). This result was achieved using two principally new approaches: *First*, the major problem proved to be expression of the P450scc system in yeast mitochondria, due to differences of P450scc topogenesis within baker's yeast from its topogenesis within mammalian mitochondria (Minenko et al. 2008; for ref. cf. Novikova et al. 2009). Duport et al. (1998) demonstrated that the P450scc system can be functionally expressed in yeast with non-mitochondrial location using cDNAs encoding mature (m) protein forms without mitochondrial targeting sequences. Despite different localization of these mature form-proteins (mP450scc is mostly plasma membrane and partially ER associated, whereas mAdR and mAdx are ER localized and cytosolic, respectively), this P450scc system was shown to be catalytically active, which indicated on non-obligatory necessity of mitochondrial surrounding for P450scc (Duport et al. 2003). *Second*, it is known that conversion of cholesterol into pregnenolone by recombinant yeast is difficult because cholesterol is not efficiently taken up by aerobically grown baker's yeast (Ness et al. 1998). To prepare steroid producing transgenic *S. cerevisiae* strains, Duport et al. (1998) had to reroute the yeast metabolism (disrupting yeast P450 gene *CYP61A* or *ERG5*, encoding sterol Δ22-desaturase, P45022DS, and expressing the plant gene encoding sterol Δ7-reductase from *Arabidopsis thaliana*) in such a way that the yeast produced during growth campesterol (ergosta-5-enol), a very close analogue of cholesterol, instead of the natural yeast sterol, i.e. ergosterol. It was shown that campesterol, which can support the vital functions of yeasts replacing ergosterol in membranes and simultaneously acts as substrate for mammalian P450scc, is transformed in vivo into pregnenolone and progesterone, thus realising the first self-sufficient steroid biosynthesis by engineered *S. cerevisiae* (Duport et al. 1998). Finally, Szczebara et al. (2003) reported the construction of transgenic *S. cerevisiae* strains that produce hydrocortisone (cortisol) from simple carbon sources (glucose, ethanol) in a single fermentation step. For metabolic engineering of these strains, the same approaches (sterol biosynthesis rerouting to campesterol and brassicasterol, directed changes in protein topogenesis, fine-tuning of gene expression) were used as described by Duport et al. (1998), including additional expression or destruction of 13 genes in the recombinant yeasts cells. Eight heterologous proteins (mature forms of P450scc, Adx, and AdR; mitochondrial forms of Adx and P450c11; cytosolic 3β-HSD; microsomal P450c17 and P450c21) of the mammalian steroidogenic pathway were simultaneously produced in a modified host (yeast genes *ATF2*, *GCY1*, *YPR1* disrupted, *ARH1* overexpressed).

The P450c11 system activity in mitochondria was relying on a partially artificial electron transfer chain combining the host-own reductase Arh1p (*a*drenodoxin *r*eductase *h*omologue 1) and the bovine Adx electron carrier (Duport et al. 2003; Szczebara et al. 2003; Dumas et al. 2006). The ER-located yeast NADPH-P450 reductase CPR (Ncp1p encoded by *ScNCP1*) supports the heterologously expressed and ER-resident mammalian P450c17 and P450c21 (Szczebara et al. 2003).

Moreover, in these and preceding studies, the major unwanted side reactions in *S. cerevisiae* were identified, like the esterification of pregnenolone (P450scc steroid product) by alcohol *O*-acetyltransferase (*ATF2*) and of campesterol (P450scc sterol substrate) by yeast steryl ester synthases (ACAT and ARE, ACAT-related enzyme, encoded by *ARE1* and *ARE2*), as well as the 20-keto reduction of the P450c17 product 17α-hydroxyprogesterone (Duport et al. 2003; Szczebara et al. 2003). Pregnenolone 3β-acetate formed by Atf2p cannot be converted by P450c17, P450c21, and 3β-HSD. The esterification (Sakaki et al. 1989; Cauet et al. 1999; Vico et al. 2002) and 20-keto reduction of steroids by yeast enzymes (Shkumatov et al. 2002, 2003, 2006) were described also by other authors. Therefore, for optimising the steroid-producing strains, overexpression of gene *TGL1*, encoding a steryl ester hydrolase, and additionally disruption of the nonessential steryl ester synthases encoding genes *ARE1* and *ARE2* (although only with minor effects) and the pregnenolone acetyltransferase gene *ATF2* (major effect) were accomplished, resulting in increased levels of free campesterol and pregnenolone (Duport et al. 2003; Szczebara et al. 2003).

The 20-keto reduction of 17α-hydroxyprogesterone into the diols 17α-hydroxy-20-dihydroprogesterone, predominantly 17α,20α-dihydroxypregn-4-ene-3-one, as demonstrated by Shkumatov et al. (2002, 2003, 2006), is assumed to be catalyzed by a concerted action of *GCY1* (galactose-inducible crystallin-like yeast protein) and *YPR1* (aldo–keto reductase) gene products (mainly Gcy1p functions as yeast 20-HSD or 20-steroid reductase), indicated by their sequence homology with bovine 20α-HSD. Disruption of both genes was leading to a loss of NADPH-keto-reductase activity with 17α-hydroxyprogesterone in vitro and of 17α,20α-dihydroxypregn-4-ene-3-one formation in vivo (Szczebara et al. 2003). As a result, *S. cerevisiae* recombinant strains were capable of self-sufficient production of hydrocortisone (11-deoxycortisol and corticosterone as main by-products) when growing in glucose- or ethanol-containing media. The amount of produced steroids during 72 h cultivation was increased from 1.6 to 16.6 μg/ml (with up to 70 % hydrocortisone of total steroids) by strain engineering.

Thus, Szczebara et al. (2003) summarised the long-term project resulting in the first production of "biohydrocortisone" by engineering of a highly complex mammalian biosynthetic pathway into baker's yeast as microbial host, which includes several coupled membrane enzymes. This comprehensive "metabolic engineering turned a unicellular microorganism into a drug-synthesizing yeast cell factory" (as titled in review) and can be considered a major breakthrough for using P450-catalyzed bioconversion in an industrial process. Such a process would yield hydrocortisone in a single fermentation step from simple carbon sources, replace a multistep chemical synthesis, and reduce the environmental impact by reducing the consumption of solvents, energy, and catalysts (Dumas et al. 2006; Brocard-Masson and Dumas 2006).

3 Transgenic *Yarrowia lipolytica* Strains in Steroid 17α-Hydroxylation

In this section, an overview on different aspects of the heterologous expression of bovine steroidogenic P450c17 in *Y. lipolytica* will be presented, describing the first functional expression of a mammalian P450 in this yeast, obtained in the end of the 1990s by Juretzek and Mauersberger at the MDC in Berlin-Buch (cf. Sect. 2.1, Table 1). Additionally, results on coexpression of P450c17 with the side-chain cleavage P450scc system in *Y. lipolytica* will be given in Sect. 4.

3.1 Endogenous Yeast Enzymes Involved in Steroid Biotransformations

To predict more precisely the catalytic properties of the designed whole-cell biocatalysts on the basis of transgenic yeast, the ability of yeast-own enzymes to catalyse transformation of substrates and/or products of mammalian steroidogenic P450 as unwanted side reactions has been investigated. Otherwise, the yeast cells contain enzymes which can directly support the functional activity of heterologously expressed P450 forms.

Yeast Enzymes Catalysing Unwanted Side Reactions of Heterologous Steroidogenic P450. With wild-type or recipient strains of *Y. lipolytica* (H222, B204-12A-213), *C. maltosa* (EH15, VSB779), and *S. cerevisiae* (GRF18), no hydroxylation activities in the C21-, C17-, or C11-positions were found towards progesterone, testosterone, deoxycorticosterone, or deoxycortisol, but different NAD(H)/NADP(H)-dependent hydroxysteroid dehydrogenase activities (oxidating the hydroxyl groups at positions 3β-, 17β-, 20α-, and 20β-HSD, or reducing the carbonyl group at positions C3, C17 and C20) catalyzed by cytosolic yeast enzymes were detected (Shkumatov et al. 1998, 2002, 2003, 2006), as it was described for *S. pombe* and *S. cerevisiae* (Pajic et al. 1999; Szczebara et al. 2003). Progesterone was reduced by alkane-growing wild-type strains of *Y. lipolytica* and *C. maltosa* at the C3- and C20-keto groups (product yields less than 2 %) and one metabolite identified as 3α-hydroxy-5α-pregnane-20-one. Furthermore, compounds corresponding to references 20α(or 20β)-dihydroprogesterone (0.6 % yield), testosterone (0.4 %), and androstenedione (0.5 %) were detected after 24 h incubation of radioactively labelled progesterone with *Y. lipolytica* and *S. cerevisiae*. Obviously, these compounds were formed as a result of 20-reduction of progesterone

(20α,20β-HSD) and a Baeyer–Villiger conversion of dihydroprogesterone to testosterone followed by 17β-oxidation to androstenedione (Shkumatov et al. 1998, 2006). A similar conversion was reported for *Aspergillus ochraceus* (Dutta et al. 1993). Testosterone and 4-androstene-3,17-dione were interconverted by in vivo biotransformations (17β-HSD activity) using alkane-grown cells of *Y. lipolytica* and *C. maltosa* (Shkumatov et al. 1998).

Thus, in contrast to several filamentous fungi, like *Aspergillus* spp., *Curvularia lunata*, *Cochliobolus lunatus*, or *Rhizopus nigricans*, which contain endogenous steroid hydroxylating enzymes, including P450 systems, and are in particular used in industrial microbiological steroid bioconversion steps (Sedlaczek 1988; Pajic et al. 1999), all yeast species tested so far do not perform any steroid hydroxylation reactions. Nevertheless, several yeast-own enzymes may have negative effects on desired P450-catalyzed steroid biotransformation reactions when expressing heterologous P450, because they perform unwanted side reactions of desired products or applied substrates. Some of these undesirable side reactions (e.g. 20-HSD activities, encoded by *GCY1* and *YPR1*, pregnenolone 3β-acetylation, *ATF2*) were shown to be diminished by corresponding gene deletions in *S. cerevisiae* (Szczebara et al. 2003; Dumas et al. 2006; cf. Sect. 2.2). Otherwise, in special cases, one can take advantage from the side reactions occurring in yeast and create new steroid biosynthetic paths with high yield of interesting steroid products, combining both highly specific reactions catalyzed by heterologous P450c17 and yeast 20α,β-HSD-like enzyme activities with chemical synthesis (Shkumatov et al. 2003; cf. Sect. 3.5).

Yeast Enzymes Supporting Heterologous P450. The yeasts *Y. lipolytica*, *C. maltosa*, *S. cerevisiae*, and *S. pombe* and other tested and completely sequenced species (cf. Fukuda and Ohta 2013; Mauersberger 2013) contain one or two own microsomal (ER-resident) NADPH-P450 reductases (encoded by *CPR* or *NCP1*), cytochrome b_5 and NADH-b_5 reductases as natural electron transfer partners for endogenous P450 (at least two P450s, 51A1 and 61A1, involved in ergosterol biosynthesis present in all yeast species, additionally up to 15 P450 genes detected species dependent, including members of the CYP52 gene family; cf. Mauersberger 2013). These enzymes can also support the function of heterologous microsomal P450 (P450c17, P450c21; first demonstrated by Sakaki et al. 1989, 1991 in *S. cerevisiae*, for other yeast species cf. Table 1 and Sect. 3 for *Y. lipolytica*) and even mitochondrial P450 (P450c27, Sakaki et al. 1996), although with different efficiency dependent on the expressed heterologous P450 (reviews by Urban et al. 1994; Pompon et al. 1996, 1997; Sakaki and Inouye 2000; Szczebara et al. 2003; Dumas et al. 2006).

Furthermore, mitochondrial P450 systems of type I expressed in yeast (e.g. P450c11) are partially supported by the host-own NADPH-*a*drenodoxin *r*eductase *h*omologues 1 protein Arh1p and the bovine adrenodoxin Adx electron carrier (Duport et al. 2003; Szczebara et al. 2003; Dumas et al. 2006). In *S. cerevisiae*, both *CPR* and *ARH1* are essential genes. Electron transfer proteins encoding Sc*ARH1* homologous genes were also found in *Y. lipolytica* and *S. pombe* genome. In fission yeast both heterologous and homologous redox chains, SpArh1p-etp1^{fd}, SpArh1p-Adx, AdR-etp1^{fd}, and AdR–Adx, can function with the heterologous P450c18, P450c11, or P450scc (Bureik et al. 2002; Schiffler et al. 2004; Ewen et al. 2008). The adrenodoxin homologous yeast gene *YAH1* (encoding a mitochondrial matrix iron–sulphur protein, *S. cerevisiae* ferredoxin Yah1p; Barros and Nobrega 1999) is involved in the biogenesis of iron–sulphur proteins and heme A synthesis (for rev. see Schiffler et al. 2004; Ewen et al. 2008). This Adx homolog Yah1p is highly conserved in fungi, plant, and animals, and it is contained in *Y. lipolytica*, *P. pastoris*, and in most yeast and filamentous fungi (cf. Mauersberger 2013). However, ScYah1p could not substitute Adx in reconstitution of steroid hydroxylation systems in vivo (Dumas et al. 1996). Contrarily, a single ferredoxin ScYah1p-like encoding gene was not found in fission yeasts *S. pombe* and *S. japonicus*, which contain *ETP1*, encoding the adrenodoxin-like mitochondrial *e*lectron *t*ransfer *p*rotein *1* (etp1), a fusion protein consisting of the N-terminal COX15 etp1^{cd} (functions in cytochrome oxidase COX complex assembly) and the carboxy-terminal ferredoxin etp1^{fd} (ferredoxin-like [2Fe-2S]-cluster, with high homology to the ferredoxin family) domains (Bureik et al. 2002; Schiffler et al. 2004). In contrast to the *S. cerevisiae* ferredoxin Yah1, the closely related iron–sulphur protein etp1^{fd} can replace Adx in the interaction with its redox partners AdR and P450. Therefore, etp1^{fd} resembles Adx more than yeast ferredoxin Yah1 in its structural and functional features. SpEtp1p-like fusion proteins were not detected in *Y. lipolytica*, *S. cerevisiae*, and other yeast and filamentous fungi. Therefore, the appearance of etp1 fusion proteins containing the adrenodoxin-like etp1^{fd} domain, which is after cleavage from the COX15 etp1^{cd} domain in mitochondria functional in electron transfer, is obviously restricted to fission yeasts.

3.2 *Recombinant* Yarrowia lipolytica *Strains Expressing Cytochrome P450c17 and CPR*

Functional expression of the bovine ER-resident and steroid transforming P450c17 (encoded by *CYP17A1* cDNA) in *Y. lipolytica* was performed to test the putative advantages of this alkane-utilising yeast as a host for heterologous P450, catalysing biotransformation reactions with hydrophobic substrates (Shkumatov et al. 1998, 2003, 2006; Juretzek 1999; Juretzek et al. 1999, 2000b; Table 1; cf. Sect. 2.1). For comparison, P450c17 was expressed in the commonly used yeast *S. cerevisiae* (host-vector system according to Schunck et al. 1991: strain GRF18 and high-copy replicative vector YEp5117α; cf. Fig. 1a) under control of the very strong galactose-inducible and glucose-repressible promoter pGAL10 (Shkumatov et al. 1998, 2002; Juretzek et al. 2000b). This comparison demonstrated the high potential of *Y. lipolytica* to perform P450-dependent biotransformation of hydrophobic steroid substrates.

Table 2 Comparison of the cytochrome P450c17-catalyzed biotransformation activity of progesterone in the yeasts *Yarrowia lipolytica* and *Saccharomyces cerevisiae*

Recipient strain and integrative or replicative vector (vector type: copy number)	Carbon source for growth	Specific activity (nmol product/OD600 × h)
Yarrowia lipolytica	Glucose	0.2
B204-12A-213/pIC17α	Ethanol	2.0
(lc *ARS-CEN*: 1–2)	Hexadecane	3.3
Yarrowia lipolytica	Glucose	1.3
PO1d(p67IC17) T4	Ethanol	7–8
(mc integrative: 10–12)	Hexadecane	9–12
Yarrowia lipolytica	Glucose	1.0–2.2
PO1d(p67IC17)/pIC17α	Glycerol	1.9–2.8
(mc integrative: 10–12 + lc *ARS-CEN*: 1–2)	Ethanol	3.0–8.5
	Hexadecane	9.0–10.0
Yarrowia lipolytica	Glucose	1.0–1.9
A15T4	Glycerol	1.7–2.7
(diploid, mc integrative: 5–6)	Ethanol	2.6–3.6
	Hexadecane	4.0–5.0
Saccharomyces cerevisiae	Glucose	0
GRF18/YEp5117α	Galactose	5.0
(*ARS* high copy: ~50–100)		

Yeast strains were grown in minimal medium with 1 % of the indicated carbon sources for 18–24 h; biotransformation assays were performed in 20-ml open glass vessels with 100 μM ^{3}H-labelled progesterone in 1 or 2 ml cell suspensions, containing normally 2–4 OD_{600} yeast cells (approximately 2–5 × 10^7 cells/ml, 5–20 OD_{600} in case of small activity) pre-grown on different substrates in fresh minimal medium with 1 % carbon source. Assays were started by adding 10 μl progesterone stock solution (10 mM in ethanol) to the cell suspensions and shaken vigorously at 200 rpm and 28–30 °C for 1–3 h. Bioconversion assays were stopped and extracted with 2 ml chloroform or dichloromethane, organic phase evaporated to dryness, and the residue dissolved in 100 μl dichloromethane or chloroform. Aliquots were separated by TLC (silica gel 60 F254 plates, Merck) using chloroform/ethyl acetate (3:1) as organic solvent and the distribution of radioactivity determined using a Berthold TLC radio-scanner. The main product formed from progesterone was 17α-hydroxyprogesterone (cf. Figs. 4 and 5). Under these assay conditions, product formation was almost linear within the first 1–3 h of incubation. For description of yeast transformants with low-copy (lc, *ARS-CEN* replicative vector), multicopy (mc, integrative) or high-copy (*ARS* replicative) vectors and derived diploid strains, see Fig. 2, and Fig. 1 for expression vectors. B204-12A-213 (*MATB leu2-17 ura3-12*). Copy numbers of vectors per haploid genome were estimated from Southern blots or given for replicative vectors according to Juretzek et al. (2001)

First Functional Expression of P450c17 in *Y. lipolytica* by Low-Copy Replicative Vectors. New host–vector systems (using replicative *ARS*/*CEN* low-copy and integrative multicopy or single-copy expression vectors) were developed for heterologous P450 expression in *Y. lipolytica* under control of the strong and regulated isocitrate lyase promoter (pICL1), which is strongly inducible during growth on ethanol, hydrocarbons, or fatty acids, repressed on glucose to a low basic expression level (5–10 %, in contrast to pGAL10 used for expression in *S. cerevisiae*) and only slightly derepressed on glycerol (Juretzek et al. 1997, 2000a, 2001; Table 2). The pICL1 is of comparable strengths as other available strong and regulated *Y. lipolytica*

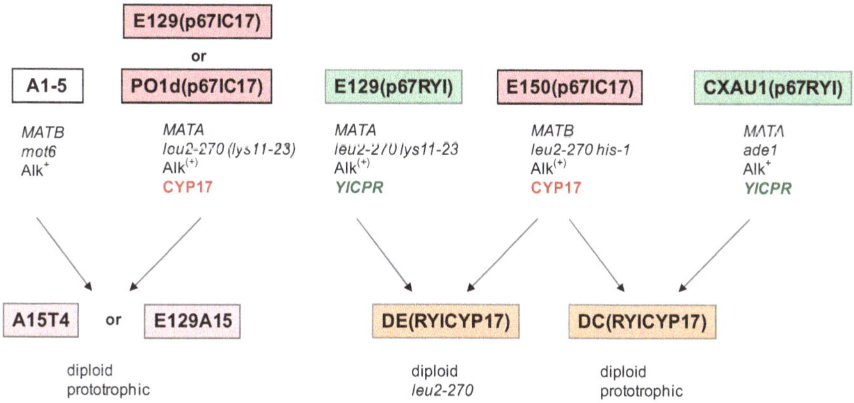

- CYP17	YlCPR CYP17	YlCPR CYP17	Multicopy expression
			Growth on:
++	+	++	Ethanol
++	(+)	++	Alkane (C16)
++	+	++	Olei cacid (C18)

Fig. 2 Recombinant haploid and diploid *Yarrowia lipolytica* strain lines tested in biotransformation of steroids after heterologous expression of bovine cytochrome P450c17 (CYP17) and overexpression of the host-own NADPH-dependent cytochrome P450-reductase (*YlCPR*) under control of promoter pICL1 Prototrophic diploid strains, like A15T4 and E129A15, were obtained by crossing the haploid CYP17 multicopy transformants PO1d(p67IC17) T4 (*MATA leu2-270 xpr2-322*) or E129(p67IC17) (*MATA leu2-270 lys211-23 xpr2-322*)—both resulting from integrative transformation with the *ura3d4*-based multicopy vector p67IC17 (Fig. 1) of the recipient strains PO1d (*MATA leu2-270 ura3-302 xpr2-322*) or E129 (*MATA leu2-270 lys211-23 ura3-302 xpr2-322*), respectively—with the wild-type-derived auxotrophic strain A1-5 (*MATB met6*). The diploid strain lines DE(RYlCYP17)—strain DE (auxotrophic, due to the presence of *leu2-270*) and DC(RYlCYP17)—strains DC1 to DC5 (prototrophic) were obtained by crossing of the haploid CYP17 multicopy transformant E150(p67IC17) with the *YlCPR* multicopy transformants E129 (p67RYl) or CXAU1(p67RYl)—resulting from integrative transformations of the recipient strains E150 (*MATB leu2-270 his-1 ura3-302 xpr2-322*), E129 (*MATA leu2-270 lys211-23 ura3-302 xpr2-322*), or CXAU1 (*MATB ade1 ura3*) with one *ura3d4*-based multicopy vector p67IC17 or p67RYl (Fig. 1), respectively. These recombinant diploids allowed high-level coexpression of bovine P450c17 and the host-own NADPH-P450 reductase during growth on ethanol, alkanes, or fatty acids (induction of pICL1). The diploid strains of the DE line are characterised by a reduced growth in minimal medium M (YNB-like mineral salt medium with ammonium sulphate as nitrogen source) due to the *leu2* auxotrophy, in contrast to prototrophic strains. This growth defect could be overcome by using E129L (*leu2* marker complemented by transformation with pINA62 containing *LEU2*) instead of E129 as recipient strain for transformation with the multicopy plasmids and subsequent diploidisation. Additionally, DE strains exhibited a significant delay in alkane utilisation (Alk$^{(+)}$), characteristic for all strains of the E-line (French inbreeding line, like E150 and E129; Barth and Gaillardin 1996) and strains directly derived from the French wild-type strain W29 (like PO1d), in contrast to the CXAU1 (derived from an American wild-type) and the German wild-type strains A1 and H222 exhibiting normal growth on alkanes (Alk^{+}, Mauersberger et al. 2001). Thus, in diploids of type DC and derived from A15, these growth defects were complemented

promoters, e.g. pPOX2, pPOT1, pALK1, or pXPR2 (Juretzek et al. 2000a). The pICL1 has the advantage to be inducible by hydrophobic substrates (alkanes, fatty acid, or triglycerides) as well as by the hydrophilic substrates ethanol or acetate.

The first functional expression of heterologous P450c17 in *Y. lipolytica* was demonstrated using the *ARS/CEN* low-copy expression plasmid pIC17α (Fig. 1b, Table 2). The expressed P450c17 was found to be functionally active in whole cells and derived microsomal membrane fractions, indicated by the highly sensitive in vivo and in vitro biotransformation assays with (radiolabelled) progesterone into 17α-hydroxyprogesterone as the major product, (Juretzek 1999; Juretzek et al. 1999, 2000b; Shkumatov et al. 1998, 2003; Table 2; cf. Sects. 3.4 and 3.5 for more details). Yeast transformants with the replicative vectors pIC17α or YEp5117α grown in minimal media containing appropriate inducers (ethanol or alkanes for *Y. lipolytica*, galactose for *S. cerevisiae*) exhibited steroid biotransformation activities (17α-hydroxylase converting steroids into 17α-hydroxy-derivatives, Table 2), indicating the functional integrity of heterologously expressed P450c17 in yeast ER and its efficient interaction with the host-own NADPH-P450 reductase (Juretzek 1999; Juretzek et al. 1999, 2000b; Shkumatov et al. 1998, 2002, 2003), as repeatedly demonstrated with different P450s (Table 1).

The P450c17-catalyzed progesterone biotransformation activity in the low-copy transformant B204-12A-213/pIC17α was induced by the growth on ethanol or hexadecane and repressed on glucose, although, in contrast to pGAL10 used for expression in *S. cerevisiae*, not completely, in accordance with the carbon-source-dependent induction of pICL1-controlled gene expression in *Y. lipolytica* (Juretzek et al. 1997, 2000a, 2001). Interestingly, the biomass-specific progesterone biotransformation activity (q_{HP}) of the ethanol- or alkane-induced *Y. lipolytica* low-copy transformant B204-12A-213/pIC17α was already in the same range with that of the galactose-induced high-copy transformant *S. cerevisiae* GRF18/YEp5117α (Table 2).

The P450c17 content in B204-12A-213 or PO1d transformants with the *ARS/CEN* low-copy vector pIC17α (Fig. 1b) was with maximally 4–5 pmol/mg cell protein hardly detectable by CO-difference spectra (COD) of ethanol-grown whole cells and therefore calculated from the P450c17 content of 14 or 10–20 pmol/mg protein, detected by COD or estimated from Western blots using microsomal fractions of these cells. The P450c17-content determination in alkane-grown B204-12A-213/pIC17α cells by COD interfered with the host-own alkane-induced P450s (cf. contribution Mauersberger 2013). The microsomal P450c17 content (about 10–20 pmol/mg microsomal protein) was therefore estimated from Western blots in comparison with microsomal P450c17 content of ethanol-grown cells. Despite low P450c17 content in this low-copy transformant, detection of P450c17 activity was possible with whole cells using the sensitive assay with radioactively labelled progesterone as substrate (Table 2). Based on above estimations of its cellular content P450c17 expressed in *Y. lipolytica*, B204-12A-213/pIC17α exhibited high specific progesterone 17α-hydroxylase activity, especially in alkane-grown whole cells (124 nmol/nmol P450 × min), compared to ethanol cells (74 nmol/nmol P450 × min) and to galactose-induced cells of *S. cerevisiae* GRF18/YEP51α (11–19 nmol/nmol P450 × min) with significant higher P450c17 content of 70–80 pmol/mg cell protein (Shkumatov et al. 1998; Juretzek et al. 2000b). The differences may reflect the different molar ratios CPR/P450c17 detected in the microsomal fractions of these cells ranging from 8.1 (hexadecane cells) and 2.6 (ethanol cells) for *Y. lipolytica* B204-12A-213/pIC17α to 0.12 for galactose-induced cells of *S. cerevisiae* GRF18/YEP51α (calculation of the CPR/P450 ratio made according to CPR data from *C. maltosa*: NADPH-P450 reductase purified of 79 kDa, 60 U/mg protein, 1 U = 210 pmol CPR; cf. Förster 2001). Otherwise, the specific P450c17 activities of the microsomal fractions of these cells were not significantly different (9.8, 8.1, and 7.1 nmol/nmol P450 × min) in the same order. Thus, the 1.7-fold higher specific P450c17 activity found in whole-cell progesterone biotransformation assays with hexadecane-grown cells is probably caused by the 3.3-fold increased NADPH-P450 reductase expression (175–580 mU/mg microsomal protein) as well as by the other specific attributes of yeast cells adapted to utilisation of the hydrophobic substrate alkane compared to the hydrophilic substrate ethanol. The comparison of the functional P450c17 expression in *Y. lipolytica* and *S. cerevisiae* demonstrates therefore the high potential of *Y. lipolytica* to perform P450-dependent biotransformation of steroids even when using low-copy replicative vectors (Table 2).

Multicopy Integrative Expression Vectors. To increase the steroid biotransformation capacity of *Y. lipolytica* cells, a pICL1-controlled and gene-dose-dependent high-level functional expression of both P450c17 and the homologous NADPH-P450 reductase (*YlCPR* gene) in haploid multicopy transformants and their coexpression in derived diploid strains was obtained using integrative multicopy vectors (Fig. 1c, d), because high-copy replicative vectors are not available for *Y. lipolytica*. In order to increase the copy number of expression cassettes in *Y. lipolytica*, a series of multicopy integrative plasmids, including self-cloning plasmids, with the defective, promoter-truncated *ura3d4* gene as multicopy selection marker, rDNA or LTR *zeta* of Ylt1 as integration targeting sequences, and pICL1/ICL1t-controlled expression cassettes for bovine P450c17 (*CYP17* cDNA in p64IC17 or p67IC17, Fig. 1d; Juretzek et al. 2000b) or the host-own NADPH-P450 reductase encoding gene *YlCPR* (p67RYl, Fig. 1c) was developed and applied according to comparable integrative vectors (p64IP- and p67IP-series and derived expression vectors for *lacZ* or *ICL1*) described by Juretzek et al. (2001).

The haploid *Y. lipolytica* strains E129, E150, PO1d, and CXAU1 were used for integrative transformation (according to Barth and Gaillardin 1996) with the *Not*I-linearised multicopy plasmids p67IC17 or p67RYl, respectively (Figs. 1 and 2). Transformants with a normal growth rate carry at least 8–12 copies of expression vectors per cell due to the used multicopy selection marker *ura3d4*. The function of this type of integrative plasmids was evaluated using the *lacZ* reporter gene of *E. coli* under pICL1-control in p64IL43 and p67IL43. The expression level of β-galactosidase in *Y. lipolytica* correlated with the copy number of integrated cassettes and increased up to 13 times in comparison with the low-copy *ARS/CEN* plasmid pIL43 contained in averaged 1.6 copies per cell (Juretzek et al. 2000b, 2001).

High copy numbers of the integrated expression vectors p67IC17 or p67RYl, ranging from 8 to 25 (rarely up to 35), were detected in these multicopy *Y. lipolytica* transformants as estimated from Southern blots (Table 2) in accordance with Juretzek et al. (2001). These multicopy vectors integrated most probably in one (maximally two) cluster at one site of integration (at least 8–12 copies totally), predominantly in tandem (head to tail) or rarely in inverse tandem (head to head) orientation as described below in Sect. 4.2. Thus, by the same approach, multicopy transformant strains of the opposite mating type expressing high levels of the heterologous bovine P450c17 or the homologous NADPH-P450 reductase (*YlCPR* gene) under pICL1 control (Fig. 1c, d) were constructed (Fig. 2; Juretzek 1999; Juretzek et al. 1999, 2000b; Gerber 1999; Förster 2001; Shkumatov et al. 1998, 2003).

Subsequently, several diploid *Y. lipolytica* strain lines (A15T4, E129A15, DE, and DC) were obtained by crossing respective multicopy transformants of the opposite mating type or with the wild-type derived *Y. lipolytica* strain A1-5. Whereas diploid strains A15T4 and E129A15 contain multiple expression cassettes only for P450c17, the diploid strains of lines DE(RYlCYP17) and DC(RYlCYP17) contain multiple expression cassettes for both P450c17 (CYP17) and NADPH-P450 reductase (CPR, indicated by RYl, for reductase of *Y. lipolytica*) originating from different haploid multicopy transformants of the recipient strains E129, E150, or CXAU1 (Fig. 2).

The integrative multicopy transformants with vector p67IC17 exhibited compared to low-copy transformants with pIC17α (in average 1.6 copies per cell, Juretzek et al. 2001) significantly increased P450c17 content and specific biotransformation activity (Table 2) in correlation with corresponding copy numbers (Juretzek 1999; Juretzek et al. 1999, 2000b). The P450c17-catalyzed biotransformation activity of progesterone increased 3.3- to 11-fold in multicopy compared to low-copy transformants, and the carbon-source-dependent induction of the pICL1-controlled P450c17 expression (induction by ethanol or alkane, slight derepression by glycerol, no complete repression by glucose) was detectable. The biomass-specific progesterone biotransformation activity (q_{HP}) of the ethanol- or alkane-induced *Y. lipolytica* multicopy transformant PO1d(p67IC17) T4 or its diploid derivative A15T4 was in the same range or up to two times higher compared with the galactose-induced high-copy transformant *S. cerevisiae* GRF18/Yep5117α (Table 2; Juretzek 1999; Juretzek et al. 2000b).

P450c17 Content. The increase of the P450c17 expression cassette copy number to 10–25 enabled the determination of the P450c17 content directly in ethanol-induced whole yeast cells when using the modified method for CO-difference spectrum (COD) measurement with cytochrome oxidase masking by the presence of 2 mM KCN and progesterone in the assay (cf. Fig. 1 in Mauersberger 2013). The maximal P450c17 content in selected haploid multicopy transformants of types E129(p67IC17), E150(p67IC17), or PO1d(p67IC17) was 40–100 pmol/mg cell protein (30–90 pmol/10^8 cells) after 18–40 h growth in minimal medium with 1 % ethanol (pICL1-induction conditions) using two different cultivation regimes, with and without minimal medium (M) change prior to the induction (medium change: from preculture MG with 1 % glucose to ME with 1 % ethanol or from MY with 1 % glycerol to ME; without medium change: from MG 0.5 % to ME or MY 0.5 % to ME; Juretzek et al. 2000b; Förster 2001). In strain PO1d(p67IC17) T4 (copy number 12–14; cf. Fig. 2) a strongly increased P450c17 content of 200 pmol/10^8 cells was detected after prolonged induction by ethanol (30–40 h and longer) under special culture conditions (repeated feeding with ethanol). Therefore, the P450c17 expression level in multicopy transformants (at least 10–12 copies) significantly increased due to gene-dose effect compared with the low P450c17 content in B204-12A-213 or PO1d transformants with the *ARS/CEN* low-copy vector pIC17α containing approximately 3–5 pmol/mg cell protein (not detectable by COD of whole cells, calculated from Western blots). The P450c17 expression level in the diploid strains DE(RYlCYP17) or DC(RYlCYP17) was with 50–60 % considerably lower than in the haploid multicopy transformants in accordance with the decreased copy numbers per haploid genome. The maximal P450c17 content detected in DE(RYlCYP17) diploids was 25–30 pmol/mg cell protein (35–40 pmol/10^8 cells) after 25–45 h growth in minimal medium with 1 % ethanol. Thus, the heterologous P450c17 expression levels in all tested multicopy integrative *Y. lipolytica* transformants (ethanol-inducible pICL1: 40–100 pmol/mg protein, from 30 up to 200 pmol/10^8 cells) were in the same range as in the high-copy *S. cerevisiae* GRF18/Yep5117α (galactose-inducible pGAL10: 50–100 pmol/10^8 cells; Juretzek et al. 2000b) and were comparable with the content of the host-own CYP52 P450s induced during growth on alkanes (in total 60–130 pmol/mg cell protein; cf. Sect. 2 and Mauersberger 2013).

NADPH-P450 Reductase (CPR) Overexpression. Gene-dose-dependent high-level expression of the homologous NADPH-P450 reductase gene *YlCPR* under pICL1-control was demonstrated using multicopy transformants (10–25 copies of p67RYl) and derived diploid strains of lines DE and DC (Fig. 2; Gerber 1999; Förster 2001). The CPR activity determined as NADPH-cytochrome c reductase (NCCR) activity in the cell-free extract (supernatant S_3) of ethanol-, fatty acid (oleic acid)-, or alkane (hexadecane)-grown cells of the haploid multicopy transformant E129(p67RYl) T6 (maximally 6.0, 5.6, and 4.4 U/mg cell protein, respectively) or CXAU1(p67RYl) T39 (3.9, 3.8, and 3.3 U/mg cell protein) increased significantly compared to the recipient strains E129 or CXAU1 (25–53 and 80–120 mU/mg cell protein in glucose, ethanol, or oleic acid cells and in hexadecane cells). The CPR activity of both multicopy transformants reached a maximum between 30 and 50 h of growth on 1 or 2 % of the inducing carbon sources and was even increased 27–30-fold in glucose-grown cells (1.1 and 1.05 U/mg protein for T6 and T39) due to the gene-dose effect (at least 10–14 copies) and the pICL1-controlled CPR expression compared with expression from one genomic *YlCPR* copy with its own promoter in the recipient strains. The maximal levels of CPR activities were in the multicopy transformants T6 and T39 approximately 70–220 times higher than in the recipient strains, except glucose- and alkane-grown cells with a nearly 30-fold increase, and the pICL1-controlled CPR expression in the multicopy transformants resulted in a 3.6- to 5.3-fold induction on ethanol, oleic acid, and alkane compared to glucose (Förster 2001; Förster and Mauersberger, unpublished).

Comparable results were obtained with selected diploid strains DE(RYlCYP17) or DC(RYlCYP17) constructed for pICL1-controlled coexpression of P450c17 and CPR (Fig. 2). The maximal CPR activities expressed in the diploid strains DE (0.6, 4.0, 3.8, and 3.7 U/mg cell protein on 2 % glucose, 1 % ethanol, 2 % oleic acid, and 2 % hexadecane, respectively) and DC1 (0.7, 2.8, 3.1, and 3.1 U/mg cell protein, on the same substrates) were 10–30 % lower than in respective haploid multicopy transformants due to decreased copy numbers per haploid genome, although the CPR activities were not so strongly reduced as observed for the P450c17 expression. This might be connected with the prevalence of expression cassettes for CPR over P450c17 in the diploid strains DE (10:2, although both parental strains contained almost equal copies of cassettes) and DC (DC1 10:2; DC3 6:4; DC5 8:2) as evidenced in Southern blots (Förster 2001, unpublished).

Thus, gene-dose-dependent high-level functional expression of bovine P450c17 (*CYP17* cDNA) and homologous NADPH-P450 reductase (*YlCPR*) in haploid multicopy transformants of *Y. lipolytica* and their coexpression in the derived DC or DE diploid strain lines was obtained under pICL1-control (Fig. 2). Heterologous P450c17 expression and coexpression of YlCPR in both diploid strain lines DE and DC were similar. The striking difference of DE and DC strains is the wild-type-like growth of DC strains on all substrates including alkanes (Alk^{+}) supplied by the CXAU1 derivative, which is in contrast considerably delayed (Alk$^{(+)}$) in DE strains due to a growth defect observed in its both parental strains, multicopy transformants of E150 and E129 (Fig. 2; cf. Mauersberger et al. 2001).

Simultaneous overexpression of P450c17 and CPR in DE and DC diploid strains resulted in only moderately increased steroid bioconversion rates (Gerber 1999; Förster 2001; Mauersberger et al. 2002; Mauersberger et al. unpublished results; Shkumatov et al. 2003, 2006; cf. Sect. 3.5). The gene-dose-dependent very high-level CPR activity (up to 100–150-fold increase compared to CPR in wild-type strains, CPR:P450 molar ratios of 10–30:1) in the diploid cells DE and DC may be too high and therefore resulted in some negative effects (NADPH depletion, uncoupled P450 reaction cycle) on the P450c17 activity. Contrarily, an assumed moderate increase of the CPR activity (single-copy expression cassette under pICL1- or pPOX2-control) in *Y. lipolytica* expressing P4501A1 (single-copy or multicopy integrated cassettes) significantly stimulated the P4501A1 activity (whole-cell biotransformation of 7-ethoxyresorufin to resorufin) from 2- to 13-fold (Nthangeni et al. 2004). A moderate 2–3-fold increase of the CPR activity in low-level P450c17 expressing *Y. lipolytica* cells (low-copy pIC17α, CPR:P450 molar ratios of 2.6–8.1:1) after growth on hexadecane compared to ethanol as carbon source also increased the specific P450c17 activity of these cells in progesterone biotransformation 1.5- to 2-fold (Shkumatov et al. 1998; Juretzek 1999; Juretzek et al. 2000b; Table 2). When p67IC17-multicopy transformants or derived diploids with high-level P450c17 expression and only wild-type CPR expression level (A15T4 or E129A15; CPR:P450 molar ratios of 0.1–0.2:1) were tested, the stimulating effect of the CPR increase in hexadecane-grown cells was less evident

(Table 2; Fig. 2). Therefore, the probable stimulating effect of a moderately increased CPR expression using a single-copy integrative vector (Nthangeni et al. 2004) together with multicopy P450c17 expression on the steroid biotransformation capacity of *Y. lipolytica* cells has to be elucidated by further studies. Additionally, as shown in Sect. 3.3, in both types of multicopy transformants for P450c17 or CPR overexpression and in derived diploid strains DE and DC for coexpression, a strong proliferation of different types of ER (subcompartments) was observed (Förster 2001; Mauersberger et al. 2002, unpublished), which might give negative effects on the steroid biotransformation capacity of the recombinant *Y. lipolytica* cells due to different localisation of CPR and P450c17.

The comparison of the P450c17-catalyzed steroid bioconversion by recombinant *Y. lipolytica* and *S. cerevisiae* cells revealed the significant advantages of the alkane-assimilating yeast *Y. lipolytica* in P450-dependent biotransformation of the hydrophobic steroid substrates (Juretzek 1999; Juretzek et al. 1999, 2000b; Shkumatov et al. 2003, 2006; cf. Sects. 3.4 and 3.5), a process important for industrial application (Szczebara et al. 2003; Dumas et al. 2006). The stable high-level expression of P450 together with its NADPH-P450 reductase opens new perspectives for further improvement of the efficiency of biotransformation reactions with recombinant *Y. lipolytica* cells, a system which seems to be useful especially for bioconversion of hydrophobic substrates (Fickers et al. 2005; Beopoulos et al. 2009, 2011; Bankar et al. 2009; Coelho et al. 2010).

3.3 *Overexpression of Cytochrome P450c17 and YlCPR Induces ER Proliferation in* Yarrowia lipolytica

The NADPH-P450 reductase and most P450 forms in fungi are integral membrane proteins co-located in the endoplasmic reticulum (ER). It was repeatedly demonstrated that overexpression of P450 forms, P450 reductase, or other membrane proteins resulted in a strong proliferation of ER in yeasts like *S. cerevisiae* or *C. maltosa*, a phenomenon called inducible membranes (Schunck et al. 1991; Wright 1993; Menzel et al. 1997; Sandig et al. 1999). Therefore, it was tested whether overexpression of P450c17 and NADPH-P450 reductase (CPR) is leading also to a strong ER proliferation in *Y. lipolytica*.

For that purpose, pICL1-controlled and gene-dose-dependent high-level functional expression of bovine P450c17 and homologous NADPH-P450 reductase (*YlCPR* gene) in haploid multicopy transformants and their coexpression in derived DC- or DE-type diploid strains of *Y. lipolytica* were obtained as shown above in Sect. 3.2 (cf. Fig. 2). Indeed, the high-level expression of P450c17 and NADPH-P450 reductase in ethanol-growing cells resulted in a strong proliferation of different ER types in *Y. lipolytica* (Fig. 3). Overexpression of P450c17 only induced the formation of a mostly tubular network of ER membranes in various parts of cytoplasm and of plasma membrane associated ER, but no karmellae-like

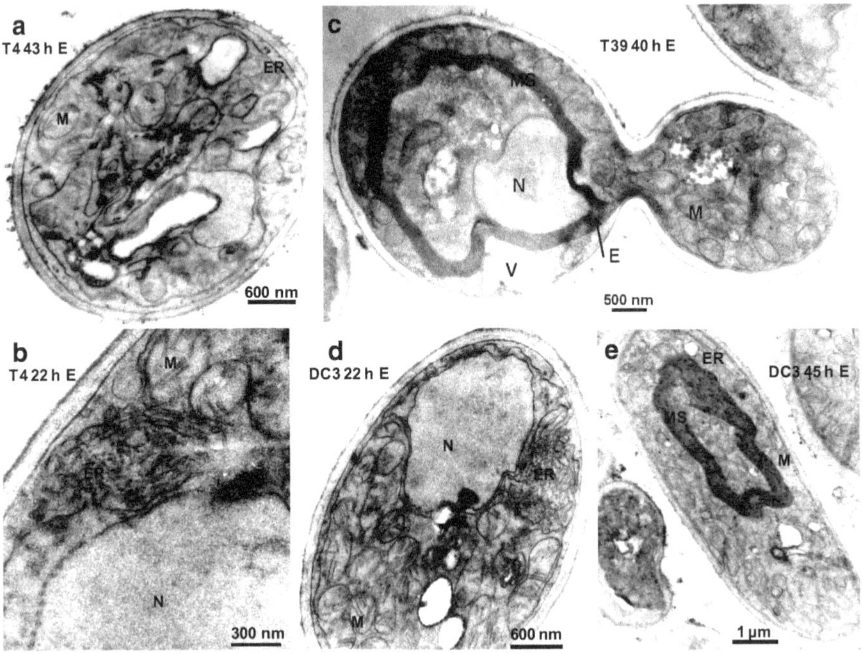

Fig. 3 Different types of ER proliferation in the *Yarrowia lipolytica* haploid multicopy transformant strains PO1d(p67IC17) T4 (**a**, **b**) overexpressing the bovine cytochrome P450c17 under pICL1 control, CXAU1(p67RYl) T39, (**c**) overexpressing the host-own NADPH-cytochrome P450 reductase YlCPR, and in the derived diploid strain DC3 (RYlCYP17), and (**d**) coexpressing both cytochrome P450c17 and YlCPR during growth on 1 % ethanol at different cultivation times. Abbreviations: *N* nucleus, *M* mitochondria, *ER* endoplasmic reticulum, *MS* karmellae-like membrane stacks, *V* vacuole, *E* ethanol, the *arrow* in **c** indicates the beginning transfer of membrane stacks to the daughter cell

membrane stacks were observed in these cells (Fig. 3a, b). Contrary, overexpression of the homologous NADPH-P450 reductase was leading to a special type of ER proliferation, forming karmellae-like membrane stacks (MS) of up to 25 membrane layers, mainly in close vicinity to the nucleus, but partially also extending into cytoplasm (Fig. 3c). In non-overexpressing cells of the *Y. lipolytica* recipient strains, ER membranes appeared normal (not shown). In P450c17 and CPR coexpressing diploid cells DE, both forms of proliferated ER were evident (Fig. 3d, e). Thus, for first time, a strong proliferation of ER in *Y. lipolytica* was shown after high-level expression of the integral membrane proteins P450c17 and NADPH-P450 reductase. Like *S. cerevisiae*, *Y. lipolytica* also exhibited different types of proliferated ER membranes depending on the expressed protein. Expression of P450c17 resulted in proliferation of a mainly tubular membrane network or plasma membrane associated ER, and during high-level expression of *YlCPR* mainly karmellae-like membrane stacks were formed.

Interestingly, there are culture time differences concerning the formation of P450c17- and CPR-depending ER proliferation types. Mostly a tubular network of ER membranes proliferated during P450c17 overexpression and in the early phase of coexpression in diploid strains (Fig. 3a, b, d), the CPR-dependent karmellae-like membrane stacks were observed almost in the later phase (after 40 h growth on ethanol) of high-level P450-reductase expression or in coexpressing diploid cells (Fig. 3e). The observed strong proliferation of different ER subcompartments, depending on which protein was overexpressed, can be used as a model to study ER proliferation and its regulation in yeast cells. The question whether these subcompartments contain predominantly the yeast P450-reductase or the heterologous P450 has to be elucidated. On the other site, if a very high-level expression of the P450 reductase and/or P450 is leading to a different localisation of both P450 system components, this might be a putative drawback for the efficiency of the P450-catalyzed biotransformation capacity of the recombinant yeast cell factory and should be investigated by further studies.

3.4 *Functional Activity of Genetically Modified* Yarrowia lipolytica *in Steroid 17α-Hydroxylation*

As shown in Sect. 3.2, recombinant haploid and diploid *Y. lipolytica* strains capable to catalyse 17α-hydroxylation of progesterone or pregnenolone upon different levels of functional expression of bovine P450c17 (*CYP17*) and host-own NADPH-P450 reductase (Yl*CPR*) have been constructed and tested for P450c17-catalyzed steroid biotransformation (Figs. 2 and 4, Table 2; Shkumatov et al. 1998, 2003, 2006; Juretzek 1999; Juretzek et al. 1999, 2000b). The expressed P450c17 was found to be functionally highly active in ethanol-grown and especially in alkane-grown cells, catalysing the biotransformation of progesterone into 17α-hydroxyprogesterone as main product (Table 2, Fig. 4).

Steroid Bioconversion Capacity of P450-Expressing Yeast Cells. As shown in Sect. 3.2 (Table 2), the values for the biomass-specific progesterone biotransformation activity (q_{HP}, 17α-HP formation rate as nmol product/OD_{600} × h), determined externally in 1 ml assays with 100 nmol labelled progesterone (100 μM) of the ethanol- or alkane-induced *Y. lipolytica* low-copy replicative and high-copy integrative transformants (like B204-12A-213/pIC17α and PO1d(p67IC17) T4) and derived diploids strains (A15T4), were in the same range with that of the galactose-induced high-copy transformant *S. cerevisiae* GRF18/YEp5117α. Using these external determined q_{HP} rates, one can calculate the volumetric productivities (Q_{HP}, 17α-HP formation rate as μM/h or mg product/l × h) of these bioreactor or shaking flask cultures. For galactose-induced *S. cerevisiae* GRF18/YEp5117α cultures, Q_{HP} of 44.0 μM/h (14.5 mg 17α-HP/l × h, cells of a 5-l-bioreactor culture, 13.3 final OD_{600}) was calculated, being in good agreement with the product formation rate of 42.5 μM/h determined in 50-ml shaking flask cultures (Shkumatov et al. 2006) and exceeding significantly the Q_{HP} rate of 6.3–11.0 μM/h reported by Sakaki et al. (1989) [11.0 μM/h for 7 × 10^7 cells/ml, extrapolated from 1.1 μM/h for

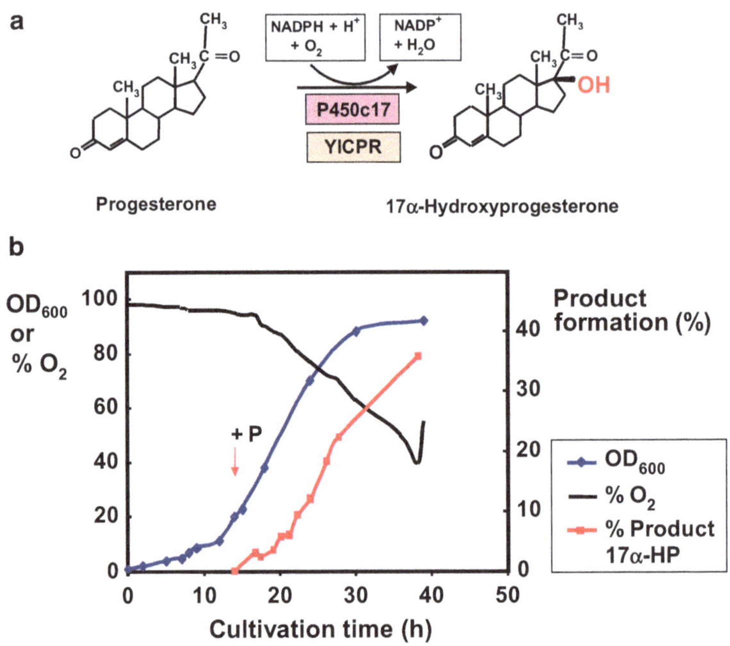

Fig. 4 Biotransformation of progesterone to 17α-hydroxyprogesterone in a 1 L bioreactor culture by hexadecane-growing cells of *Yarrowia lipolytica* A15T4 expressing the ER-resident bovine cytochrome P450c17. (**a**) Reaction scheme of progesterone biotransformation, catalyzed by bovine P450c17 (P450c17, *CYP17A*) expressed in *Y. lipolytica*; electron transfer from NADPH to heterologous P450c17 by functional interaction with the host-own microsomal NADPH-P450 reductase (YlCPR). (**b**) P450c17-catalyzed progesterone biotransformation by alkane-induced A15T4 cells during growth on hexadecane (Juretzek 1999; Juretzek et al. 2000b). Recombinant *Y. lipolytica* diploid strain A15T4, derived by crossing of PO1d(p67IC17) T4 x A1-5, containing multiple copies of the expression cassette pICL1-*CYP17A* cDNA-ICL1t (cf. Fig. 2); preculture on glucose; 1-l-bioreactor culture on 1 % hexadecane (C16, induction of pICL1); after 14 h cultivation 0.5 g progesterone (P) in 10 ml dimethylformamide and 2 % C16 were added. Samples were taken and extracted and product formation (17α-HP, 17α-hydroxyprogesterone as main product) from progesterone was analysed by HPLC

7×10^6 cells/ml] and Ness et al. (1998) [6.3–6.7 μM/h for approximately 10×10^7 cells/ml] for progesterone biotransformation with *S. cerevisiae* expressing bovine P450c17. For ethanol- or hexadecane-induced cultures of the *Y. lipolytica* low-copy replicative transformant B204-12A-213/pIC17α, Q_{HP} of 17.8 or 34.9–39.6 μM/h (cells of 5-l-bioreactor cultures, 13.4 or 15.5 final OD_{600}) were calculated, respectively (Juretzek 1999). According to Table 2, the Q_{HP} values of the ethanol- or alkane-induced multicopy transformant and derived diploid strain cultures (like *Y. lipolytica* PO1d(p67IC17) T4 or A15T4) might be even up to threefold higher, provided optimal biotransformation conditions in bioreactors without oxygen limitations occurring in shaking flask cultures, although some growth limitations were observed with E129- and PO1d-based multicopy transformants and derived diploid DE strains (cf. Fig. 2).

The bioconversion capacity of *Y. lipolytica* expressing bovine P450c17 was experimentally demonstrated with selected diploid strains, like A15T4 (normal growth on alkanes, high level expression of P450c17 due to 5–6 vector copies per haploid genome, alkane-induced CPR level; Fig. 2, Table 2), using a 1-l-bioreactor culture with hexadecane as carbon source and pICL1-inducing substrate, permitting sufficient oxygen supply ($pO_2 > 40$ %) and high biomass concentrations (OD_{600} approximately 100, 5–10 × 10^8 cells/ml) of up to 150 g/l yeast wet weight (Fig. 4; Juretzek 1999; Juretzek et al. 2000b). After adding progesterone (0.5 g in 10 ml dimethylforamide, i.e. 1,590 μM compared to 100 μM used in usual external biotransformation assays; cf. Table 2) to alkane-growing (P450c17-induced) cells, at 14 h of cultivation, the biotransformation to 17α-hydroxyprogesterone (17α-HP) as main product started immediately and continued in the stationary growth phase reached due to hexadecane exhaustion. The main part of substrate and product were found in the cell-free culture medium. The overall biotransformation rate Q_{HP} was 23.3 μM/h (maximal rate Q_{HPmax} 26.8 μM/h) under these conditions (Fig. 4). Comparable results were obtained with ethanol-induced cells of diploid strain DE13 (coexpression of P450scc system and P450c17; cf. Sect. 4) exhibiting a productivity Q_{HP} of 26.7 μM/h in shaking flasks when using 100 μM progesteron.

Compared with other recombinant yeasts, these results for biotransformation of progesterone to 17α-hydroxyprogesterone with *Y. lipolytica* and *S. cerevisiae* are the highest values published to date for P450-catalyzed steroid biotransformations with yeasts. For the shark P450c17 expressed in *Pichia pastoris*, a progesterone biotransformation rate of 1.3 μM/h (10^9 cells/ml) was described (Trant 1996). For *Schizosaccharomyces pombe* expressing human P450c11 (CYP11B1, 11β-hydroxylase), conversion of 11-deoxycortisol to cortisol with a rate of 8.4–9.4 μM/h (4–5 × 10^7cells/ml) was reported (Drăgan et al. 2005), whereas expression of human P450c18 (CYP11B2, aldosterone synthase) resulted in maximally 0.03 μM/h corticosterone formation from 11-deoxycorticosterone (Bureik et al. 2002, 2004). The conversion of progesterone to 11-deoxycorticosterone and that of 17α-hydroxyprogesterone to 11-deoxycortisol occurred in *S. pombe* expressing P450c21 (CYP21, 21-hydroxylase) with maximal rates of 0.9 and 10.8 μM/h, respectively (Drăgan et al. 2006). Biotransformations with recombinant baker's yeast expressing bovine CYP21 had comparable or lower steroid conversion rates of 17α-HP to 11-deoxycortisol (Sakaki et al. 1990, 1991; 5.1 μM/h with 1 × 10^7 cells/ml, extrapolated to 25.5 μM/h with 5 × 10^7 cells/ml, Wu et al. 1991; 4.8–5.7 μM/h, cell density not given, Szczebara et al. 2003), but the experimental setup (cell density, substrate concentration) was not entirely comparable.

These values were calculated from literature data to compare the steroid bioconversion capacity of the recombinant yeasts, although different cell densities were applied for bioconversion assays, e.g. *P. pastoris*, 10^9 cells/ml; *S. pombe* 10^7 cells/ml (Bureik et al. 2002, 2004; Drăgan et al. 2005, 2006); and *Y. lipolytica* 10^7–10^8 cells/ml (external biotransformation assay; cf. Table 2; bioreactor culture; cf. Fig. 4). For comparison, with recombinant *S. cerevisiae* strains, self-sufficient production of hydrocortisone (70 %, with 11-deoxycortisol and corticosterone as by-products) from glucose or ethanol with an overall steroid product formation rate of 0.27 μM/h (with 0.18 μM/h for hydrocortisone) was reported (Szczebara et al. 2003; cf. Sect. 2.2), whereas for progesterone biosynthesis, the product formation rate Q_P was 1.91 μM/h (Duport et al. 1998).

Substrate Specificity of P450c17 Expressed in Yeast. The first activity of mammalian P450c17 is to catalyse hydroxylation of pregnenolone and progesterone at C17 position (17α-hydroxylase) to generate 17α-hydroxypregnenolone and 17-α-hydroxyprogesterone. The second enzymatic activity follows in cleavage of the C_{17}–C_{20} bond (17,20-lyase) of either 17α-hydroxypregnenolone or 17-α-hydroxyprogesterone to form dehydroepiandrosterone (DHEA) and androstenedione, respectively (Lieberman and Warne 2001). The ratio of 17,20-lyase to 17α-hydroxylase activities in mammalia is regulated by the CPR/P450c17 ratio, presence of cytochrome b_5, or serine/threonine phosphorylation of P450c17.

Biotransformation of the steroids progesterone, pregnenolone and related derivatives was demonstrated with recombinant yeast cells of *Y. lipolytica* and *S. cerevisiae* expressing bovine P450c17 (Shkumatov et al. 1998, 2002, 2003, 2006; Juretzek 1999; Juretzek et al. 2000b; Novikova et al. 2009). Among several substrates tested, progesterone was hydroxylated by P450c17 expressed in *S. cerevisiae* GRF18/Yep5117α with the highest activity (Shkumatov et al. 2002, 2006). The 17α-hydroxylase activity of P450c17 towards tested substrates decreased in the sequence progesterone >> 11α- > 11β- >> 19- or 21-hydroxyprogesterone (11-deoxycorticosterone, no activity detected for the two last compounds), whereas the 20α-ketoreduction (yeast 20α-HSD) was observed for 17α- > 21- >> 19-hydroxyprogesterone (no activity) and a minor 20α-HSD activity was also detected for progesterone resulting in the formation of 20α-dihydroprogesterone (Shkumatov et al. 2002, 2003, 2006). Although besides progesterone and pregnenolone the steroid derivatives 11β-, 11α-, 21-, or 19-hydroxyprogesterone were not tested for P450c17 expressed in *Y. lipolytica*, one can assume a comparable substrate specificity in this case. These in vivo experiments in yeasts did not reveal other P450c17 activity towards progesterone, in particular, 16α-hydroxylase activity (Shkumatov et al. 2006). Because no significant androstenedione (the product of the 17,20-lyase reaction catalyzed by P450c17) formation was detected, bovine P450c17 expressed in both yeast *S. cerevisiae* or *Y. lipolytica* exhibited only the 17α-hydroxylase activity and lacked the 17,20-lyase activity (Shkumatov et al. 2002, 2006), as it was previously described for P450c17 expressed in *S. cerevisiae* (Sakaki et al. 1989).

Side-Product Formation. With the recombinant diploid *Y. lipolytica* E129A15 (multiple copies of cassette pICL1-CYP17-ICL1t for P450c17 expression; see Fig. 2 for its construction), the conversion of progesterone (I) into the target main product 17α-hydroxyprogesterone (II) was found already at the initial stage (2–6 h) of biotransformation, whereas steroid side products III and IV (17α,20(α or β)-diols) were accumulated later (12–30 h) after nearly complete conversion of I to II (Fig. 5, see also Sect. 4 for strain DE13; Shkumatov et al. 2003, 2006). For identification, side products were prepared in large quantities after prolonged

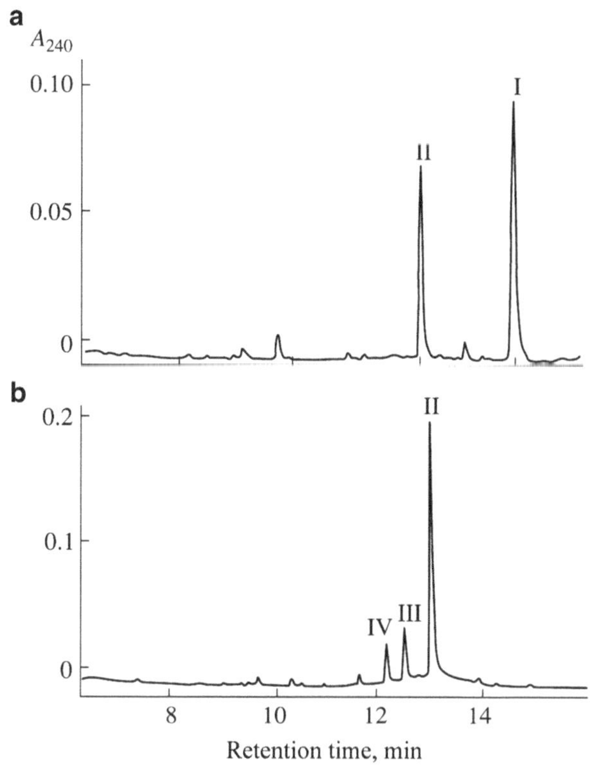

Fig. 5 HPLC separation of steroid products formed upon progesterone biotransformation with ethanol-induced cells of the recombinant diploid *Yarrowia lipolytica* strain E129A15 expressing cytochrome P450c17 under pICL1 control. Cultivation and steroid biotransformation were carried out in 250-ml Erlenmeyer flasks (culture volume of 50 ml) at 28–29 °C, pH 5–6, under aeration conditions (200 rpm). The strain *Y. lipolytica* E129A15 (cf. Fig. 2 for its construction) was cultivated in YPD medium (Difco) containing 1 % yeast extract, 2 % peptone and 2 % glucose. The inducing substrate ethanol (up to 1 % final concentration) was added to the flasks at 24 h, after complete consumption of *D*-glucose by the culture. After 6 h, 1 % ethanol was once more added, and after 18 h cultivation, the cells were centrifuged and transferred to YPE medium (YP-ethanol medium containing 1 % ethanol) and 100 μM progesterone was added (medium change to biotransformation medium). From the biotransformation assay 1 ml samples were taken at 3 and 18 h after addition of steroid substrates, the cells were separated by centrifugation, the supernatant was extracted twice with 2 ml ethyl acetate and the combined organic phase was evaporated on a rotary evaporator. The dry residue was dissolved in 500 μl methanol and analysed by HPLC. Column Kromasil 100-C18, 5 μm, 125 × 4 mm; elution with a gradient of solution B (acetonitrile) in a solution A (H_2O): 5 min, 20 % B; 5–10 min, 20 → 80 % B; 10–17 min, 80 % B; 17–20 min, 80 → 20 % B; flow rate 1 ml/min; spectrophotometric registration in the range of 220–340 nm; ordinate axis: absorption A at 240 nm; Peak I, starting substrate progesterone; peak II, 17α-hydroxyprogesterone (the major product); peaks III and IV, steroid by-products, detected after 3 h (**a**) and 18 h (**b**) incubation of induced cells with progesterone. Figure was taken and modified from Shkumatov et al. (2003)

progesterone biotransformation with *Y. lipolytica* E129A15 and purified by a combination of TLC and HPLC. The data of ^{1}H NMR spectroscopy, mass spectrometry, and HPLC for the steroid side products III and IV and their comparison with the corresponding data for chemically synthesised 17α,20β-dihydroxypregn-4-ene-3-one (Kovganko et al. 2001) and with those for the isolated progesterone side metabolites of *S. cerevisiae* GRF18/YEp5117α expressing P450c17 (Shkumatov et al. 2002) allowed to ascribe them the structures of 17α,20β-dihydroxypregn-4-ene-3-one and 17α,20α-dihydroxypregn-4-ene-3-one, respectively.

The reduction of 17α-hydroxyprogesterone (II) to 17α,20β- (III) and 17α,20α-dihydroxypregn-4-ene-3-one (IV) is due to the functioning of the corresponding yeast 20β- and 20α-dihydroxysteroid dehydrogenase (20-steroid reductase) activities (20β- and 20α-HSD). It is worth mentioning some distinctive features of progesterone bioconversion by recombinant yeasts expressing bovine P450c17, indicating potential advantages of *Y. lipolytica* compared to *S. cerevisiae*. In case of *S. cerevisiae* GRF18/YEp5117α transformants, the major side product was the diol IV (III:IV was about 1:50), whereas in case of *Y. lipolytica* E129A15, the ratio was close to 1:1 with a small prevalence of diol III (Fig. 5). In addition, these both side compounds did not exceed 40–45 % in the latter case even after 48–96 h biotransformation in the YP-ethanol medium (Shkumatov et al. 2003, 2006). At the same time, in the case of *S. cerevisiae* GRF18/YEp5117α, the yield of steroid IV can exceed 60–80 % of the starting progesterone after 24–96 h biotransformation in the nonselective YP-galactose medium (Shkumatov et al. 2002, 2003, 2006). This was due to D-galactose induction of biosynthesis of both bovine P450c17 and yeast Gcy1p protein (crystallin-like protein), an ortholog of mammalian 20α-HSD (Szczebara et al. 2003).

Interestingly, when 20α- or 20β-dihydro derivatives of progesterone (pregn-4-ene-(20α,β)-ol-3-ones, V and VI) were used as starting substrates for bioconversion to study the conjugation between the activities of 17α-hydoxylase and 20-dehydrogenase of recombinant *Y. lipolytica* and *S. cerevisiae* cells, the following sequence of reactions was observed: first oxidation of 20-dihydroprogesterones (by the yeast 20-HSD back reaction or by an oxidase activity of P450c17 towards V and VI) to progesterone (I) followed by its 17α-hydoxylation to II (by P450c17) and final stereoselective 20α- or 20β-reduction (yeast 20-HSDs) to III and IV, indicated by the appearance of I, II, III, and IV as intermediates and products of V or VI bioconversion (Shkumatov et al. 2003, 2006; Novikova et al. 2009).

Side-Product Reduction by Chemical Oxidation. The formation of the reduced derivatives III and IV (17α,20(α,β)diols by HSD-like yeast enzymes) as side products along with the main product 17α-hydroxyprogesterone (II) during progesterone biotransformation by P450c17 expressed in recombinant strains makes the yield of the target product II lower and hampers its isolation from the mixture. Therefore, the Jones oxidation with chromic acid ($H_2Cr_2O_7$) in acetone of steroid mixtures II to IV formed during progesterone biotransformation by *Y. lipolytica* E129A15 or *S. cerevisiae* GRF18/YEp5117α cultured on nonselective media with the addition of the corresponding inducers (ethanol or hexadecane or galactose, respectively) has been studied (Shkumatov et al. 2003). In all the cases,

17α-hydroxyprogesterone (II) was not oxidised. At the same time, 17α,20β-diol (III) and 17α,20α-diol (IV) were almost completely converted into androst-4-ene-3,17-dione, the second natural product of P450c17 (due to 17,20-lyase activity), which was although almost not formed by bovine P450c17 expressed in yeasts. The yield of 17α hydroxyprogesterone was only slightly increased due to partial conversion of the side product IV (17α,20α-diol) into II, whereas the main part of III and IV was oxidized to androstenedione (Shkumatov et al. 2003). These results extend the possibilities of chemical and enzymatic steroid synthesis. A combination of biotechnological methods using various recombinant microorganisms producing mammalian P450c17 and chemical oxidation methods allows the preparation of the target steroid set (II to IV and androstenedione) from progesterone as the only starting compound, which can be useful as raw materials for the production of steroid medicines and other applications (Shkumatov et al. 2003).

Identification of 20α-, 20β-, and 17β-Steroid Reductases in *Y. lipolytica*. The main side products of progesterone biotransformation by recombinant yeast cells expressing P450c17 were 17α,20α- and 17α,20β-dihydroxypregn-4-ene-3-ones (Fig. 5, Table 3; Shkumatov et al. 2003, 2006), indicating the existence of yeast enzymes acting as mammalian 20α -, 20β-, and 17β-hydroxysteroid dehydrogenases (HSDs). The proteins Gcy1p and Ypr1p belonging to the aldo–keto reductase superfamily AKR may be analogues of 20α-HSD in *S. cerevisiae* (Szczebara et al. 2003; cf. Sect. 2.2), whereas analogues of 17β-HSD in *S. cerevisiae* may be Ayr1p (1-acylhydroxyacetone-phosphate reductase) for reduction and Fox2p (hydroxyacyl-CoA-dehydrogenase) for oxidation (Vico et al. 2002). A search for steroid side product generating 20α/β- and 17β-steroid reductase analogues (HSD) performed in the *Y. lipolytica* genome database (Dujon et al. 2004) revealed at least six proteins with statistically significant similarity to the 20α-/17β-HSD members of the AKR superfamily (Shkumatov et al. 2006). The function of these proteins in relation to the formation of side products occurring during 17α-hydroxyprogesterone conversion remains to be elucidated by gene disruption studies. The 20α-HSD was purified from *Y. lipolytica* E129A15 cells. The substrate specificity (reduction of 17α-hydroxyprogesterone and progesterone, oxidation of 17α,20α-dihydroxypregn-4-ene-3-one, and 20α-dihydroprogesterone), the cofactor dependence and molecular weight of 39 kDa relate the *Y. lipolytica* enzyme to 20α-HSD isoenzyme I of mammalia (Shkumatov et al. unpublished results).

Liquidation of Side Products by Selective Inhibition. In *S. cerevisiae ATF2* encoded *O*-acetyltransferase catalyses acetylation of 3β-OH-groups of pregnenolone, 17α-hydroxypregnenolone, or DHEA. Because pregnenolone 3β-acetate cannot be converted by P450c17, Atf2p activity decreases significantly the P450c17 biotransformation efficiency for pregnenolone, and *ATF2* gene disruption was successfully used to avoid undesirable 3β-O-acetylation of pregnenolone (Cauet et al. 1999; Vico et al. 2002; Szczebara et al. 2003; cf. Sect. 2.2). It was also shown that Atf2p can catalyse formation of acetyl esters of isoamyl alcohol and some other aliphatic alcohols (Vadali et al. 2004) opening an alternative approach to diminish the side-product pregnenolone 3β-acetate formation. Addition of such aliphatic alcohols to cultural medium may

Table 3 Progesterone biotransformation by recombinant diploid *Yarrowia lipolytica* strains expressing bovine cytochrome P450c17 (CYP17) and overexpressing the host-own NADPH-dependent cytochrome P450-reductase (YlCPR) under control of the pICL1 promoter

Strain	Medium for biotrans-formation	Change of medium	Yield (%) I Pro	II 17αHP	III 17α,20β DHPre	IV 17α,20α DHPre	V 20βDH Pro	VI 20αDH Pro
Y. lipolytica E129A15	YP-ethanol	+	2.0	58.2	21.0	18.7	<0.5	<0.5
	YP-ethanol	–	5.1	80.4	8.2	6.2	<0.5	<0.5
	YNB-ethanol	–	4.1	86.0	5.0	4.8	<0.5	<0.5
Y. lipolytica DE	YNB-ethanol	+	13.8	80.7	2.5	3.0	<0.5	<0.5
Y. lipolytica DC3	YP-ethanol	+	0	47.2	21.8	19.1	5.6	6.3
	YP-ethanol	–	19.0	60.0	10.2	9.8	<0.5	<0.5
	YNB-ethanol	–	8.0	71.8	11.5	7.7	<0.5	<0.5
Y. lipolytica DC5	YP-ethanol	–	6.0	92.0	<0.5	<0.5	<0.5	<0.5
	YP-ethanol	+	0.8	94.7	2.0	1.5	<0.5	<0.5

The product yield during long-term progesterone biotransformation by the diploid strains E129A15 (P450c17 expression), DE (line DE(RYlCYP17) strain, coexpression of YlCPR and P450c17), and DC3 and DC5 (both line DC(RYlCYP17) strains, coexpression of YlCPR and P450c17) was determined. Description of these diploid strains and their parental haploid multicopy transformants was given in Fig. 2. Biotransformation assays (after or without change to fresh YP- or YNB-ethanol media) were conducted during 24 h, using 100 μM progesterone (Pro, I) as substrate and ethanol-induced cells, cultivated in nonselective rich medium YP-ethanol (YPD, 1 % ethanol as carbon source instead of 2 % glucose) or in minimal medium YNB-ethanol (YNB, 1 % ethanol as carbon source) after pre-cultivation in YPD or YNBD (2 % or 1 % glucose). Cultivation, enzyme induction, and biotransformation conditions were the same as shown below. Main and side products detected by HPLC were II, 17α-hydroxyprogesterone (17αHP, the main product) and the side III, 17α,20-β-dihydroxypregn-4-ene-3-one (17α,20βDHPre), IV, 17α,20α-dihydroxypregn-4-ene-3-one (17α,20αDHPre), V, 20β-dihydroprogesterone (20βDHPro), and VI, 20α-dihydroprogesterone (20αDHPro)

decrease pregnenolone acetylation in *S. cerevisiae* due to competition of steroid with non-steroid alcohols for interaction with Atf2p. Cells of transgenic strain *S. cerevisiae* GRF18/YEp5117α expressing P450c17 were used to carry out pregnenolone biotransformation in rich YP-galactose medium. Formation of 17α-hydroxypregnenolone, pregnenolone 3β-acetate, and 17α-hydroxypregnenolone 3β-acetate was detected. Concentration ratio of 17α-hydroxypregnenolone to pregnenolone 3β-acetate was increased up to 5.3 (0.8 in the control) in the case of addition of 0.5 % vol. isoamyl alcohol (Faletrov et al. 2008a).

Interestingly, genome analysis of *Y. lipolytica* revealed the absence of *S. cerevisiae* Atf2p homologues although other *O*-acetyltransferases are present. Indeed, during pregnenolone biotransformation using *Y. lipolytica* transgenic strain DE54 (DE5-type; see Sect. 4.2) expressing P450c17 under pICL1-control, the steroid was quantitatively converted to 17α-hydroxypregnenolone, and formation of 3β-*O*-acetates of 17-α-hydroxypregnenolone or pregnenolone was not detected (Faletrov et al. 2008a, b).

3.5 Optimisation of Progesterone to 17α-Hydroxyprogesterone Bioconversion

It was established that chemical Jones oxidation of the purified side products 17α,20(α,β)diols (III and IV) (cf. Sect. 3.5) is not the only effective way to decrease the number and change the nature of side products formed during progesterone biotransformation with transgenic *Y. lipolytica* strains. The following representative members of the recombinant diploid strain lines E129A15, DC, and DE of *Y. lipolytica* (cf. Fig. 2 for their construction) were investigated to compare their abilities as catalysts of progesterone 17α-hydroxylation and 20-oxidation/reduction: E129A15 (pICL1-controlled P450c17 overexpression from multiple expression cassettes, basic YlCPR expression from one genomic copy, CPR: P450c17 ratio nearly 0.33:1), DC3, DC5 (both strains of DC line), and DE (one strain of DE line; DC and DE strains with pICL1-controlled coexpression of P450c17 and YlCPR from multiple expression cassettes, CPR:P450c17 ratio 10–15:1). With these diploid strains, the catalytic characteristics of P450c17 expressed in *Y. lipolytica* with and without overexpression of the host-own NADPH-P450 reductase (YlCPR) and the product yield of progesterone biotransformations were compared. Ethanol-induced cells (18–24 h induction in YP-ethanol after pre-growth in YPD as described in Fig. 5, Shkumatov et al. 2006) were used for short- (for kinetic data) and long-term (for the influence of induction and biotransformation conditions on the main- and side-product yield, Table 3) progesterone biotransformation assays and the expression levels of P450c17 ($3.3–4.2 \times 10^4$ P450c17 molecules/cell, determined by CO-difference spectra with whole cells) and of CPR (24- to 36-fold increased CPR expression in strains DC5 and DE compared to E129A15, measured as NADPH-cytochrome c reductase activities) were determined in these cells. The general activity (multiplication of P450c17 molecule number expressed per cell with the Vmax values of the biotransformation reaction) was somewhat higher for strain E129A15 (47) compared to DE (39) and DC5 (43). Concerning their catalytic efficiency (*kcat*/Km, $\text{min}^{-1} \times 10^6\ \text{M}^{-1}$), the recombinant diploid strains can be ranked as DC5 (14.3) > DE (13.1) > E129A15 (10.0). Thus, the very high-level coexpression of CPR in DE and DC strains did not lead to a significant increase in the catalytic efficiency of the expressed P450c17. This might be connected with the observed strong proliferation of different ER membrane subcompartments in P450c17 and CPR overexpressing *Y. lipolytica* cells (cf. Sect. 3.3).

During exhaustive progesterone biotransformation with the *Y. lipolytica* diploid strains E129A15, DE, DC3, and DC5, a strain- and culture-condition-dependent formation of main and side products was observed (Table 3). Long-term progesterone biotransformation over 24 h by *Y. lipolytica* strain E129A15 grown in rich medium YP-ethanol (cells pre-grown 24 h in YPD or till glucose exhaustion, followed by 18–24 h ethanol-induced P450c17 expression and subsequent biotransformation in YP-ethanol) with medium change prior to biotransformation (as indicated in Fig. 5; Shkumatov et al. 2006) was associated with relatively high diols (III and IV, 17α,20β- and 17α,20α-dihydroxypregn-4-ene-3-one) formation as side products (21.0 % and 18.7 %, respectively) with a moderate main product II (17α-hydroxyprogesterone) yield of 58.2 % (Table 3). Significant higher main product II yields were achieved in YP-ethanol or YNB-ethanol medium without medium change prior to biotransformation (80.4 % and 86.0 % for E129A15), and this was associated with lower side-product III and IV formation. Comparable results were obtained with diploid strain DC3 (pICL1-controlled overexpression of P450c17 and YlCPR), although with lower main product yield. With strain DC3, the formation of small amounts (5.6 and 6.3 %) of the minor products V and VI (20β-and 20α-dihydroprogesterone) was observed when medium change prior biotransformation was applied, obviously arising from low activities of the yeast 20α- and 20β-HSD enzymes towards progesterone (Shkumatov et al. 2006), although these side products were almost not found with the strains E129A15, DE or DC5.

The diploid strain DE exhibited a main steroid product II yield of 80.7 % in selective minimal medium YNB with low side product III and IV levels of 2.5–3 % even when using medium change to fresh YNB-ethanol prior to biotransformation (Table 3). The maximal main product II yield of up to 94.7 % was achieved by using strain DC5 in YP-ethanol medium (with and without medium change), which also exhibited low residual substrate and side-product III and VI (2.0 % and lower) levels (Table 3). This practically full side-product absence (total <4.5 %) is the major advantage of the *Y. lipolytica* diploid strain DC5 because it allowed not destroying 20α- and 20β-HSD analogues genes for gaining efficient biotransformation of progesterone (Rudaya et al. 2006; Rudaya 2007; Shkumatov et al. 2007a; Novikova et al. 2008). Contrarily, long-term progesterone biotransformation over 24 h with *S. cerevisiae* GRF18/YEp5117α in YP-galactose medium led to the appearance of yeast 20α-hydroxysteroid dehydrogenase activity (20α-HSD) resulting in considerable formation of mainly 17α,20α-dihydroxypregn-4-ene-3-one (IV) of 60 % up to 86 %. Otherwise, side-product III to VI formation with *Y. lipolytica* strains E129A15 and DC3 did not exceed 35–52 % in YP-ethanol medium with medium change (Table 3; Shkumatov et al. 2002, 2003, 2006), and this amount was finally reduced below 4.5 % by strain selection and optimisation of the culture, induction and biotransformation conditions.

Additionally, for strains *Y. lipolytica* E129A15, DE, and DC5, the repressive effect of initial glucose concentration in the growth media on the pICL1-dependent P450c17 expression was studied. The decrease of initial glucose concentration in YPD medium for growth from 2 to 0.4 % caused an increased specific rate of progesterone 17α-hydroxylation due to earlier carbon source shift from pICL1-repressing glucose to pICL1-inducing ethanol or hexadecane (according to Juretzek et al. 2000a, b, 2001) and the complete absence of residual glucose in the induction medium. In particular, this and simultaneous inducer and progesterone addition (shown below) resulted in decrease of time for 50 % progesterone to 17α-hydroxyprogesterone bioconversion from approximately 10–2 h.

Furthermore, simultaneous addition of the steroid substrate (progesterone) and the P450c17-inducer (ethanol or hexadecane) decreased the time needed for more than 90 % bioconversion of progesterone to 17α-hydroxyprogesterone to 4–8 h (Rudaya 2007; Shkumatov et al. 2007a; Novikova et al. 2008). Absence of unproductive oxygen redox cycles catalyzed by P450c17 without substrate(s) leading to its inactivation can be a reason for increased 17α-hydroxyprogesterone yield. Besides, during long-time cultivation in rich YPD medium, induction of 20-HSD orthologues (like the proteins Gcy1p and Ypr1p in *S. cerevisiae*) is possible, especially at high sugar concentration (Chang et al. 2007).

Thus, *Y. lipolytica* strains and conditions were selected with very low side-product content allowing efficient progesterone biotransformation into 17α-hydroxyprogesterone without prior destroying 20α- and 20β-HSD analogue genes. The highly selective production of 17α-hydroxyprogesterone with *Y. lipolytica* strains like DC5 was achieved by the selected conditions for pICL1-controlled P450c17 expression and biotransformation on ethanol or *n*-alkanes (absence of galactose-induced protein Gcy1p or Ypr1p, which in case of pGAL10-controlled P450c17 expression in *S. cerevisiae* on galactose led to strongly increased formation of 17α,20α-dihydroxypregn-4-ene-3-one as the main side-product IV), as well as by the rapid excretion of the desired product 17α-hydroxyprogesterone by *Y. lipolytica* cells. This was indicated by a lower rate of reconsumption of 17α-hydroxyprogesterone (which is more hydrophilic than progesterone and obviously does not induce yeast 20α,β-HSD) and the observed faster consumption of 20α- or 20β-dihydroprogesterone by ethanol-induced *Y. lipolytica* cells compared to galactose-induced *S. cerevisiae* cells expressing P450c17 (Shkumatov et al. 2006, 2007a).

Thus, as shown in Sects. 3.4 and 3.5, distinctive features of P450c17-expressing *Y. lipolytica* strains are the easily selectable conditions for low activity of 20(α,β)-hydroxysteroid dehydrogenases and the absence of side reactions of 3-*O*-acetylation for C21-3β-OH-Δ5-steroids (pregnenolone or 17α-hydroxypregnenolone). This was leading to a significant reduced extent of side products of progesterone or pregnenolone bioconversion obtained by *Y. lipolytica* strain selection and phenotypical optimisation without the necessity for 20α,20β-HSD analogue gene disruption and indicates several advantages of *Y. lipolytica* as alternative host–vector systems for heterologous expression of steroidogenic P450s compared to *S. cerevisiae* and other yeasts.

4 Transgenic *Yarrowia lipolytica* Strains for Sterol Side-Chain Cleavage and 17α-Hydroxylation

After successful functional expression of bovine P450c17, the mammalian sterol (cholesterol) side-chain cleavage P450scc (CYP11A1) system was selected for coexpression with P450c17 in *Y. lipolytica* to generate a multistep steroid bioconversion chain in this yeast. Whereas the expressed bovine P450c17 protein was shown to be functionally active with the host-own NADPH-P450 reductase (CPR) for functional expression of P450scc, additional coexpression of the two electron transfer proteins NADPH-adrenodoxin reductase (AdR) and adrenodoxin ([2FE-2S] ferredoxin, Adx) will be necessary, as indicated by previous in vitro reconstitution experiments with selected purified P450s (mitochondrial P450scc and P450c11, microsomal P450c21 and P450Coh) using purified or partially purified yeast NADPH-P450 reductases from *C. maltosa* or *Y. lipolytica*, respectively, showing only minor activity with P450scc or P450c11 (0.2–1.4 %), but significant activities with ER-resident P450c21 or coumarin hydroxylating P450Coh (42–85 %) in comparison with their natural electron transfer components (Shkumatov and Smettan 1991; Novikova et al. 2009; Shkumatov et al. unpublished results).

4.1 Substrate Specificity of Cytochrome P450scc

The first reaction leading to the biosynthesis of all the steroid hormones in mammalian is generally considered to be that concerned with the conversion of the sterol precursor, cholesterol, to the C21-steroid, pregnenolone. The widely accepted pathway for side-chain cleavage can be summarized as follows: cholesterol → (22R)-22-hydroxycholesterol → (20R,22R),20,22-dihydroxy-cholesterol → pregnenolone and aldehyde. The process requires 3 mol of NADPH, 3 mol of molecular oxygen, and three proteins, P450scc (CYP11A1), and electron-transferring proteins, ferredoxin reductase (in the adrenal, adrenodoxin reductase—AdR) and ferredoxin (in the adrenal, adrenodoxin—Adx). It was considered earlier that P450scc has narrow substrate specificity. Some modifications, especially 22,23-double bond in structures of ergosterol or some phytosterols, play critical role for the ability of sterols to be substrates for P450scc. Specificity of P450scc towards cholesterol being the only physiological substrate established is considered to be rather strict (Pikuleva 2006).

Phytosterols as Substrates for P450scc. The interaction of P450scc with cholesterol and plant oil phytosterols (β-sitosterol, campesterol, brassicasterol) was studied in vitro using a spectrophotometric titration approach and the reconstituted P450scc system from purified AdR, Adx, and P450scc. Evidence was given that selected plant oil phytosterols can serve as substrates for the side-chain cleavage P450scc system. Both cholesterol and β-sitosterol caused a type I spectral change typical for sterol substrate-P450 interaction, and β-sitosterol was converted by a reconstituted P450scc system into pregnenolone with approximately 40 % of the rate of cholesterol (Shkumatov and Smettan 1991; Novikova et al. 2009; Shkumatov et al. unpublished). With a phytosterol mixture from the sterol fraction of rape oil distillate (containing 40 % β-sitosterol, 31 % campesterol, 29 % brassicasterol, as well as tocopherol and a fatty acids mixture) as substrate, it was shown that besides cholesterol also β-sitosterol and campesterol (but not brassicasterol) were converted to pregnenolone by the reconstituted P450scc system, confirming that additional ethyl or methyl groups at C24 of a potential substrate do not alter significantly the P450scc substrate specificity. It was also established that brassicasterol (campesterol analogue with Δ23-24 bond) can act as competitive inhibitor of P450scc. Intracellularly produced campesterol was used as substrate by the P450scc expressed in *S. cerevisiae* (Duport et al. 1998, 2003; Szczebara et al. 2003; cf. Sect. 2.2).

Uptake of Exogenously Added Sterols and Their Intracellular Modifications. Accessible sterols (sitosterol, campesterol, cholesterol) can represent itself substrates for production of steroid hormones. Transgenic microorganisms, which express the mammalian P450scc-dependent sterol side-chain cleavage system, can be used as biocatalysts for one-stage microbiological synthesis of pregnenolone and other pregnanes (C21-steroids, including mineralo- and glucocorticoids). This could be useful to avoid advanced chemical stage(s) for addition of the C20–C21 fragment to 17-carbon of androgens used in the classical technology.

Some principal features of yeasts restrict their potential as the basis of such biocatalysts. These are (1) the inability of wild-type yeast strains, like *S. cerevisiae*, to uptake exogenously supplied sterols from media under aerobic conditions, the so-called phenomenon of aerobic sterol exclusion (Crowley et al. 1998; Ness et al. 1998; Wilcox et al. 2002; Duport et al. 2003; Szczebara et al. 2003; Raychaudhuri and Prinz 2006), and (2) the formation of steryl esters inside yeast cells of *S. cerevisiae* or *Y. lipolytica* (Czabany et al. 2007; Athenstaedt et al. 2006; Beopoulos et al. 2009, 2011). The aerobic sterol exclusion phenomenon has been thoroughly studied with *S. cerevisiae* in connection with ergosterol biosynthesis (inhibitors, mutants, or gene disruption on squalene synthase or HMG-CoA reductase) to improve uptake of exogenously added sterols, but comparable studies with the strictly aerobic yeast *Y. lipolytica* are required. The established steryl ester (SE) formation with fatty acids from sterols in yeast endoplasmic reticulum can compete with their conversion by expressed P450scc. Otherwise, SE can be hydrolyzed to maintain intracellular pool of free sterols, which is under strict control. Interestingly, for sterols without Δ7- or Δ22-bond or the 24β-methyl smaller uptake rate and larger SE formation degree were observed (Taylor and Parks 1981). In *S. cerevisiae*, SE formation is catalyzed by the steryl ester synthase (acyl-CoA sterol acyltransferase) enzymes Are1p and Are2p (ER-located), and their hydrolysis by the steryl ester hydrolases (sterol esterases) Yeh1p, Tgl1p (both in lipid particles), and plasma membrane located Yeh2p (Czabany et al. 2007). Slight differences (*YlARE1—ScARE1/ARE2*-like, YlARE2—different from *ScARE1/ARE2, YlTGL1, YlYEH* genes involved in SE metabolism) were found for *Y. lipolytica.* Interestingly, only small amounts (2–5 %) of SE and mainly triacylglycerols (TAG) are contained in *Y. lipolytica*, whereas the SE and TAG fractions account each for 50 % of the storage lipids in *S. cerevisiae*, located mainly in lipid particles (Czabany et al. 2007; Athenstaedt et al. 2006; Beopoulos et al. 2009, 2011). The problems connected with the undesirable 3β-*O*-acetylation (acetyltransferase Atf2p encoded by *ATF2* gene) of sterol-derived steroid products like pregnenolone in the yeast *S. cerevisiae* were discussed above (Sects. 2.2 and 3.1), although they were obviously not significant in *Y. lipolytica* due to the absence of Atf2p-like acetyltransferase in this yeast (cf. Sect. 3.5; Faletrov et al. 2008a, b).

4.2 Strain Construction for Coexpression of the Cytochrome P450scc System and Cytochrome P450c17

For testing the functional expression of the cholesterol side-chain cleavage P450scc system, the construction of recombinant *Y. lipolytica* strains has been performed containing pICL1-controlled expression cassettes with cDNAs for the mature (m) forms of all three individual protein components of this enzyme system, namely, NADPH-adrenodoxin reductase (human mAdR, shortly AdR), adrenodoxin (human mAdx, shortly Ax), and P450scc (mP450scc, CYP11A1, human or bovine, short names Ph or Pb, respectively), or alternatively the fusion protein mAdx-mP450scc (human, shortly AxP). The decision to express mature protein forms of the P450scc system in *Y. lipolytica* was based on previous studies demonstrating a catalytically active P450scc system when mature forms of P450scc, Adx, and AdR were expressed in *S. cerevisiae* (Duport et al. 1998, 2003; Szczebara et al. 2003) and overcame serious problems for the expression of mitochondrial localised P450scc in yeasts (cf. Sect. 2.2).

Several new *Y. lipolytica* expression vectors were constructed on the basis of the multicopy integrative plasmids p64PT and p67PT, which are comparable to p64IP and p67IP (only part of pICL1D to *Bam*HI site in front of the *ICL1* intron, ICL1i, contained), described by Juretzek et al. (2001), except they contain the complete pICL1-*Sph*I-ICL1t unit for insertion of a cDNA or gene to be expressed into the *Sph*I site (cf. p67RYl and p67IC17 in Fig. 1 and Sect. 3.2). Respectively, integrative expression vectors of the p64-series (p64AdR, p64AxP, p64Pb, p64Ph) and of the p67-series (p67AdR, p67Ax, p67AxP, p67Pb, p67Ph) for components of the cholesterol side-chain cleavage system (cDNAs for mature forms) have been constructed and used for integrative transformation into selected *Y. lipolytica* strains (Yovkova 2006; Yovkova et al. 2007; Novikova et al. 2008, 2009; Fig. 6; cf. Fig. 1). To combine expression cassettes for all three components of the P450scc system and P450c17 in one recombinant yeast, integrative multicopy transformants of the haploid *Y. lipolytica* strains E129L (*MATA LEU2 lys11-23 ura3-302 xpr2-322*) and E150 (*MATB leu2-270 his-1 ura3-302 xpr2-322*) were obtained *by* simultaneous integrative transformation of mixtures of two to four *ura3d4*-based multicopy expression vectors of the same p64- or p67-series, respectively, flanked by rDNA- or LTR- *zeta*-sequences after linearisation prior to transformation but containing different expression cassettes, e.g. p64AdR and p64AxP or p67AdR, p67Ax, p67Pb and p67IC17 for strain E150, and p67Ax, p67AxP, p67IC17 or p67AdR, p67Ax, p67Pb and p67IC17 for strain E129L.

By Southern blotting (Fig. 6c), evidence was given for simultaneous integration in total of at least 10–12 copies per haploid genome of up to three different pICL1-controlled expression cassettes containing vectors into one multicopy transformant, such as E129L(p67Ax p67AxP p67IC17), shortly E129L(Ax Axp P17) T2 or T3, representing two strains with three integrated vectors out of 10 transformants tested, which was first time demonstrated for *Y. lipolytica*. In other E129L transformants of this assay, at least 10–12 copies of one (T4) or two vectors (seven transformants:

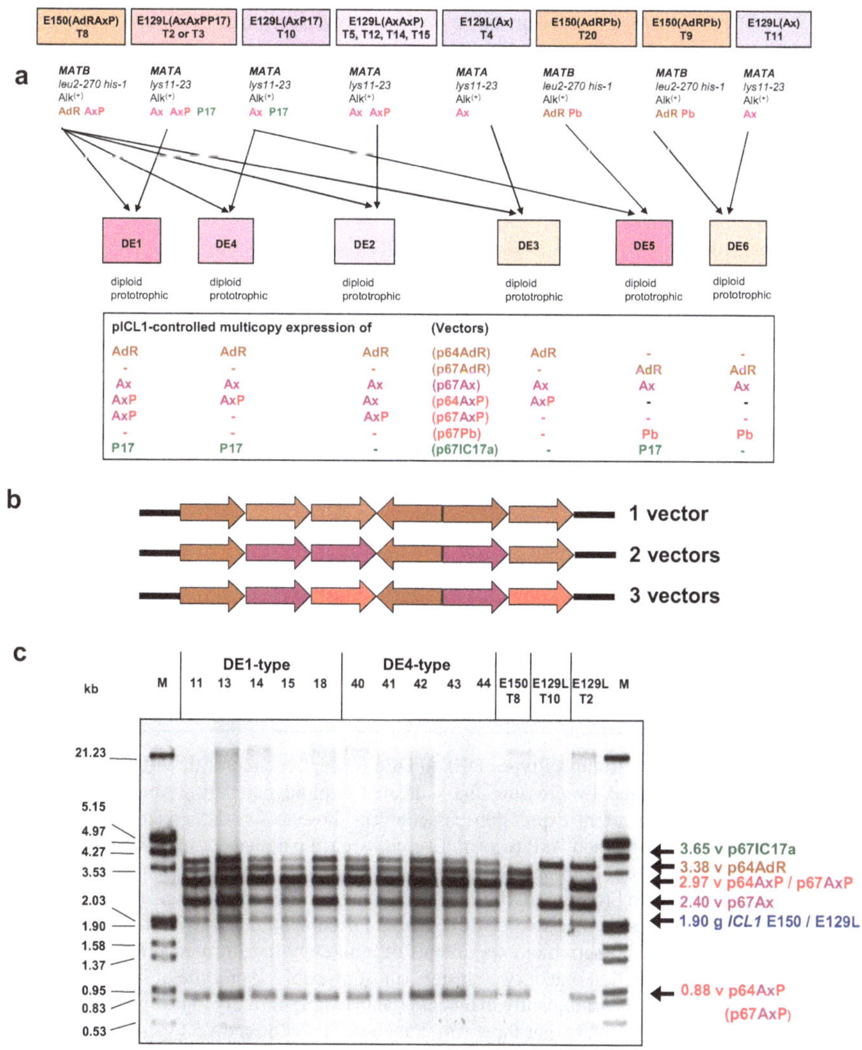

Fig. 6 Construction of recombinant *Yarrowia lipolytica* strains of types DE1 to DE6 for pICL1-controlled expression of the cytochrome P450scc system components and of cytochrome P450c17 by diploidisation using haploid multicopy transformants (**a**), schematic diagram of vectors organisation in the multicopy transformants (**b**), and Southern blot of DNA from selected haploid parental and derived diploid strains for testing the presence of multiple copies of different expression cassettes (**c**). (**a**) Description of multicopy transformants of strains *Y. lipolytica* E150 and E129L, selected as parental haploid strains for diploidisation, e.g. E150(AdRAxP) T8 (shortly E150 T8), for transformant T8 of strain E150 with vector pair p64AdR and p64AxP simultaneously (both vectors integrated), or E129L(AxAxPP17) T2, E129L(AxPP17) T10 (shortly E129L T2, E129L T10), and E129L(Ax) T4, for transformants T2, T10, and T4 of strain E129L with vectors p67Ax, p67AxP, and p67IC17 simultaneously, in which three, two, or only one vector was found integrated in multiple copies (in total at least 10–12), respectively. Designation of other transformants followed the same rules. For the parental transformants selected for the creation of DE1- and DE4-type diploids, evidence for integration of two (E150 T8, E129L T10) or three (E129L T2) expression vectors is provided in Southern blot (**c**). Almost all tested individual

T1, T5-T15) were integrated (Fig. 6a, c). Other transformants with two integrated vectors, such as E150(AdR AxP) T8 (E150 T8, shortly for E150(p64AdR p64AxP) T8), transformed with two p64-series vectors (p64AdR, p64AxP) and E129L(Ax P17) T10 (E129L T10, shortly for E129L(p67AxP p67IC17) T10), transformed with three p67-series vectors (p67Ax, p67AxP, p67IC17), both used as parental strains for subsequent diploidisation are shown together with E129L(Ax Axp P17) T2 as examples in Fig. 6c.

The copy number ratios of two or three integrated multicopy vectors varied in individual transformants, from rare dominance of one vector (E150 T8) to predominantly nearly equal distribution of two or three vectors (E129L T2 and T10, Fig. 6c). The initially determined total vector copy numbers of these multicopy transformants were from 6 to 32 with a majority from 8 to 13. Therefore, three vectors were found integrated in at least three to four copies each, considering that most probably in total 10–12 vector copies integrated and stabilised due to the *ura3d4* multicopy selection marker used, although initially lower and higher copy numbers up to 30–40 were found less frequent with these vector types, and upon further cultivation, it stabilised at 10–12 copies in average (amplification and deamplification, observed by Juretzek et al. 2001).

Thus, individual transformants contained one, two, or three of the vectors p67Ax, p67AxP, and p67IC17 used for E129L transformation, p64AdR and p64AxP used for E150 transformation, and p67AdR, p67Ax, p67Pb, or p67IC17

Fig. 6 (continued) diploid strains of types DE1 to DE6 (except one out of three DE5-type strains tested, not shown) obtained by crossing the indicated haploid parental strains contained the expected three to four different expression cassettes (in three to five integrated vectors; DE1- and DE2-type diploids contained AxP twice delivered by each parental strain containing p64AxP or p67AxP) originating from both parental haploid transformants as summarised in principle (**a**) and evidenced by Southern blotting (**c**). (**b**) Principles of integrated vector organisation in clusters, containing multiple copies of one, two or three vectors used simultaneously for transformation, evidenced by additional Southern blot experiments of multicopy transformants (not shown). The multicopy vectors integrated in one or two clusters (most probably at one site of integration) of at least 10–12 copies totally, predominantly in tandem (head to tail) or rarely in inverse tandem (head to head) orientation. Therefore, two or three simultaneously transformed vectors of the same p64- or p67-type but containing different expression cassettes were found in the same cluster of multiple vector copies, as illustrated in **b**. (**c**) Southern blot for individual diploid strains of type DE1 and DE4 in comparison with the respective parental strains. All tested diploids contained the expected four expression cassettes reflecting the content provided by both parental strains with slight differences in copy number ratios. Chromosomal DNA of all strains was digested by *Eco*RV. The selected probe pICL1-AxP (*Sal*I fragment from vector p67AxP) detected simultaneously the *ICL1* gene (1.9 kb band, one genomic copy per haploid genome, as internal standard for comparison with the multiple copies of integrated vectors) and, due to the presence of pICL1, all four different expression cassettes (multiple copies, visible as fragments of different size from 2.40 to 3.65 kb, with significantly higher signal intensities compared to the 1.90 kb *ICL1* fragment). Abbreviations: Alk$^{(+)}$, delay in growth on alkanes, compared wild-type strains like H222 or CXAU1 (Mauersberger et al. 2001; cf. Fig. 2); P17, P450c17; M, molecular weight standard λ-DNA *Eco*RI-*Hin*dIII digested; 11–18, DE1-type diploids DE11 to DE18; 40–44, DE4-type diploids DE40 to DE44; g, band from the one genomic copy of *ICL1*; v, multicopy bands from the expression cassettes of integrated vectors

used for E150 and E129L transformations, respectively (Fig. 6a, c). Although simultaneous integrative transformations of mixtures of up to four vectors (p67AdR, p67Ax, p67Pb, and p67IC17) were also tested, maximally three vectors were found integrated in one transformant up till now.

As evidenced in additional Southern blots, these vectors integrated in multiple copies in *tandem* (head to tail, dominant) or *inverse tandem* (head to head, rare) orientation mostly at one site of integration in the genome. Vectors of the same type (p67- or p64-type) with up to three different expression cassettes were found integrated in the same cluster of multiple vector copies as shown in the schematic diagram in Fig. 6b. This simultaneous integration in clusters of up to three expression vectors (each in at least three to four copies per haploid genome) of the p64 or p67 series was first time demonstrated for *Y. lipolytica* during these studies.

Subsequently, by diploidisation, new recombinant strains of *Y. lipolytica* were obtained from selected multicopy transformants of the opposite mating types *MATA* or *MATB* of the haploid strains E150 and E129L (containing one to three expression vectors), as shown in principle in Fig. 6 for the DE1 to DE6 types of recombinant diploid strains (Yovkova 2006; Yovkova et al. 2007; Novikova et al. 2008, 2009). These diploid strains and the additionally constructed control strains of type DE7 to DE10 contain several combinations of up to four expression cassettes (or up to five different vectors) for the P450scc system and P450c17 under control of the inducible promoter pICL1. On the one hand, diploid strains of types DE1 to DE4 all contain the AxP fusion as the P450scc component (vectors p64AxP, p67AxP), whereas the DE5-, DE6-, DE9-, and DE10-type strains contain single Pb (p67Pb) instead of the AxP fusion (Fig. 6). Otherwise, two series of recombinant diploid strains were constructed containing up to four different expression cassettes (each in at least two to five copies per haploid genome) for several cDNAs encoding either all three protein components of the cholesterol side-chain cleavage P450scc system (AdR, Ax, and Pb in DE6-type or fusion protein AxP in DE2- and DE3-type strains) or cDNAs encoding the three P450scc system components as well as P450c17 for coexpression (DE5 with Pb, or DE1 and DE4 with AxP, respectively, Fig. 6). Additionally, control strains without cDNAs for some of these four proteins (DE7 lacking P450scc—Pb, DE8 containing P450c17 and lacking Pb, DE9 lacking Adx, DE10 lacking AdR) were constructed to test the possible contribution of host-own proteins in the formation of the functional P450scc enzyme system, such as the presence of a functional adrenodoxin reductase-like or adrenodoxin-like proteins in *Y. lipolytica*.

For selected recombinant diploid *Y. lipolytica* strains of types DE1 to DE10 the ethanol- or alkane-induced heterologous expression of all three components of the P450scc system (human AdR, Adx, fusion protein AxP or bovine Pb) was demonstrated by Western blotting with anti-AdR, -Adx, or -P450scc antibodies (Agalarov 2008; Novikova et al. 2008). Heterologous protein expression was detected from 6 to 72 h after addition of the pICL1-inducing carbon sources ethanol or hexadecane to glucose-grown cells. These Western blot results were in good agreement with the results concerning the expression cassettes present in these strains shown by Southern blotting (Fig. 6). Additionally, in strains of type DE1, DE4 and DE5 functional coexpression of bovine P450c17 with the human P450scc system

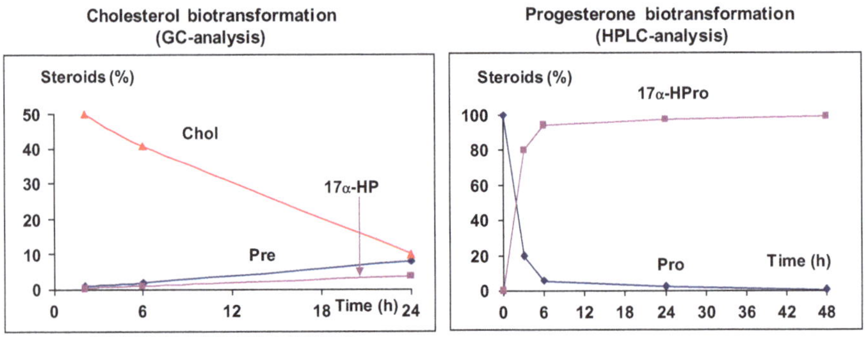

Time (h)	Pre (%)	17α-HP (%)	Chol (%)
2	1	0.4	50
6	2	1	41
24	8	4	10

Abbreviations:
Chol - cholesterol
Pre - pregnenolone
17α-HP - 17α-hydroxypregnenolone
Pro - progesterone
17α-HPro -17α-hydroxyprogesterone

Fig. 7 Cholesterol and progesterone biotransformation by the recombinant diploid *Yarrowia lipolytica* strain DE13 after ethanol-induced coexpression of the cholesterol side-chain cleavage cytochrome P450scc system and cytochrome P450c17. The DE1-type diploid strain *Y. lipolytica* DE13(AdR Ax AxP P17) was derived from the parental haploid strains E150(p64AdR p64AxP) T8 and E129L(p67Ax p67AxP p67IC17) T3 and contains multiple copies of five expression cassettes for mature forms of human AdR, Ax, and AxP (from both vectors p64AxP and p67AxP) and bovine P450c17 (P17) under control of the isocitrate lyase promoter pICL1 (Fig. 6). Western blot results demonstrated the ethanol-induced heterologous expression of the P450scc system proteins AdR, Ax, and AxP (human Ax-P450scc fusion protein). Functional expression of P450c17 was demonstrated by its enzymatic activity in progesterone biotransformation using whole cells. Cultivation and induction of heterologous protein expression was performed in shaking flasks in YPD with 0.4 % glucose for 24 h (to exhaustion of glucose). At 24 h and second time at 30 h, 1 % ethanol was added as inducing carbon sources to start the pICL1-controlled heterologous protein expression of the cholesterol side-chain cleavage system proteins AdR, Ax, and AxP and of P450c17. After glucose exhaustion at 24 h to glucose pre-grown cells, 50 μM cholesterol or 100 μM progesterone was added together with the inducing substrate 1 % ethanol to test their biotransformation; thus, induction and biotransformation were performed simultaneously. Culture samples were taken 2, 3, 6, 24, and 48 h after biotransformation substrate addition, and organic extracts were analysed by GC or HPLC

proteins was shown under comparable cultivation and ethanol-induction conditions, evidenced by the nearly complete P450c17-catalyzed biotransformation of progesterone into 17α-hydroxyprogesterone during 6–10 h (cf. Fig. 7 for DE13).

4.3 *Sterol and Steroid Bioconversion with Recombinant* Yarrowia lipolytica *Coexpressing AdR, Adx, and Cytochromes P450scc and P450c17*

For selected recombinant diploid *Y. lipolytica* strains of types DE1 to DE10, heterologous expression of the mammalian P450scc system and its coexpression with bovine P450c17 were demonstrated by Western blotting (cf. Sect. 4.2), and functional expression of human AdR, Adx, and the fusion protein AxP as well as of bovine P450scc (Pb) and P450c17 (P17) was tested by steroid or sterol biotransformation studies (Fig. 7). To confirm the functional coexpression of the P450scc system and of P450c17 in recombinant *Y. lipolytica* DE strains, the bioconversion of cholesterol or β-sitosterol into pregnenolone (P450scc system) and further to 17α-pregnenolone (P450c17) or of progesterone into 17α-hydroxyprogesterone (P450c17) was studied using selected DE1 to DE6-type diploid strains. The diploid strains D13 (containing expression cassettes for AdR, Adx, AxP twice, P17), D31 (AdR, Adx, AxP), D40 (AdR, Adx, AxP, P17), DE51 or DE54 (AdR, Adx, Pb, P17), and DE61(AdR, Adx, Pb) were used (cf. Figs. 6 and 7 for detailed strain descriptions and results with DE13).

Commercially available pure cholesterol or β-sitosterol was used as substrates for P450scc to be transformed into pregnenolone. Heterologous protein expression was induced by adding to glucose-grown *Y. lipolytica* cells (24-h shaking flask cultivation in YPD with 0.4 % glucose to reach glucose exhaustion, because glucose should be completely consumed at the moment of inducer addition; cf. Sect. 3.5) either 1 % (v/v) of ethanol (Fig. 7) or 1 % hexadecane at 24 h of cultivation (beginning of induction phase) and second time at 30 h of cultivation. Higher glucose content in the YPD preculture (1–2 %) will significantly delay heterologous protein expression. These cultivation and expression conditions were confirmed by Western blotting and tested with strains coexpressing P450c17 (DE13, DE40, DE51, DE54), for which the P450-catalyzed progesterone or pregnenolone biotransformation activity was easily detectable (see the cultivation and biotransformation conditions in Figs. 5 and 7). The bioconversion substrate (50 μM cholesterol, 50 μM β-sitosterol, or up 100 μM progesterone) was added together with the inducing carbon source at 24 h of cultivation. Biotransformation substrate addition made first 24 h after the start of induction (48 h cultivation) was less efficient, as discussed in Sect. 3.5 for progesterone biotransformation.

With ethanol-induced cells of strain DE13 (coexpression of P450c17 and P450scc system, i.e. AxP, Adx, and AdR) a time-dependent formation of small amounts of pregnenolone (up to 8 % in 24 h) and 17α-hydroxypregnenolone (4 %) relatively to faster decreasing quantities of cholesterol was detected (Fig. 7). The P450scc activity of these cells was rather low compared with their high P450c17-catalyzed biotransformation capacity of progesterone into 17α-hydroxyprogesterone (Fig. 7) or of pregnenolone into 17α-hydroxypregnenolone (without formation of 3β-*O*-acetylated side products, due to the absence of *ScATF2* homologues encoded *O*-acetyltransferase activity in *Y. lipolytica*) with strain DE54 (Faletrov et al. 2008b; cf. Sects. 3.5 and 4.1). Thus, under these conditions, strain DE13 was catalysing at least partial biotransformation of cholesterol into pregnenolone, which was subsequently converted to 17α-hydroxypregnenolone by coexpressed P450c17, whereas with β-sitosterol no significant product formation was found. Contrarily, β-sitosterol and campesterol were shown to be converted in vitro by a reconstituted P450scc system to pregnenolone (cf. Sect. 4.1). Further conversion of 17α-hydroxypregnenolone into dihydroepiandrosterone could not be detected, obviously due to the low rate of sterol conversion by P450scc and the very low 17,20-lyase activity of bovine P450c17 expressed in yeast.

The results exemplified for strain DE13 in Fig. 7 provided evidence for efficient biotransformation of progesterone (strains DE13, DE40, DE51) or of pregnenolone (strain DE54, Faletrov et al. 2008a, b) into their 17α-hydroxy products, whereas under these conditions, the bioconversion of cholesterol (or β-sitosterol) to pregnenolone by the same recombinant yeast cells was less efficient. The reasons for the low cholesterol and very low β-sitosterol bioconversion rates of the tested DE-type strains (DE13 and others) are currently under investigation and perspectives for their elimination should be elucidated (as discussed in Sect. 4.1). The low sterol bioconversion abilities of these *Y. lipolytica* DE strains are probably mainly due to (1) the very low sterol uptake of *Y. lipolytica* cells under the applied cultivation conditions (possibly due to its potential toxicity and sterols planar structure) versus long-chain *n*-alkanes serving as very good growth substrates and/or (2) the occurring sterol or product modifications by the yeast cells, i.e. conversion of intracellular pool of the sterol and/or its P450scc product(s) to substances, which are not able to be exported from the cells. Contrarily, under the same culture conditions DE1, DE3, or DE4-type strains are able to perform efficiently P450c17-catalyzed biotransformation of the steroids progesterone and pregnenolone and to excrete the steroid products into the medium.

Interestingly, there is a striking decrease of the cholesterol content not correlating with the observed slow product formation rate (Fig. 7). Additionally, when analysing cell-free supernatant fractions, it was found that about 50 % of cholesterol or β-sitosterol were taken up by the cells and/or absorbed on cell surface 6 h after addition and then stayed practically constant in cases strains exhibited no bioconversion. Whether cholesterol uptake by yeast cells can be stimulated using addition of atorvastatin (known to block partially the in situ ergosterol biosynthesis) or by moderate ultrasound exposure is recently under investigation.

Otherwise, whereas, in contrast to *S. cerevisiae*, 3-*O*-acetylation of at least pregnenolone and derivates as undesirable metabolic way may be not significant in *Y. lipolytica* due to the absence of acetyltransferase activity (no *ScATF2* homologues gene) in this yeast, the formation of steryl esters (SE) from sterols and fatty acids by acyl-CoA sterol acyltransferases (*YlARE1*, *YlARE2*) and their deposit in lipid particles (Beopoulos et al. 2009, 2011) may contribute to the observed low activity of the expressed P450scc system in the DE strains of *Y. lipolytica*. This has to be investigated by further studies, including disruption of *YlARE1/2* (or application of inhibitors) and overexpression of steryl ester hydrolases (*YlTGL1* and *YlYEH1/2*).

Thus, despite all difficulties in detection of cholesterol or β-sitosterol bioconversion to pregnenolone, which was the expected crucial point of these studies (cf. Sects. 3.1 and 4.1), bioconversion activity of both P450scc and P450c17 enzymes was demonstrated in principle with ethanol-induced cells of the strain DE13 coexpressing AdR, Ax, and AxP as well as P450c17, evidenced by the conversion of cholesterol into pregnenolone and subsequent formation 17α-hydroxypregnenolone. Additionally, these cells of DE13 as well as other strains, like DE40, DE51, or DE54, exhibited a high P450c17-catalyzed biotransformation capacity of progesterone or pregnenolone into 17α-hydroxy products (Fig. 7). These promising results clearly demonstrated first time the coexpression of a functionally active cholesterol side-chain cleavage P450scc system and of P450c17 in *Y. lipolytica* cells, thus catalysing the coupled bioconversion from cholesterol (and probably also from β-sitosterol) into pregnenolone followed by subsequent 17α-hydroxylation of pregnenolone to 17α-hydroxypregnenolone. Because P450c17 expressed in yeast showed no or very low 17,20-lyase activity towards 17α-hydroxy products (17α-hydroxyprogesterone and 17α-hydroxypregnenolone), formation of dihydroepiandrosterone was not detected.

5 Conclusions

The growing demand for steroid pharmaceutical preparations generates a need for increasing level of their production. Present-day steroid industry couples the chemical and biological approaches taking advantage of the best aspects of each. New technologies of steroid drug synthesis are based on the use of transgenic microorganisms capable of reproducing some processes in steroidogenic organs or tissues of mammals.

Prospects for increasing the productivity of transgenic microorganisms are the genetic manipulations, directed to alteration of intracellular metabolism with the rise of target steroids and decrease of producing unwanted by-products which are result of host-own enzymatic systems functioning («switching off», activation or additional expression of definite genes). Besides, genetic manipulations can be directed to the improvement of expression level and/or activity of mammalian proteins (especially P450 systems) in the microorganisms (e.g. mutagenesis of native cDNAs resulting in more effective transcription or/and translation; modification, substitution, or removal of the inherent N-terminal addressing amino acid presequences, which ensure their targeting into subcellular compartments; construction of new vectors including more effective promoters; increasing of heterologous genes copy number in the host cell; and coexpression of heterologous proteins with their own native partners, involved in certain pathway).

The yields and activity of heterologous P450s can be improved owing to optimisation of cultivation parameters (e.g. the use of cultural media able to provide for necessary level of cofactor synthesis and heme availability in the cell, using small- or large-scale cultivation conditions of recombinant microorganisms, selection of optimal inducers for P450 protein expression). Moreover, studies on structural–functional features of heterologous P450 proteins, which are expressed in different microorganisms (exhibiting distinct peculiarities and advantages), can generate useful information for selection of suitable host organisms.

The productivity of transgenic microorganisms can be improved, also owing to application of various approaches to increase the cell permeability for hydrophobic P450 substrates, because limited solubility of substrates in water (or cultural media) imposes a limit on the extent of steroid hydroxylated. For example, use of organic solvents, two-phase systems, detergents, cyclodextrins, inhibitors of ergosterol synthesis and action of ultrasound are among these approaches (Fernandes et al. 2003; Lu et al. 2007; Manosroi et al. 2008). In this respect promising results were recently obtained for application of two-phase systems for steroid biotransformation with *Y. lipolytica* strains expressing heterologous P450s (Braun et al. 2012).

The results presented here on heterologous P450 expression in the yeast *Y. lipolytica* provide evidence the high potential of this yeast as host for biotransformation of hydrophobic substrates. Comparison of functional P450c17 expression in the alkane-utilising yeast *Y. lipolytica* and in the commonly used yeast *S. cerevisiae* showed significant advantages of alkane-assimilating yeast cells for P450 catalyze biotransformations of hydrophobic substrates. New integrative multicopy plasmids for *Y. lipolytica* were used to increase the copy number (at least 8–12) of P450 expression cassettes, to optimise the electron transfer to P450 by coexpression of heterologous or host-own NADPH-P450 reductases (CPR) and to increase the biotransformation capacity of the recombinant "yeast cell factory". *Yarrowia lipolytica* strains and conditions were selected with very low side-product content (total <4.5 %) allowing to perform efficient progesterone bioconversion without prior destroying 20α- and 20β-HSDs analogues genes. In contrast to *S. cerevisiae*, there is no 3-*O*-acetyltransferase in *Y. lipolytica*. Multicopy integration vectors and subsequent diploidisation were used for strain construction expressing the three components of the P450scc system (AdR, AdX, P450scc) and of P450c17 in *Y. lipolytica*. Improvement of bioconversion of cholesterol (and β-sitosterol) to pregnenolone is under investigation.

In a recent study, functional coexpression of human CYP2D6 or CYP3A4 with hCPR in *Y. lipolytica* was demonstrated, and the potential of these recombinant yeast cells for bioconversion of hardly soluble hydrophobic steroids (testosterone, 17α-testosterone, progesterone) was tested two-liquid biphasic culture systems. Especially organic solvent phases which can be efficiently taken up and metabolised by the cell (like ethyl oleate) enable a more efficient bioconversion as compared to aqueous systems (Braun et al. 2012).

The data described in this review are not comprehensive, but they demonstrate the wide possibilities of yeast for biotransformation of different steroid substrates. The construction of suitable new strains of *Y. lipolytica* and the study of their biocatalytic potential can open perspectives for creating of new technologies in this field. The results on steroidogenic P450 expression in the yeasts *Y. lipolytica* and *S. cerevisiae* could be used in the following directions: (1) elaboration of new biotechnological approaches for synthesis of pharmaceutically active steroid hormones by using the recombinant yeasts *Y. lipolytica* and *S. cerevisiae* (impact on technology and on public health; Faletrov et al. 2008b) and (2) application of new recombinant *Y. lipolytica* and *S. cerevisiae* yeast cells as test systems for primary screening of potentially active compounds inhibiting the P450c17 enzyme system, which is involved in steroidogenesis and the development of hormonal carcinogenesis (impact on the science in this field and on public health; Shkumatov et al. 2007a–c; Faletrov et al. 2008a, b; Novikova et al. 2009).

Acknowledgements Parts of this work were supported by the Russian Foundation for Basic Research (projects 05-04-48049a and 08-04-00842a), the Belorussian Foundation for Basic Research (project BO4MC-030), the Bundesministerium für Bildung und Forschung of Germany (BMBF, former Bundesministerium für Forschung und Technologie, BMFT, projects 0310257 and 0310654), and the INTAS project CA 03-51-4366. The authors would like to thank the PhD and diploma students which have been involved in theses studies (since the middle of the 1990s). We thank Wolf-Hagen Schunck (Berlin) for his support and fruitful discussions during the first steps of this research, and we gratefully acknowledge the permanent interest and support of the late (deceased) Valentin N. Luzikov (Moscow), Jean-Marc Nicaud (Thiverval-Grignon) and of Gerold Barth (Dresden) to this research.

References

Agalarov JA (2008) Coexpression of the cholesterol hydroxylase/lyase system components in *Yarrowia lipolytica* cells. Diploma thesis, Biochemistry Department, Biological Faculty, Lomonosov Moscow State University

Ahmad S, Garg SK, Johri BN (1992) Biotransformation of sterols: selective cleavage of the side chain. Biotechnol Adv 10:1–67

Andersen MD, Møller BL (2002) Use of methylotrophic yeast *Pichia pastoris* for expression of cytochromes P450. Methods Enzymol 357:333–342

Andersen MD, Busk PK, Svendsen I, Møller BL (2000) Cytochromes P-450 from cassava (*Manihot esculenta* Crantz) catalyzing the first steps in the biosynthesis of the cyanogenic glucosides linamarin and lotaustralin. Cloning, functional expression in *Pichia pastoris*, and substrate specificity of the isolated recombinant enzymes. J Biol Chem 275:1966–1975

Athenstaedt K, Jolivet P, Boulard C, Zivy M, Negroni L, Nicaud J-M, Chardot T (2006) Lipid particle composition of the yeast Yarrowia lipolytica depends on the carbon source. Proteomics 6:1450–1459

Bankar AV, Kumar AR, Zinjarde SS (2009) Environmental and industrial application of *Yarrowia lipolytica*. Minireview. Appl Microbiol Biotechnol 84:847–865

Barros MH, Nobrega FG (1999) *YAH1* of *Saccharomyces cerevisiae*: a new essential gene that codes for a protein homologous to human adrenodoxin. Gene 233:197–203

Barth G, Gaillardin C (1996) *Yarrowia lipolytica*. In: Wolf K (ed) Nonconventional yeasts in biotechnology. Springer, Heidelberg, pp 313–388

Beopoulos A, Cescut J, Haddouche R, Uribelarrea JL, Molina-Jouve C, Nicaud JM (2009) *Yarrowia lipolytica* as a model for bio-oil production. Prog Lipid Res 48:375–387

Beopoulos A, Nicaud JM, Gaillardin C (2011) An overview of lipid metabolism in yeasts and its impact on biotechnological processes. Appl Microbiol Biotechnol 90:1193–1206

Bernhardt R (2006) Cytochrome P450 as versatile biocatalysts. J Biotechnol 124:128–145

Bourel G, Nicaud J-M, Nthangeni B, Santiago-Gomez P, Belin J-M, Husson F (2004) Fatty acid hydroperoxide lyase of green bell pepper: cloning in *Yarrowia lipolytica* and biogenesis of volatile aldehydes. Enzyme Microb Technol 35:293–299

Boyle SM, Popp MP, Smith WC, Greenberg RM, James MO (1998) Expression of CYP2L1 in the yeast *Pichia pastoris*, and determination of catalytic activity with progesterone and testosterone. Mar Environ Res 46:25–28

Braun A, Geier M, Buehler B, Schmid A, Mauersberger S, Glieder A (2012) Steroid biotransformation in biphasic systems with *Yarrowia lipolytica* expressing human liver cytochrome P450 genes. Microb Cell Fact 11:106. doi:10.1186/1475-2859-11-106

Brocard-Masson C, Dumas B (2006) The fascinating world of steroids: *S. cerevisiae* as a model organism for the study of hydrocortisone biosynthesis. Biotechnol Genet Eng Rev 22:213–252

Bureik M, Schiffler B, Hiraoka Y, Vogel F, Bernhardt R (2002) Functional expression of human mitochondrial CYP11B2 in fission yeast and identification of a new internal electron transfer protein, etp1. Biochemistry 41:2311–2321

Bureik M, Hübel K, Drăgan CA, Scher J, Becker H, Lenz N, Bernhardt R (2004) Development of test systems for the discovery of selective human aldosterone synthase (CYP11B2) and 11-ß-hydroxylase (CYP11B1) inhibitors. Discovery of a new lead compound for the therapy of congestive heart failure, myocardial fibrosis and hypertension. Mol Cell Endocrinol 217:249–254

Cauet G, Degryse E, Ledoux C, Spagnoli R, Achstetter T (1999) Pregnenolone esterification in *Saccharomyces cerevisiae*. A potential detoxification mechanism. Eur J Biochem 261:317–324

Chang Q, Griest TA, Harter TM, Petrash JM (2007) Functional studies of aldo-keto reductases in *Saccharomyces cerevisiae*. Biochim Biophys Acta 1773:321–329

Coelho MAZ, Amaral PFF, Belo I (2010) *Yarrowia lipolytica*: an industrial workhorse. In: Méndez-Vilas A (ed) Current research, technology and education topics in applied microbiology and microbial biotechnology, vol 2. Formatex, Badajoz, pp 930–944

Cornelissen S, Julsing MK, Schmid A, Bühler B (2012) Comparison of microbial hosts and expression systems for mammalian CYP1A1 catalysis. J Ind Microbiol Biotechnol 39:275–287

Crowley JH, Leak FW, Shianna KV, Tove S, Parks LW (1998) A mutation in a purported regulatory gene affects control of sterol uptake in *Saccharomyces cerevisiae*. J Bacteriol 180:4177–4183

Czabany T, Athenstaedt K, Daum G (2007) Synthesis, storage and degradation of neutral lipids in yeast. Biochim Biophys Acta 1771:299–309

Dietrich M, Grundmann L, Kurr K, Valinotto L, Saussele T, Schmid RD, Lange S (2005) Recombinant production of human microsomal cytochrome P450 2D6 in the methylotrophic yeast *Pichia pastoris*. Chembiochem 6:2014–2022

Drăgan CA, Zearo S, Hannemann F, Bernhardt R, Bureik M (2005) Efficient conversion of 11-deoxycortisol to cortisol (hydrocortisone) by recombinant fission yeast Schizosaccharomyces pombe. FEMS Yeast Res 5:621–625

Drăgan CA, Hartmann RW, Bureik M (2006) A fission yeast-based test system for the determination of IC50 values of anti-prostate tumor drugs acting on CYP21. J Enzyme Inhib Med Chem 21:547–556

Dujon B, Sherman D, Fischer G et al (2004) Genome evolution in yeasts. Nature 430:35–44

Dumas B, Cauet G, Lacour T, Degryse E, Laruelle L, Ledoux C, Spagnoli R, Achstetter T (1996) 11 beta-hydroxylase activity in recombinant yeast mitochondria. In vivo conversion of 11-deoxycortisol to hydrocortisone. Eur J Biochem 238:495–504

Dumas B, Brocard-Masson C, Assemat-Lebrun K, Achstetter T (2006) Hydrocortisone made in yeast: metabolic engineering turns a unicellular microorganism into a drug-synthesizing factory. Biotechnol J 1:299–307

Duport C, Spagnoli R, Degryse E, Pompon D (1998) Self-sufficient biosynthesis of pregnenolone and progesterone in engineered yeast. Nat Biotechnol 16:186–189

Duport C, Schoepp B, Chatelain E, Spagnoli R, Dumas B, Pompon D (2003) Critical role of the plasma membrane for expression of mammalian mitochondrial side chain cleavage activity in yeast. Eur J Biochem 270:1502–1514

Dutta TK, Datta J, Samanta TB (1993) Onset of new catalytic activity in immobilized spores of *Aspergillus ochraceus* TS due to in situ germination: C17-C20 lysis accompanies 11 alpha-hydroxylation of steroid. Biochem Biophys Res Commun 192:119–123

Ehmer PB, Bureik M, Bernhardt R, Müller U, Hartmann RW (2002) Development of a test system for inhibitors of human aldosterone synthase (CYP11B2): screening in fission yeast and evaluation of selectivity in V79 cells. J Steroid Biochem Mol Biol 81:173–179

Ewen KM, Schiffler B, Uhlmann-Schiffler H, Bernhardt R, Hannemann F (2008) The endogenous adrenodoxin reductase-like flavoprotein arh1 supports heterologous cytochrome P450-dependent substrate conversions in Schizosaccharomyces pombe. FEMS Yeast Res 8:432–441

Faletrov YV, Frolova NS, Rudaya EV, Radyuk VG, Shkumatov VM (2008a) Inhibition of pregnenolone and related steroids 3β-acetylation in recombinant yeasts *Saccharomyces cerevisiae* GRF18/YEp5117α by aliphatic alcohols. In: Kilchevskii AV (ed) Molecular and applied genetics: Proceedings, vol 7, Minsk, pp 45–48

Faletrov YV, Frolenkov KA, Novikova LA, Luzikov VN, Mauersberger S, Barth G, Shkumatov VM (2008b) Transgenic yeasts for pharmacological investigations. In: 9th International symposium on cytochrome P450 biodiversity and biotechnology, Nice, France, 08–12 June 2008, Abstracts

Fallon RD, Payne MS, Picataggio SK, Wu S (1999) Transformed yeast strains and their use for the production of monoterminal and diterminal aliphatic carboxylates. WO9904014

Fallon RD, Payne MS, Picataggio SK, Wu S (2004) Transformed yeast strains and their use for the production of monoterminal and diterminal aliphatic carboxylates. US2004146999

Fernandes P, Cruz A, Angelova B, Pinheiro HM, Cabral JMS (2003) Microbial conversion of steroid compounds: recent developments. Enzyme Microb Technol 32:688–705

Fickers P, Benetti PH, Wache Y, Marty A, Mauersberger S, Smit MS, Nicaud JM (2005) Hydrophobic substrate utilisation by the yeast *Yarrowia lipolytica*, and its potential applications. FEMS Yeast Res 5:527–543

Förster A (2001) Untersuchungen zur heterologen Expression von steroidwandelnden Cytochrom P450 Systemen in der Hefe *Yarrowia lipolytica*. Diploma thesis, Technische Universität Dresden, Germany

Fukuda R, Ohta A (2013) Utilization of hydrophobic substrates by *Yarrowia lipolytica*. In: Barth G (ed) Yarrowia lipolytica, vol 1. Springer, Heidelberg

Gerber J (1999) Untersuchungen zur Optimierung des Elektronentransportsystems für die Cytochrom P450 katalysierte Biotransformation von Steroiden in *Yarrowia lipolytica*. Diploma thesis, Technische Universität Dresden

Gonzalez FJ, Korzekwa KR (1995) Cytochromes P450 expression systems. Annu Rev Pharmacol Toxicol 35:369–390

Hakki T, Zearo S, Drăgan CA, Bureik M, Bernhardt R (2008) Coexpression of redox partners increases the hydrocortisone (cortisol) production efficiency in CYP11B1 expressing fission yeast *Schizosaccharomyces pombe*. J Biotechnol 133:351–359

Juretzek T (1999) Entwicklung von Wirts-Vektor-Systemen zur heterologen Expression von Proteinen in der nichtkonventionellen Hefe *Yarrowia lipolytica* und ihre Anwendung für die Cytochrom P450-katalysierte Stoffumwandlung. PhD thesis, Technische Universität Dresden

Juretzek T, Prinz A, Schunck W-H, Barth G, Mauersberger S (1997) Expressionskassetten zur heterologen Expression von Proteinen in der Hefe *Yarrowia lipolytica* unter Kontrolle des regulierbaren Promotors der Isocitratlyase. Patent DE19525282

Juretzek T, Mauersberger S, Schunck W-H, Nicaud J-M, Barth G (1999) Functional expression of heterologous cytochrome P450 in the yeast *Yarrowia lipolytica* and its use in biotransformation of steroids. Curr Genet 35(3–4):307

Juretzek T, Wang H-J, Nicaud J-M, Mauersberger S, Barth G (2000a) Comparison of promoters suitable for regulated overexpression of heterologous genes in the alkane-utilizing yeast *Yarrowia lipolytica*. Biotechnol Bioprocess Eng 5:320–326

Juretzek T, Mauersberger S, Barth G (2000b) Recombinant haploid or diploid *Yarrowia lipolytica* cells for the functional heterologous expression of cytochrome P450 systems. Patent WO0003008, DE19932811

Juretzek T, Le Dall MT, Mauersberger S, Gaillardin C, Barth G, Nicaud J-M (2001) Vectors for gene expression and amplification in the yeast *Yarrowia lipolytica*. Yeast 18:97–113

Katsumata T, Hasegawa A, Fujiwara T, Komatsu T, Notomi M, Abe H, Natsume M, Kawaide H (2008) *Arabidopsis* CYP85A2 catalyzes lactonization reactions in the biosynthesis of 2-deoxy-7-oxalactone brassinosteroids. Biosci Biotechnol Biochem 72:2110–2117

Ko KW, Lin F, Katsumata T, Sugai Y, Miyazaki S, Kawaide H, Okada K, Nojiri H, Yamane H (2008) Functional identification of a rice *ent*-kaurene oxidase, OsKO2, using the *Pichia pastoris* expression system. Biosci Biotechnol Biochem 72:3285–3288

Kolar NW, Swart AC, Mason JI, Swart P (2007) Functional expression and characterisation of human cytochrome P45017α in *Pichia pastoris*. J Biotechnol 129:635–644

Kovganko NV, Kashkan ZN, Shkumatov VM (2001) Simple synthesis of 17α,20-β-dihydroxypregn-4-en-3-one. Chem Nat Comp 37:55–56

Lieberman S, Warne PA (2001) 17-Hydroxylase: an evaluation of the present view of its catalytic role in steroidogenesis. J Steroid Biochem Mol Biol 78:299–312

Lu W, Du L, Wang M, Guo Y, Lu F, Sun B, Wen J, Jia X (2007) A novel substrate addition method in the 11ß-hydroxylation of steroids by *Curvularia lunata*. Food Bioprod Process 85:63–72

Manosroi J, Saowakhon S, Manosroi A (2008) Enhancement of 17α-hydroxyprogesterone production from progesterone by biotransformation using hydroxypropyl-β-cyclodextrin complexation technique. J Steroid Biochem Mol Biol 112:201–204

Masuda Y, Park SM, Ohkuma M, Ohta A, Takagi M (1994) Expression of an endogenous and a heterologous gene in *Candida maltosa* by using a promoter of a newly-isolated phosphoglycerate kinase (PGK) gene. Curr Genet 25:412–417

Matsuzaki F, Wariishi H (2005) Molecular characterization of cytochrome P450 catalyzing hydroxylation of benzoates from the white-rot fungus *Phanerochaete chrysosporium*. Biochem Biophys Res Commun 334:1184–1190

Mauersberger S (2013) Cytochrom P450 of the alkane-utilizing yeast *Yarrowia lipolytica*. In: Barth G (ed) Yarrowia lipolytica. Springer, Heidelberg

Mauersberger S, Kärgel E, Matyashova RN, Müller H-G (1987) Subcellular organization of alkane oxidation in *Candida maltosa*. J Basic Microbiol 27:565–582

Mauersberger S, Ohkuma M, Schunck W-H, Takagi M (1996) Chapter 12. *Candida maltosa*. In: Wolf K (ed) Non-conventional yeasts in biotechnology. A handbook. Springer, Berlin, pp 411–580

Mauersberger S, Wang H-J, Gaillardin C, Barth G, Nicaud J-M (2001) Insertional mutagenesis in the *n*-alkane-assimilating yeast *Yarrowia lipolytica*: generation of tagged mutants in genes involved in hydrophobic substrates utilization. J Bacteriol 183:5102–5109

Mauersberger S, Förster A, Gerber J, Juretzek T, Barth G (2002) Functional expression of heterologous cytochrome P450 in *Yarrowia lipolytica* and its use in biotransformation of steroids. In: Third *Yarrowia lipolytica* international meeting (TYLIM2002), Dresden, Germany, 17–20 July 2002, Abstracts p 114 (Poster 41)

McCormick SP, Alexander NJ, Proctor RH (2006) Heterologous expression of two trichothecene P450 genes in *Fusarium verticillioides*. Can J Microbiol 52:220–226

Menke HH, Nan LM, Selten GCM, Smaal EB, Slijkhuis H (1990) Cloning and co-expression of bovine adrenal cytochrome P450scc and adrenodoxin in *Kluyveromyces lactis*. In: 20th FEBS-Meeting, Budapest, Hungary, 19–24 August 1990, program p 215, poster P-TH-381

Menzel R, Vogel F, Kärgel E, Schunck W-H (1997) Inducible membranes in yeast: relation to the unfolded-protein-response pathway. Yeast 13:1211–1229

Minenko AN, Novikova LA, Luzikov VN, Kovaleva IE (2008) Import of hybrid forms of CYP11A1 into yeast mitochondria. Biochim Biophys Acta 1780:1121–1130

Ness F, Achstetter T, Duport C, Karst F, Spagnoli R, Degryse E (1998) Sterol uptake in *Saccharomyces cerevisiae* heme auxotrophic mutants is affected by ergosterol and oleate but not by palmitoleate or by sterol esterification. J Bacteriol 180:1913–1919

Novikova LA, Yovkova V, Faletrov YV, Agalarov JA, Luzikov VN, Shkumatov VM, Barth G, Mauersberger S (2008) Functional expression of the cholesterol side-chain cleavage cytochrome P450scc system and of P450c17 in the yeast *Yarrowia lipolytica*. In: 9th International symposium on cytochrome P450 biodiversity and biotechnology, Nice, France, 08–12 June 2008, Abstracts p 68

Novikova LA, Faletrov YV, Kovaleva IE, Mauersberger S, Luzikov VN, Shkumatov VM (2009) From structure and functions of steroidogenic enzymes to new technologies of gene engineering. Review. Biochemistry (Moscow) 74:1482–1504

Nthangeni MB, Urban P, Pompon D, Smit M, Nicaud J-M (2004) The use *Yarrowia lipolytica* for the expression of human cytochrome P450 CYP1A1. Yeast 21:583–592

Oeda K, Sakaki T, Ohkawa H (1985) Expression of rat liver cytochrome P-450MC cDNA in *Saccharomyces cerevisiae*. DNA 4:203–210

Ohkuma M, Kobayashi K, Kawai S, Hwang CW, Ohta A, Takagi M (1995a) Identification of a centromeric activity in the autonomously replicating TRA region allows improvement of the host-vector system of *Candida maltosa*. Mol Gen Genet 249:447–455

Ohkuma M, Park S-M, Zimmer T, Menzel R, Vogel F, Schunck W-H, Ohta A, Takagi M (1995b) Proliferation of intracellular membrane structures upon homologous overproduction of cytochrome P-450 in *Candida maltosa*. Biochim Biophys Acta (Biomembranes) 1236: 163–169

Ortiz de Montellano PR (ed) (2005) Cytochrome P450: structure, mechanism and biochemistry, 3rd edn. Plenum, New York

Pajic T, Vitas M, Zigon D, Pavko A, Kelly S, Komel R (1999) Biotransformation of steroids by the fission yeast *Schizosaccharomyces pombe*. Yeast 15:639–645

Park SM, Ohkuma M, Masuda Y, Ohta A, Takagi M (1997) Galactose-inducible expression systems in *Candida maltosa* using promoters of newly-isolated *GAL1* and *GAL10* genes. Yeast 13:21–29

Peters FT, Drăgan CA, Wilde DR, Meyer MR, Zapp J, Bureik M, Maurer HH (2007) Biotechnological synthesis of drug metabolites using human cytochrome P450 2D6 heterologously expressed in fission yeast exemplified for the designer drug metabolite 4′-hydroxymethyl-alpha-pyrrolidinobutyrophenone. Biochem Pharmacol 74:511–520

Peters FT, Drăgan CA, Schwaninger AE, Sauer C, Zapp J, Bureik M, Maurer HH (2009a) Use of fission yeast heterologously expressing human cytochrome P450 2B6 in biotechnological synthesis of the designer drug metabolite N-(1-phenylcyclohexyl)-2-hydroxyethanamine. Forensic Sci Int 184:69–73

Peters FT, Drăgan CA, Kauffels A, Schwaninger AE, Zapp J, Bureik M, Maurer HH (2009b) Biotechnological synthesis of the designer drug metabolite. 4′-hydroxymethyl-alpha-pyrrolidinohexanophenone in fission yeast heterologously expressing human cytochrome P450 2D6 – a versatile alternative to multistep chemical synthesis. J Anal Toxicol 33:190–197

Picataggio S, Rohrer T, Deanda K, Lanning D, Reynolds R, Mielenz J, Eirich LD (1992) Metabolic engineering of *Candida tropicalis* for the production of long-chain dicarboxylic acids. Biotechnology (NY) 10:894–898

Pikuleva IA (2006) Cytochrome P450s and cholesterol homeostasis. Pharmacol Ther 112:761–773

Pompon D, Louerat B, Bronine A, Urban P (1996) Yeast expression of animal and plant P450s in optimized redox environments. Methods Enzymol 272:51–64

Pompon D, Gautier JC, Perret A, Truan G, Urban P (1997) Simulation of human xenobiotic metabolism in microorganisms. Yeast a good compromise between E. coli and human cells. Review. J Hepatol 26(Suppl 2):81–85

Prinz A (1995) Testung des *ICL1*-Promoters zur funktionellen heterologen Expression von Cytochrom P450 in der Hefe *Yarrowia lipolytica*. Universität Gesamthochschule Kassel, Fachbereich 19 (Biologie/Chemie)

Raychaudhuri S, Prinz WA (2006) Uptake and trafficking of exogenous sterols in *Saccharomyces cerevisiae*. Biochem Soc Trans 34:359–362

Renaud JP, Peyronneau MA, Urban P, Truan G, Cullin C, Pompon D, Beaune P, Mansuy D (1993) Recombinant yeast in drug metabolism. Review. Toxicology 82:39–52

Rudaya EV (2007) Catalytic efficiency and substrate specificity of recombinant yeast expressing cytochrome P450c17. PhD Thesis, September 2007, Byelorussian State University

Rudaya EV, Frolova NS, Mauersberger S, Shkumatov VM (2006) Catalytic efficiency of progesterone biotransformation by recombinant *Yarrowia lipolytica* yeasts. In: International conference on molecular, membrane and basic cellular principles of functions of biosystems (VII. congress of BOOFiB = BSSPhB), 21–23 June 2006, Minsk, Belarus, Poster I, pp 129–131

Sakaki T, Inouye K (2000) Practical application of mammalian cytochrome P450. Review. J Biosci Bioeng 90:583–590

Sakaki T, Shibata M, Yabusaki Y, Murakami H, Ohkawa H (1989) Expression of bovine cytochrome P450c17 cDNA in *Saccharomyces cerevisiae*. DNA 8:409–418

Sakaki T, Shibata M, Yabusaki Y, Murakami H, Ohkawa H (1990) Expression of bovine cytochrome P450c21 and its fused enzymes with yeast NADPH-cytochrome P450 reductase in *Saccharomyces cerevisiae*. DNA Cell Biol 9:603–614

Sakaki T, Akiyoshi-Shibata M, Yabusaki Y, Manabe K, Murakami H, Ohkawa H (1991) Progesterone metabolism in recombinant yeast simultaneously expressing bovine cytochromes P450c17 (CYP17A1) and P450c21 (CYP21B1) and yeast NADPH-P450 oxidoreductase. Pharmacogenetics 1:86–93

Sakaki T, Kominami S, Hayashi K, Akiyoshi-Shibata M, Yabusaki Y (1996) Molecular engineering study on electron transfer from NADPH-P450 reductase to rat mitochondrial P450c27 in yeast microsomes. J Biol Chem 271:2609–26213

Sandig G, Kärgel E, Menzel R, Vogel F, Zimmer T, Schunck WH (1999) Regulation of endoplasmic reticulum biogenesis in response to cytochrome P450 overproduction. Drug Metab Rev 31:393–410

Santiago-Gómez MP, Vergely C, Policar C, Nicaud J-M, Belin J-M, Rochette L, Husson F (2007) Characterization of purified green bell pepper hydroperoxide lyase expressed by *Yarrowia lipolytica*: radicals detection during catalysis. Enzyme Microb Technol 41:13–18

Schiffler B, Bureik M, Reinle W, Müller EC, Hannemann F, Bernhardt R (2004) The adrenodoxin-like ferredoxin of *Schizosaccharomyces pombe* mitochondria. J Inorg Biochem 98:1229–1237

Schunck W-H, Mauersberger S, Huth J, Riege P, Müller H-G (1987a) Function and regulation of cytochrome P-450 in alkane-assimilating yeast I. Selective inhibition with carbon monoxide in growing cells. Arch Microbiol 147:240–244

Schunck W-H, Mauersberger S, Kärgel E, Huth J, Müller H-G (1987b) Function and regulation of cytochrome P-450 in alkane-assimilating yeast II. Effect of oxygen-limitation. Arch Microbiol 147:245–248

Schunck W-H, Vogel F, Gross B, Kärgel E, Mauersberger S, Köpke K, Gengnagel C, Müller H-G (1991) Comparison of two cytochromes P-450 from *Candida maltosa*: primary structures, substrate specificities, and effects of their expression in *Saccharomyces cerevisiae* on the proliferation of the endoplasmic reticulum. Eur J Cell Biol 55:336–345

Sedlaczek L (1988) Biotransformation of steroids. Crit Rev Biotechnol 7:187–236

Shiningavamwe A, Obiero G, Albertyn J, Nicaud J-M, Smit M (2006) Heterologous expression of the benzoate para-hydroxylase encoding gene (*CYP53B1*) from *Rhodotorula minuta* by *Yarrowia lipolytica*. Appl Microbiol Biotechnol 72:323–329

Shkumatov VM, Smettan G (1991) Heterologous reconstruction of monooxygenase systems: "Substrate specific cytochrome P-450-microorganism" (Russian). In: Finogenova TV, Sharyshev AA (eds) Alkane metabolism and oversynthesis of metabolites by microorganisms. Center for Biological Research, USSR Academy of Sciences, Pushchino, Russia, pp 78–86

Shkumatov VM, Radyuk VG, Usova EV, Schunck W-H, Mauersberger S (1998) Transgenic strains of microorganisms in the biosynthesis of steroids (Russian). In: Sviridov VV (ed) Chemical problems in the creation of new materials and technologies. Minsk University Press, Minsk, pp 559–566

Shkumatov VM, Usova EV, Poljakov YS, Frolova NS, Radyuk VG, Mauersberger S, Chernogolov AA, Honeck H, Schunck W-H (2002) Biotransformation of steroids by a recombinant yeast strain expressing bovine cytochrome P-45017α. Biochemistry (Moscow) 67:456–467

Shkumatov VM, Usova EV, Radyuk VG, Kashkan ZN, Kovganko NV, Juretzek T, Mauersberger S (2003) Oxidation of 17α,20ß- and 17α,20α-dihydroxypregn-4-en-3-ones, side products of progesterone biotransformation with recombinant microorganisms, expressing cytochrome P-45017α. Russ J Bioorgan Chem 29:581–587

Shkumatov VM, Frolova NS, Rudaya EV, Faletrov YV, Mauersberger S, Barth G (2006) Range of substrates and steroid bioconversion reactions performed by recombinant microorganisms *Saccharomyces cerevisiae* and *Yarrowia lipolytica* expressing cytochrome P450c17. Appl Biochem Microbiol (Moscow) 42:472–478

Shkumatov VM, Falertov YV, Mauersberger S, Barth G, Novikova LA, Luzikov VN (2007a) Recombinant microorganisms for the synthesis of steroids and testing of medicines. EMBO Workshop The chemistry and biochemistry of catalysis by biological systems, 20–22 June 2007, EMBL Hamburg, Abstracts p 20 (invited oral presentation) and p46

Shkumatov VM, Faletrov YV, Mauersberger S, Barth G, Novikova LA, Luzikov VN (2007b) New approaches to the synthesis of steroids and testing of medical compounds. In: XVIIIth Mendeleev congress on general and applied chemistry, 23–28 Sept 2007, Moscow, Russia (invited oral presentation)

Shkumatov VM, Usova EV, Frolova NS, Barth G, Mauersberger S (2007c) Effect of steroid biosynthesis modifiers on the progesterone biotransformation by recombinant yeasts expressing cytochrome P450c17. Biochem (Moscow) Suppl Ser B Biomed Chem 1:87–94

Slijkhuis H, Selten GCM Smaal EB (1989) Process for the biochemical oxidation of steroids and genetically engineered cells used therefor. Patent EP0340878

Slijkhuis H, Smaal EB, Selten GCM (1990–2004) Process for multiple oxidation of steroids and genetically engineered cells used therein. Patent series NL88202080, EP0360361, WO8910963, WO9003428, US20040067579

Storbeck KH, Kolar NW, Stander M, Swart AC, Prevoo D, Swart P (2008) The development of an ultra performance liquid chromatography-coupled atmospheric pressure chemical ionization mass spectrometry assay for seven adrenal steroids. Anal Biochem 372:11–20

Szczebara FM, Chandelier C, Villeret C, Masurel A, Bourot S, Duport C, Blanchard S, Groisillier A, Testet E, Costaglioli P, Cauet G, Degryse E, Balbuena D, Winter J, Achstetter T, Spagnoli R, Pompon D, Dumas B (2003) Total biosynthesis of hydrocortisone from a simple carbon source in yeast. Nat Biotechnol 21:143–149

Taylor FR, Parks LW (1981) An assessment of the specificity of sterol uptake and esterification in *Saccharomyces cerevisiae*. J Biol Chem 256:13048–13054

Theron CW (2007) Heterologous expression of cytochrome P450 monooxygenases by the yeast *Yarrowia lipolytica*. M.S. thesis, University of the Free State, Bloemfontein, South Africa

Thevenieau (2006) Ingenierie metabolique de la levure *Yarrowia lipolytica* pour la production d'acides dicarboxyliques a partier d'huiles vegetales. PhD thesis, INRA, Paris-Grignon

Thevenieau F, Nicaud J-M, Gaillardin C (2009) Chapter 26. Applications of the non-conventional yeast *Yarrowia lipolytica*. In: Satyanarayana T, Kunze G (eds) Yeast biotechnology: diversity and applications. Springer, Dordrecht, pp 589–613. doi:10.1007/978-1-4020-8292-4_26

Thevenieau F, Beopoulos F, Desfougeres T, Sabirova J, Albertin K, Zinjarde S, Nicaud J-M (2010) Uptake and assimilation of hydrophobic substrates by the oleaginous yeast *Yarrowia lipolytica*. In: Timmis KN (ed) Handbook of hydrocarbon and lipid microbiology, vol 2, part 7, chap 48. Springer, Berlin, pp 1513–1527

Trant JM (1996) Functional expression of recombinant spiny dogfish shark (*Squalus acanthias*) cytochrome P450c17 (17α-hydroxylase/$C_{17,20}$-lyase) in yeast (*Pichia pastoris*). Arch Biochem Biophys 326:8–14

Urban P, Truan G, Bellamine A, Laine R, Gautier JC, Pompon D (1994) Engineered yeasts simulating P450-dependent metabolisms: tricks, myths and reality. Review. Drug Metabol Drug Interact 11:169–200

Vadali RV, Horton CE, Rudolph FB, Bennett GN, San KY (2004) Production of isoamyl acetate in ackA-pta and/or ldh mutants of *Escherichia coli* with overexpression of yeast ATF2. Appl Microbiol Biotechnol 63:698–704

Vico P, Cauet G, Rose K, Lathe R, Degryse E (2002) Dehydroepiandrosterone (DHEA) metabolism in *Saccharomyces cerevisiae* expressing mammalian steroid hydroxylase CYP7B: Ayr1p and Fox2p display 17β-hydroxysteroid dehydrogenase activity. Yeast 19:873–886

Wilcox LJ, Balderes DA, Wharton B, Tinkelenberg AH, Rao G, Sturley SL (2002) Transcriptional profiling identifies two members of the ATP-binding cassette transporter superfamily required for sterol uptake in yeast. J Biol Chem 277:32466–32472

Wilson CR, Craft DL, Eirich LD, Eshoo M, Maduri KK, Cornett CA, Brenner AA, Tang M, Loper JC, Gleeson M (2001) Cytochrome P450 monooxygenase and NADPH cytochrome P450

oxidoreductase genes and proteins related to the omega-hydroxylase complex of *Candida tropicalis* and methods relating thereto. Patent US6331420
Wright R (1993) Insights from inducible membranes. Curr Biol 3:870–873
Wu DA, Hu MC, Chung BC (1991) Expression and functional study of wild-type and mutant human cytochrome P450c21 in Saccharomyces cerevisiae. DNA Cell Biol 10:201–209
Yamazaki S, Sato K, Suhara K, Sakaguchi M, Mihara K, Omura T (1993) Importance of the proline-rich region following signal-anchor sequence in the formation of correct conformation of microsomal cytochrome P-450s. J Biochem 114:652–657
Yasumori T (1997) Expression of a human cytochrome P450 in *Schizosaccharomyces pombe*: comparison with the expression in *Saccharomyces cerevisiae*. In: Giga-Hama Y, Kumagai H (eds) Foreign gene expression in fission yeast *Schizosaccharomyces pombe*. Springer, Berlin, pp 111–121
Yasumori T, Chen LS, Li QH, Ueda M, Tsuzuki T, Goldstein JA, Kato R, Yamazoe Y (1999) Human CYP2C-mediated stereoselective phenytoin hydroxylation in Japanese: difference in chiral preference of CYP2C9 and CYP2C19. Biochem Pharmacol 57:1297–1303
Yovkova V (2006) Construction of recombinant *Yarrowia lipolytica* strains for the heterologous expression of the cholesterol-site chain cleaving cytochrome P450 system (German). Diploma Thesis, May 2006, Institute of Microbiology, Dresden University of Technology
Yovkova V, Novikova LA, Luzikov VN, Barth G, Mauersberger S (2007) Expression of the cholesterol side-chain cleavage cytochrome P450scc system in *Yarrowia lipolytica*. In: IV International symposium “intracellular protein catabolism”, VAAM (Association for General and Applied Microbiology) Annual meeting, Osnabrueck, Germany, 1–4 April 2007, Abstracts PB078, pp 103–104, Biospektrum Sonderausgabe Tagungsband 2007
Zimmer T, Schunck W-H (1995) A deviation from the universal genetic code in *Candida maltosa* and consequences for heterologous expression of cytochromes P450 52A4 and 52A5 in *Saccharomyces cerevisiae*. Yeast 11:33–41
Zöllner A, Drăgan CA, Pistorius D, Müller R, Bode HB, Peters FT, Maurer HH, Bureik M (2009) Human CYP4Z1 catalyzes the in-chain hydroxylation of lauric acid and myristic acid. Biol Chem 390:313–317
Zöllner A, Parr MK, Drăgan CA, Dräs S, Schlörer N, Peters FT, Maurer HH, Schänzer W, Bureik M (2010) CYP21-catalyzed production of the long-term urinary metandienone metabolite 17ß-hydroxymethyl-17α-methyl-18-norandrosta-1,4,13-trien-3-one: a contribution to the fight against doping. Biol Chem 391:119–127

Cytochromes P450 of the Alkane-Utilising Yeast *Yarrowia lipolytica*

Stephan Mauersberger

Abstract Since the first studies in 1980s, P450 has been shown to be induced during growth of *Y. lipolytica* on alkanes, fatty alcohols and fatty acids and evidenced to be involved in terminal hydroxylation of alkanes and ω-hydroxylation of fatty acids without differentiation between P450 isoforms. The alkane-hydroxylating P450 exhibit an extraordinary high in vivo activity with turnover numbers up to 3,000/min.

Altogether *Y. lipolytica* contains 17 P450 genes and electron transfer protein genes, encoding for NADPH-P450 reductase, NADH-b_5 reductase and cytochrome b_5. The multiple paralog P450 genes *ALK1* to *ALK12*, classified into the CYP52 family of the P450 supergene family CYP, were detected by gene cloning and deduced from the *Y. lipolytica* genome sequence and represent predominantly alkane- or fatty acid-inducible genes. The multiplicity in the CYP52 family reflects an adaptation to the utilisation of different hydrocarbons and fatty acids and is assumed to be a result of gene duplications and divergent evolution from an ancestral gene. Additionally, P450 genes were deduced, encoding for P450 14DM (*CYP51F*) and P450 22DS (*CYP61A*) being involved in ergosterol biosynthesis, and moreover three putative P450 genes ($P450_1$ to $P450_3$) were detected, which function remains to be elucidated.

The P450 *ALK* genes have obviously diversified in their inducibiltiy and regulation, and in the substrate, chain-length and regioselectivity of encoded P450. However, functional analyses have been performed only for a subset of P450s of *Candida* spp. and *Y. lipolytica*, and the in vivo function of most individual P450ALK remains to be studied.

S. Mauersberger (✉)
Institute of Microbiology, Technische Universität Dresden, Mommsenstrasse 13, 01062 Dresden, Germany
e-mail: Stephan.Mauersberger@tu-dresden.de

G. Barth (ed.), *Yarrowia lipolytica*, Microbiology Monographs 25, 227
DOI 10.1007/978-3-642-38583-4_8, © Springer-Verlag Berlin Heidelberg 2013

Contents

1 Introduction 228
2 Function and Regulation of Alkane-Induced Cytochromes P450 in *Yarrowia lipolytica* 231
 2.1 Induction of Cytochrome P450 in *Yarrowia lipolytica* by Alkanes 231
 2.2 Localisation and Regulation of Alkane-Induced Cytochrome P450 in *Yarrowia lipolytica* 235
3 Presence of Multiple Cytochrome P450 Genes in Fungi 240
 3.1 Multiple Cytochrome P450 *ALK* Genes in Yeasts 240
 3.2 Multiple Cytochrome P450 Genes in Other Fungi 244
4 Functional Analysis of Yeast Cytochrome P450 *ALK* Genes 245
 4.1 Analysis of *Yarrowia lipolytica* Cytochrome P450ALK Function and Regulation. 245
 4.2 Electron Transport Proteins to Cytochrome P450 in Yeasts 252
5 Conclusions 255
References 257

Abbreviations

aa	Amino acids
AHA	Alkane hydroxylase activity
AMOS	Alkane monooxygenase system
C.	*Candida*
ER	Endoplasmic reticulum
FADH	NAD(P)$^+$-dependent fatty alcohol dehydrogenases (*ADH* genes)
FAHA	Fatty acid ω-hydroxylase activity
FALDH	NAD(P)$^+$-dependent fatty aldehyde dehydrogenases (*ALD* genes)
FAO	Fatty alcohol oxidases (*FAO* genes)
P450	Cytochrome P450
S.	*Saccharomyces*
Y.	*Yarrowia*

Throughout this section *n*-alkanes and fatty acids (FA) of defined chain lengths will be referred to as C10 (decane), C12 (dodecane), C14 (tetradecane), hexadecane (C16), dodecanoic or lauric acid (C12FA), palmitic acid (C16FA), etc., in accordance with the number of carbon atoms they contain

1 Introduction

Among alkane-assimilating yeasts *Yarrowia* (*Y.*) *lipolytica* stands out as the organism which has been most thoroughly studied at the biochemical and genetic levels, mainly due to its efficient utilisation of a wide range of hydrophobic substrates (alkanes, triglycerides, fatty acids), metabolic product formation (citric acid, α-ketoglutaric acid, dicarboxylic acids) and the high protein secretion (proteases, lipases) capability (Barth and Gaillardin 1996; Fickers et al. 2005; Thevenieau et al.

2009, 2010; Coelho et al. 2010). The complete genome sequence of *Y. lipolytica* is available since 2004 (Dujon et al. 2004; http://www.genolevures.org/). Because of these characteristic features and qualification as GRAS (Generally Regarded As Save) organism by FDA (US Food and Drug Administration) for citric acid production, *Y. lipolytica* has been a target of intensive studies for biotechnological application (for reviews, see Fickers et al. 2005; Coelho et al. 2010; Thevenieau et al. 2009, 2010; Beopoulos et al. 2010a, b, 2011).

In alkane-assimilating yeasts, like *Candida (C.) maltosa*, *C. tropicalis* and *Y. lipolytica*, several cytochromes P450 (P450 or CYP) encoded by alkane-inducible paralog *ALK* genes are involved in the oxidation of alkanes to the corresponding fatty acids (FA, monoterminal or primary alkane oxidation), and of fatty acids to dicarboxylic acids (DCA, diterminal or ω-oxidation), catalysing the terminal hydroxylation of alkanes (alkane monooxygenase system, AMOS or alkane hydroxylase activity, AHA) and the ω-hydroxylation of fatty acids (fatty acid ω-hydroxylase activity, FAHA), respectively (previous review and reports on *C. maltosa* and *C. tropicalis*, Mauersberger et al. 1996; Zimmer et al. 1998; Craft et al. 2003; Eschenfeldt et al. 2003; and reviews on *Y. lipolytica*, Barth and Gaillardin 1996; Fickers et al. 2005; Coelho et al. 2010; Thevenieau et al. 2009, 2010; Beopoulos et al. 2010a, b, 2011; Takai et al. 2012; Fukuda and Ohta 2013). These multiple paralogs of P450 *ALK* genes are classified into the CYP52 family of the P450 supergene family CYP, comprising thousands of individual members (in 2011 approximately 18,500; Nelson 2009; http://drnelson.utmem.edu/CytochromeP450.html).

The P450 enzymes are distributed widely throughout all organisms, from viruses (Lamb et al. 2009) and archaebacteria to humans. These P450 monooxygenases are heme-thiolate enzymes involved in the oxidative metabolism of great variety of endo- and xenobiotic chemicals. Despite differences in phylogenetic origin, cellular localisation, electron donor partner and substrate specificities of these P450 enzymes, their active sites and catalytic mechanisms appear to be remarkably conserved (Ortiz de Montellano 2005).

After uptake into the cell, the metabolism of alkanes in alkane-assimilating yeasts is initiated by a P450 catalysed terminal hydroxylation (AMOS or AHA), resulting in fatty alcohol (1-alkanol) formation from alkane, the first step of primary alkane oxidation, whereas degradation of imported external fatty acids proceeds mainly after fatty acid activation to acyl-CoA compounds by acyl-CoA synthetase II (ACS II, *FAT2*) in the peroxisomal β-oxidation. Furthermore, the first step of the possible diterminal or ω-oxidation of free fatty acids in yeasts is catalysed by P450-dependent fatty acid ω-hydroxylases (FAHA), resulting in ω-hydroxy fatty acid production from fatty acid (Fickers et al. 2005; Fukuda and Ohta 2013). These class II P450s of yeasts with AHA or FAHA of yeasts are ER resident together with an NADPH-dependent P450 reductase (*CPR* or *NCP1* genes) for the electron transport. The subsequent second step of primary alkane oxidation is performed either by soluble (or ER associated) NAD^+- or $NADP^+$-dependent fatty alcohol dehydrogenases (FADH, *ADH* genes) or by peroxisomal membrane-bound hydrogen peroxide-producing fatty alcohol oxidases (FAO, *FAO* genes), which convert the terminal hydroxy groups of 1-alkanols, 1,ω-diols or ω-hydroxy fatty acids into corresponding fatty aldehydes (Kemp et al. 1990; Mauersberger et al. 1992b; Ilchenko et al. 1994; Fickers et al. 2005). The third step involves the oxidation of the fatty aldehyde to a free fatty acid catalysed by ER and/or peroxisomal NAD (P)$^+$-dependent fatty aldehyde dehydrogenases (FALDH, *ALD* genes). These oxidation steps finally result in the formation of free fatty acids from alkanes or dicarboxylic acids (DCA) from fatty acids, which are activated by ACS II (*FAT2*) and metabolised via β-oxidation in peroxisomes, as indicated above for the main substrate flux during fatty acid utilisation. Additionally, fatty acids of appropriate chain lengths (C14-C18) are activated by ACS I (*FAT1*, *FAA1*) and utilised directly for lipid biosynthesis in ER or lipid particles (Fickers et al. 2005; Thevenieau et al. 2009, 2010; Beopoulos et al. 2010a, b, 2011).

Stimulated by the use of *Y.* (*Candida*) *lipolytic*a and other *Candida* yeasts for the production of single cell protein (SCP) from alkanes in the 1960–1970s and further biotechnological applications, including citric acid production from alkanes, the biochemistry of alkane-utilising yeast developed rapidly in the 1970–1980s (Tanaka and Fukui 1989; Barth and Gaillardin 1996; Mauersberger et al. 1996; Fickers et al. 2005; Finogenova et al. 2005). In this context the alkane-mediated induction of P450 in these alkane-assimilating yeasts was revealed by CO-difference spectra, followed by biochemical investigations on these P450s in the 1970–1980s (Ilchenko et al. 1980; Mauersberger and Matyashova 1980; Mauersberger et al. 1980, 1981; Takagi et al. 1980; Schunck et al. 1987a, b; for reviews see Mauersberger et al. 1996; Barth and Gaillardin 1996; van den Brink et al. 1998; Fickers et al. 2005). Whereas these first studies did not distinguish between different alkane-induced P450 forms (presence of substantially one alkane-induced P450 was assumed), striking evidence for the presence of multiple alkane-induced P450s in one yeast organism accumulated at the beginning of 1990s. This was first demonstrated for *Candida* yeasts, including mainly *C. maltosa*, *C. tropicalis* (Sanglard and Fiechter 1989; Ohkuma et al. 1991a, b, 1995a, 1998; Schunck et al. 1991; Seghezzi et al. 1992; Mauersberger et al. 1992a, 1996) and *C. apicola* (Lottermoser et al. 1996), and later also for *Y. lipolytica* (Iida et al. 1998, 2000) and other yeasts by molecular cloning of P450 encoding *ALK* genes and by genome sequencing.

Recent progress made in the genomic era revealed the presence of huge amount of different P450 forms in fungi counting from only two to three in *Schizosaccharomyces (S.) pombe* and *Saccharomyces (S.) cerevisiae*, to 10–20 in other yeast species, including *Y. lipolytica* and *Candida* spp., and up to approximately 100–200 P450 genes in one fungal organism in the case of different filamentous fungi (*Aspergillus*, *Fusarium*, *Magnaporthe* and *Phanerochaete* spp.). These multiple P450 forms in fungi (hitherto more than 6,000 to 8,000 fungal genes coding for putative P450s are known, depending on the database, FCPD/CFGP or CYPED) are involved in primary (e.g. CYP51, CYP52, CYP56, CYP61) and secondary (e.g. mycotoxins or higher plant hormone, production) metabolism, in detoxification and/or degradation of xenobiotics, in the denitrification process and in fatty acid subterminal hydroxylation (Crešnar and Petrič 2011), and most of them wait to be characterised in detail. The isolation of genes involved in alkane, fatty acid and triglyceride utilisation by reverse and classical genetics and the analysis of complete genome sequences of *Y. lipolytica* and other alkane-assimilating yeasts revealed the presence of several multigene families (e.g. *LIP*, P450 *ALK*, *ADH*, *ALD*, *POX* genes) connected with theses metabolic pathways in yeasts (Fickers et al. 2005; Thevenieau et al. 2009, 2010; Beopoulos et al. 2010a, b, 2011; Petzsch 2012). In this contribution the present knowledge on the P450 forms of the yeast *Y. lipolytica* is summarised, including data from the first detection of alkane-induced P450 in this yeast in the 1970–1980s to the recent analysis of multiple P450 forms obtained due to gene cloning and genome sequencing in the last two decades.

2 Function and Regulation of Alkane-Induced Cytochromes P450 in *Yarrowia lipolytica*

Following the initial studies in the 1970s on *Candida* spp., induction of P450 by long-chain *n*-alkanes has been reported for a large number of yeast species being able to utilise these unconventional substrates, including species of the genera *Candida*, *Pichia*, *Debaryomyces* and *Yarrowia* (*C.*) *lipolytica* (reviewed by Mauersberger et al. 1996; van den Brink et al. 1998; Fickers et al. 2005; Crešnar and Petrič 2011).

2.1 *Induction of Cytochrome P450 in* Yarrowia lipolytica *by Alkanes*

The presence of alkane-inducible P450 in *Y.* (*Candida*) *lipolytica* has been demonstrated since the beginning of 1980s (Ilchenko et al. 1980; Mauersberger and Matyashova 1980; Marchal et al. 1982). Subsequently the function and regulation of alkane-inducible P450 of *Y. lipolytica* was studied in comparison with *C. maltosa* and other alkane-utilising yeasts (Schunck et al. 1987a, b; Mauersberger 1991; Mauersberger et al. 1991, 1992a, b; Mauersberger, Sharyshev and Schunck, unpublished results). However, its P450 systems and further enzymes (FAO, FADH, FALDH) catalysing the first oxidation steps of alkanes to the corresponding fatty acids have not been characterised in detail as of *C. maltosa* and *C. tropicalis* (Schunck et al. 1991; Mauersberger et al. 1992a, b, 1996; Ohkuma et al. 1995a, 1998; Zimmer et al. 1998).

Spectral determination of P450 using the CO-difference spectrum (COD) method can be performed using whole yeast cells. This method has been widely used to demonstrate the P450 formation in yeast cells under the influence of extracellular factors (carbon source and oxygen). Dithionite-reduced

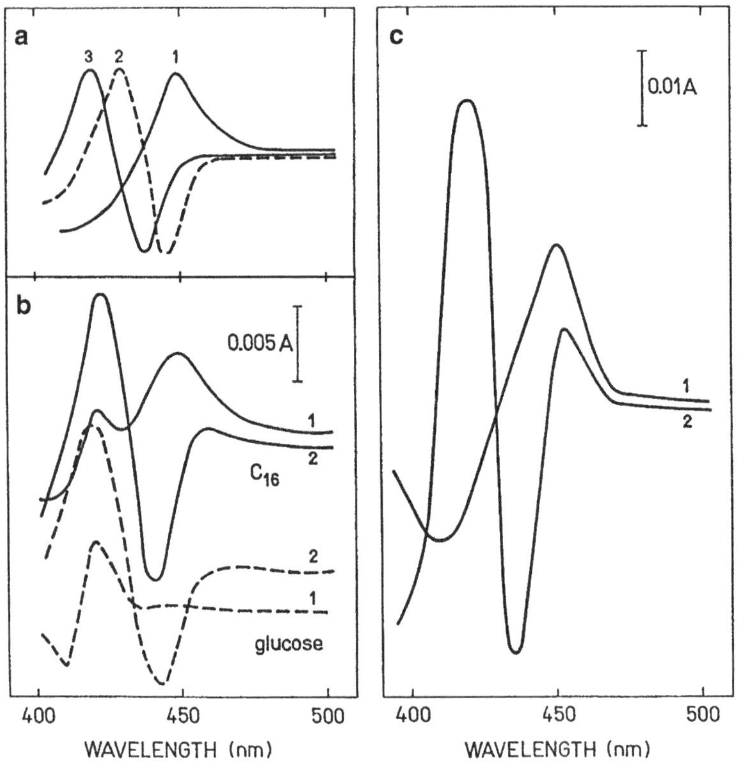

Fig. 1 Determination of cytochrome P450 in *Yarrowia lipolytica* by CO-difference spectra using a modified method masking the cytochrome oxidase. Modification of the CO-difference spectrum (COD) method for *Y. lipolytica* due to its high cytochrome oxidase content for masking cytochrome a_3 by adding 1–3 mM KCN to the assay, thus preventing its interference with P450 in COD (Schunck et al. 1987b; Mauersberger 1991; Mauersberger et al. 1996). (**a**) Spectral interference of P450 (*1*, COD of microsomal membrane fraction of alkane-grown *C. maltosa* EH15), cytochrome oxidase (*2*, COD of crude mitochondrial fraction of glucose-grown *C. maltosa*) and other CO-binding pigments (*3*, COD of cytosolic fraction of glucose-grown *C. maltosa*) during COD determination. (**b**, **c**) P450 determination by COD in whole yeast cells (**b**) of *Y. lipolytica* EH52 grown on hexadecane (*continuous line*) or glucose (*dashed line*) and in crude microsomes from alkane-grown *Y. lipolytica* (**c**). *1*, COD without dithionite addition in presence of 1 mM KCN to mask cytochrome a_3 CO complex, showing P450 reduced endogenously in whole cells (**b**), or by addition of NADPH and NADH to microsomes (**c**); *2*, COD of the same samples reduced with dithionite

CO-difference spectra yield a reliable quantitation of high total P450 content, provided that there is no severe spectral interference with mitochondrial cytochrome a_3 (Fig. 1). Due to high interfering cytochrome oxidase content in *Y. lipolytica* (250–360 pmol $a+a_3$/mg protein, Table 1) compared to *C. maltosa* (50–130 pmol $a+a_3$/mg protein) or *S. cerevisiae*, a modified method was developed to detect low P450 content (Fig. 1, Table 1; Schunck et al. 1987b; Mauersberger 1991; Mauersberger et al. 1996). It involves omitting dithionite and measurement

Table 1 Regulation of the cytochrome P450 content in *Yarrowia lipolytica* and *Candida maltosa* by the carbon sources and by additional oxygen limitation or selective partial CO inhibition of growth

		Yarrowia lipolytica		*Candida maltosa*	
		P450	a + a_3	P450	a + a_3
Growth substrates	Culture conditions	(nmol/g dry weight or 50 % of the pmol/mg protein values)			
Parex (C11-C19)	O_2 saturated	18–25	–	35–45	50–65
	O_2 limited (<5 %, red. air)	–[a]	–	120–155	–
Decane (C10)		35–47*	–	40–52*	–
Dodecane (C12)		55–58*	–	50–55*	–
Tetradecane (C14)	O_2 saturated (65–80 %)	–	–	60–65	–
	O_2 limited (<5 %, N_2)	–	–	130–140	–
Hexadecane (C16)		19–45*	–	25–50*	–
	O_2 saturated (65–80 %)	25–30	125–180	25–35	50–60
Hexadecane	O_2 limited (<5 %, N_2)	60–70	–	100–150	60–65
	CO limited (O_2 65–80 %)	110–120	–	135–160	35–40
Pristane		–	–	12–18[c]	–
1-Hexadecanol	Without pristane	4–5*	–	7–12*	–
	With 1.2 % pristane	–	–	15–20[d]	–
Palmitic acid	Without pristane	2–3*	–	4–13*	–
Glycerol		0*	–	2–4*	–
		–	–	4–8	25–30
Ethanol		0*,[b]	–	0*	–
Acetic acid		0*	–	0*	–
Glucose		0*	150–250*	0/0*	50–60*

Determination of total P450 content by CO-difference spectra (cf. Fig. 1) and of cytochrome a + a_3 content by reduced/oxidised difference spectra with whole cells after growth in minimal medium in shaking flasks (*, with oxygen limitation; P450 probably underestimated with the classical dithionite-reduced COD method) or in bioreactors without and with oxygen limitation (<5 % oxygen saturation by additional nitrogen streaming) or with selective partial CO inhibition of alkane utilisation at 65–80 % oxygen saturation (Mauersberger et al. 1980, 1984) on 1.2–3 % hexadecane or Parex (C11 to C19 alkane mixture) in fermenters (6 l or 4 l working volume, respectively), or in the small-scale fermenter system (Schunck et al. 1987a, b; Mauersberger et al. 1996; cf. Figs. 2 and 3); 0.5–1.2 % of other substrates were used; results (Mauersberger and Matyashova 1980; Ilchenko et al. 1980; Mauersberger et al. 1980, 1984; Schunck et al. 1987b; Mauersberger and Schunck, unpublished results, cf. Fig. 3) were obtained with *Y. lipolytica* H222, EH52 or Y155 (data given as 50 % of determined pmol/mg cell protein values to be comparable with the data given in nmol/g dry weight for Y155) and with *C. maltosa* EH15 (50 % of pmol/mg protein values) or NP4 (nmol/g dry weight)

[a]Not determined

[b]Under the applied growth conditions with 1 % ethanol no P450 was observed, although Ilchenko et al. (2003) described a P450 induction in *Y. lipolytica* strain 704 in the first hour's growth on 2–3 % ethanol

[c]Control cultivation with 1.2 % pristane (2,6,10,14-tetramethyl-pentadecane) in the presence of 0.1 % galactose

[d]Cultivation on hexadecane-1-ol in the small-scale fermenter led to formation of a slightly increased P450 content compared to shaking flasks' cultures without pristane, due to the inducing effect of pristane (Mauersberger et al. 1981) used as inert hydrocarbon phase to dissolve the solid substrate (Schunck et al. 1987a, b)

of endogenously or NAD(P)H-reduced CO complexes of P450 in whole cells or in membrane fractions, respectively, in the presence of potassium cyanide (1–3 mM) or antimycin A (20–60 μg/ml assay) and oxidising the cells in advance by streaming with air, to prevent reduction of the cytochrome oxidase by dithionite, thus masking the a_3-CO-complex formation and its interference with P450 in the CO-difference spectrum. By this modified method alkane-induced P450 was detected in all wild-type, mutant and laboratory strains of *Y. lipolytica* tested in different laboratory (Schunck et al. 1987b; Mauersberger 1991; Mauersberger et al. 1991, 1996; Iida et al. 1998; Takai et al. 2012). The problems in P450 detection due to spectral interference might be the reason for some contradictory results with this yeast by other authors (Peterson 1970; Baroncelli et al. 1979; Takagi et al. 1980). It should be mentioned that by the COD spectrum method, it is not possible to discriminate between different forms of P450 with different substrate specificities or between enzymatic active and inactive but spectral detectable forms of P450s.

This modified COD spectrum method for quantification of P450 in whole yeast cells was successfully applied for the detection of heterologous P450 expression in *Y. lipolytica*, provided that this P450 can be reduced endogenously without addition of potassium dithionite, what is in most cases stimulated and accelerated by the addition of a P450 substrate to the assay (Juretzek 1999; Juretzek et al. 2000; Shkumatov et al. 2002, 2006; Mauersberger et al. 2013).

As in other alkane-utilising yeasts (e.g. *C. maltosa*, *C. guilliermondii*; *C. tropicalis*, *Debaryomyces hansenii*, anamorph *C. famata*, formerly *Torulopsis candida*), P450 was found to be induced during growth of *Y. lipolytica* on *n*-alkanes, like hexadecane (constant total P450 content of 25–35 nmol/g d.w. observed during growth under oxygen saturation conditions in bioreactors), and to a smaller extent (2–5 nmol/g d.w.) also during growth on long-chain fatty alcohols (hexadecane-1-ol, 13–17 % of C16) and fatty acids (palmitic acid, 7–10 %), but not on glucose, glycerol, ethanol or acetate (Ilchenko et al. 1980; Mauersberger and Matyashova 1980; Marchal et al. 1982; Table 1; Figs. 1 and 4). The P450 content in these cells was determined as total content by the COD spectrum method without discrimination between P450 isoforms. Interestingly, Ilchenko et al. (2003) reported later the induction of significant amounts of P450 in the first 3- to 10-h growth of wild-type *Y. lipolytica* strain 704 on 1–5 % ethanol (slow or no growth on 3–5 % ethanol) and discussed the probable P450 participation in ethanol oxidation as the microsomal ethanol oxidation system (MEOS), comparable to mammalian P450 CYP2E.

Interestingly, significantly increased P450 content (up to 130–140 pmol/mg protein, compared to 68–77 pmol/mg protein of the wild-type *Y. lipolytica* strain) was also achieved after 5-h alkane-mediated P450 induction by decane (C10) or hexadecane (C16) in several Alk mutants of *Y. lipolytica* and *C. maltosa* exhibiting the alkane chain-length-dependent AlkAb ($C10^-$ $C16^+$), AlkAc ($C10^+$ $C16^-$, all mutants were C12-1-ol^+) and AlkD phenotypes (Mauersberger 1991; Alk-phenotype classification see Mauersberger et al. 2001; Mauersberger and Nicaud 2003; Thevenieau et al. 2007). Additionally, the induction of P450 (formation of spectral detectable P450) in *Y. lipolytica* and *C. maltosa* by hexadecane

Table 2 Different degree of glucose repression on cytochrome P450 induction by decane (C10) and hexadecane (C16) in the yeasts *Yarrowia lipolytica* and *Candida maltosa*

Yeast strain	Presence of 2 % glucose	Relative P450 induction (%) by the alkanes (*pmol/mg cell protein*)		
		C10	C16	Without alkane
Candida maltosa EH15	–	100 (*104*)	100 (*120*)	3
	+	52	13	0
Yarrowia lipolytica H222	–	100 (*68*)	100 (*77*)	3
	+	74	25	0

P450 determination by CO-difference spectra with whole cells (cf. Fig. 1) after 5-h incubation of glucose-grown cells (0.5 g wet weight in 100 ml minimal medium) with 0.5 % alkane in shaking flasks with or without 2 % glucose present. Determined specific P450 content is given in parentheses in italics. Results taken from Mauersberger (1991)

(C16) was more sensitive to glucose repression than induction by decane (C10) as shown in Table 2 (Mauersberger 1991). These results were considered as hints for the presence of multiple, differently regulated alkane-induced P450 forms in these yeasts (cf. Sect. 3). Furthermore, glycerol significantly repressed the P450 induction in *Y. lipolytica* (Iida et al. 2000). In contrast, glycerol and galactose are no or less repressive substrates for alkane-mediated P450 induction in *C. maltosa* (Mauersberger et al. 1981, 1996).

2.2 *Localisation and Regulation of Alkane-Induced Cytochrome P450 in* Yarrowia lipolytica

Localisation of Cytochrome P450. During subcellular fractionation of alkane-grown *Y. lipolytica* and *C. maltosa* cells, P450 and NADPH-P450 reductase were enriched in the microsomal fraction (Table 3; Delaissé et al. 1981; Marchal et al. 1982; Mauersberger et al. 1987). The P450-dependent fatty acid (lauric acid) ω-hydroxylase system (C12FAHA) was shown to be most probably localised in the endoplasmic reticulum (ER) of *Y. lipolytica* (Delaissé et al. 1981), as clearly demonstrated immunocytochemically for alkane-grown *C. maltosa* (Vogel et al. 1992). Microsomal membrane fractions of *Y. lipolytica* catalysed in vitro the NADPH-dependent and CO-sensitive terminal hydroxylation of alkanes (AHA for C12 and C16 alkanes) and the ω-hydroxylation of lauric acid (C12FAHA), with 2–4 and 3–7 nmol product/nmol P450 × min (Table 3), respectively, results being comparable to data for *C. maltosa* (Mauersberger et al. 1987, 1992a; Blasig et al. 1988; Mauersberger and Schunck, unpublished results).

Hexadecane-1-ol, palmitic acid and probably hexadecanal, intermediates of monoterminal *n*-alkane oxidation, were identified as the main products of NADPH-dependent in vitro [1-^{14}C]-hexadecane oxidation by crude microsomal membrane fractions from alkane-grown *Y. lipolytica* or *C. maltosa* cells using TLC, GC or GC-MS. Additionally, intermediates of diterminal *n*-alkane oxidation

Table 3 Cytochrome P450 content, NADPH-cytochrome P450 reductase, alkane hydroxylase (AHA) and fatty acid ω-hydroxylase (FAHA) activities in microsomal membrane fractions of alkane-grown *Yarrowia lipolytica* and *Candida maltosa*

Yeast species	P450 (nmol/mg protein)	NADPH-P450 reductase (U/mg protein)	AHA (C16) (nmol/nmol P450 × min)	FAHA (C12FA) (nmol/nmol P450 × min)
Yarrowia lipolytica	0.1–0.3	0.3–0.6	2–4	3–7
Candida maltosa	0.3–0.5	0.4–0.7	4–8	2–6

Crude microsomal membrane fractions prepared after mechanical disruption of hexadecane-grown cells of *Y. lipolytica* H222 or EH52 and *C. maltosa* EH15 or VSB779; P450 content determined by CO-difference spectra (cf. Fig. 1); hydroxylase enzyme assays with [1-^{14}C]-hexadecane (C16) for AHA and [1-^{14}C]-lauric acid (C12FA) for FAHA, NADPH-P450 reductase activity measured as NADPH-cytochrome c reductase (Mauersberger et al. 1984, 1987; Schunck et al. 1991)

(16-hydroxy-hexadecanoic acid, ω-HFA16, and hexadecanedioic acid, DCA16) were also detected in these assays (according to Mauersberger et al. 1987; Schunck et al. 1991), although they represented only about 5–10 % of the total oxidation products. These microsomal fractions of *Y. lipolytica* and *C. maltosa* catalysed also the NADPH/O_2-dependent oxidation of [1-^{14}C]-lauric acid or [1-^{14}C]-palmitic acid. Using TLC and GC, the ω-hydroxy fatty acid (ω-HFA) and 1,ω-dicarboxylic acid (DCA) were identified as main oxidation products, and ω-aldehyde of fatty acid (ω-AFA) was detected as minor product (Mauersberger and Schunck, unpublished results; Blasig et al. 1988 for *C. maltosa*). Comparable results were shown by Delaissé et al. (1981) for FAHA (C12FA) and Marchal et al. (1982) for AHA (C16) and FAHA (C12FA to C18FA) in hexadecane-grown *Saccharomycopsis (Y.) lipolytica.* Due to diminished occurrence of oxidation products by reason of inhibition with CO, it was concluded that microsomal (probably ER resident) P450-containing enzyme systems catalyse the ω-hydroxylation of alkanes and fatty acids in these yeasts. The presence of different alkane-induced P450s in the ER-derived membrane fractions was later demonstrated by purification of at least two different P450 forms from alkane-induced *C. maltosa* cells (Mauersberger et al. 1992a, 1996) and clearly evidenced by first P450 gene cloning from *C. maltosa and C. tropicalis* (Schunck et al. 1989, 1991; Sanglard and Fiechter 1989; Ohkuma et al. 1991a, b; Seghezzi et al. 1992) and later from *Y. lipolytica* (Iida et al. 1998, 2000) and will be given in detail in Sect. 3.

Selective Inhibition of Cytochrome P450 by Carbon Monoxide. The in vivo function of alkane-induced P450 in *Y. lipolytica* catalysing alkane hydroxylation was demonstrated by selective strong inhibition of alkane (hexadecane) utilisation by low CO concentrations (pO_2 85–90 %) in the culture medium, under conditions where the utilisation of hexadecane-1-ol, palmitic acid or glucose continued unaffected (Figs. 2 and 3). In particular, the uninhibited utilisation of hexadecane-1-ol (the direct product of P450-catalysed alkane hydroxylation) in contrast to complete blocking of hexadecane degradation clearly showed the high selectivity of CO action, being directed to the P450 enzyme system catalysing the first step of alkane degradation (Fig. 2). In contrast to the high CO sensitivity of hexadecane utilisation

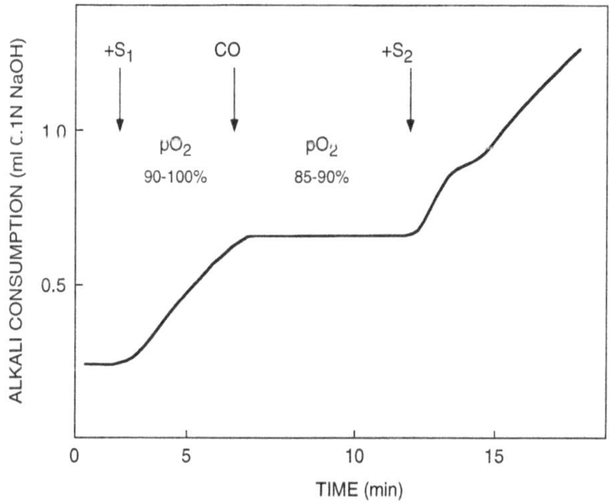

Fig. 2 Selective inhibition of the cytochrome P450-catalysed first step of alkane degradation in *Yarrowia lipolytica* through low carbon monoxide concentrations. Cultivation in 100 ml minimal medium in a small-scale fermenter system (Schunck et al. 1987a, b; Mauersberger et al. 1996); alkali consumption for pH titration, to compensate growth-related proton extrusion, is measured as growth parameter; selective inhibition of first step of hexadecane utilisation (S_1, 35 mg) in presence of 1.2 % pristane by alkane-grown *Y. lipolytica* EH52 with low carbon monoxide (*CO*) concentration in the medium, whereas utilisation of hexadecane-1-ol (S_2, 20 mg) added after complete CO inhibition of P450-catalysed alkane hydroxylation continued uninhibited. Inhibition of fatty alcohol, fatty acid or glucose utilisation was found at significantly higher CO concentrations, obviously due to CO inhibition of cytochrome oxidase (cf. Fig. 3; Schunck et al. 1987a)

(50 % of inhibition by 40–50 μM CO at pO_2 93–95 %) for the utilisation of hexadecane-1-ol, palmitic acid or glucose, a much lower sensitivity to CO streaming (50 % inhibition at 625 μM CO at pO_2 of 25 %) was observed (Fig. 3, cf. Schunck et al. 1987a). Obviously, the other hemoproteins, like cytochrome oxidase binding CO in competition to oxygen, exhibit much lower affinities towards carbon monoxide. Concerning the CO sensitivity of substrate utilisation, essentially the same results were obtained with several other alkane-assimilating yeast strains, including *C. maltosa*, *Debaryomyces hansenii* (anamorph *C. famata*, formerly *Torulopsis candida*), *Debaryomyces formicarius* (*D. vanrijiae* or *Schwanniomyces vanrijiae*) and *Pichia guilliermondii* (Schunck et al. 1987a; Sharyshev and Mauersberger, unpublished results).

In particular, under oxygen limitation conditions, P450-catalysed alkane hydroxylation appears to represent the rate-limiting step of the whole pathway (Schunck et al. 1987a; Fickers et al. 2005; Fig. 3). The relatively low oxygen affinity of P450 is indicated by the finding that the apparent Ko_2-values of substrate assimilation fall in the order hexadecane (11–13 μM) > hexadecane-1-ol (7–8 μM) > glucose (1–2 μM), probably reflecting the different affinities of oxygen-activating enzymes involved in alkane assimilation, e.g. P450, long-chain alcohol oxidase (FAO),

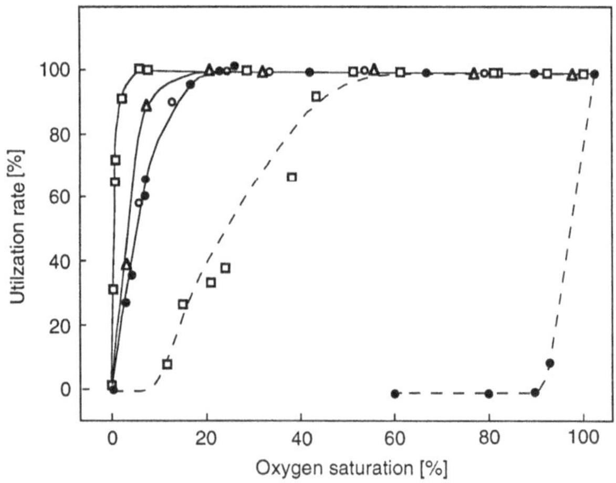

Fig. 3 Utilisation rate of different substrates by *Yarrowia lipolytica* in dependence on the pO_2 of the culture medium and CO sensitivity of alkane and glucose utilisation. Short-time cultivation of *Y. lipolytica* EH52 in a small-scale fermenter system (Schunck et al. 1987a; Mauersberger et al. 1996; cf. Fig. 2) using hexadecane (*open circles*), hexadecane (*filled circles*) and hexadecane-1-ol (*open triangles*), the last both in presence of 1.2 % of pristane, as substrates for cells pre-grown on alkane, or glucose (*open squares*) for glucose-grown cells. Oxygen saturation was varied stepwise by additional streaming with nitrogen, and the substrate utilisation rate was determined as alkali consumption (0.1 N NaOH) for pH titration. Additionally, the CO sensitivity of substrate utilisation (*dashed line*) is shown for hexadecane (*filled circles*) and glucose (*open squares*). Different fixed pO_2-values were adjusted by additional streaming of the culture medium with carbon monoxide. Comparable to glucose results on CO sensitivity were obtained with hexadecane-1-ol and palmitic acid

acyl-CoA oxidases (ACO) and cytochrome oxidase. In fact, these significant differences were found in the pO_2 dependence of utilisation of hexadecane, hexadecane-1-ol, palmitic acid and glucose by *Y. lipolytica*, *C. maltosa*, *Debaryomyces formicarius* and other yeasts, and therefore, the P450 system is distinguished by the lowest oxygen affinity among the O_2-activating enzymes involved in alkane assimilation (Fig. 3; Schunck et al. 1987a, Sharyshev and Mauersberger, unpublished results). Interestingly, the yeasts *Y. lipolytica* and *C. maltosa* exhibited different apparent K_{O_2}-values of hexadecane utilisation (11–13 μM and 25–26 μM, respectively), whereas the apparent K_{O_2}-values for utilisation of hexadecane-1-ol (7–8 μM and 10–11 μM) and glucose (1–2 μM and 2–4 μM) were only slightly different (Fig. 3; cf. Schunck et al. 1987a; Mauersberger, unpublished results).

From the growth rate on alkanes, the substrate utilisation rates with hexadecane in bioreactors under oxygen substrate saturation conditions (Fig. 2) and the total P450 content of these yeast cells, the in vivo substrate turnover numbers for alkane-hydroxylating P450 in *Y. lipolytica* and *C. maltosa* were estimated to be in the range of 1–3 μmol hexadecane hydroxylated per min and nmol P450 (Schunck et al. 1987a; Blasig et al. 1988; Mauersberger et al. 1996). Thus, in alkane-utilising yeasts, like *Y. lipolytica*, *Candida* spec. and others, the alkane-induced P450s are

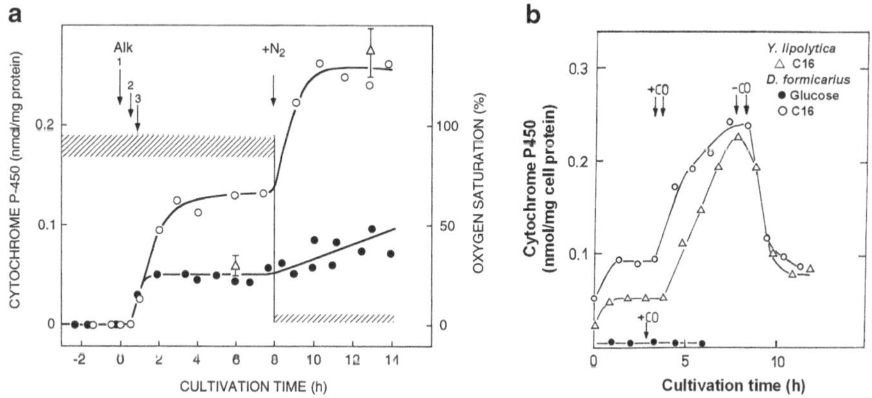

Fig. 4 Regulation of alkane-induced cytochrome P450 content in *Yarrowia lipolytica, Candida maltosa* and *Debaryomyces formicarius* by substrate induction, oxygen limitation and partial and selective CO inhibition during cultivation on alkanes. (**a**) Induction of P450 after carbon source changing from glucose to *n*-alkanes during cultivation of *C. maltosa* EH15 (C14: *open circles* or C16: *open triangles*) or of *Y. lipolytica* EH52 (C16: *filled circles*). Cultivation in a small-scale fermenter system (100 ml working volume; Schunck et al. 1987a, b; Mauersberger et al. 1996); shaking flasks preculture on 1 % glucose, 3-h fermenter cultivation on 0.2–0.5 % glucose; *1*, addition of 0.5 % hexadecane or tetradecane (C16 or C14, *Alk*, at 0 h, 0.2 % residual glucose); *2*, glucose completely consumed (0.5 h); *3*, after a short lag-phase alkane utilisation started and first P450 was detectable in COD (1 h; cf. Fig. 1); alkane concentration kept 0.25–0.5 % due to alkali consumption for pH titration; alkane chain-length-dependent constant P450 levels found at 80–90 % of oxygen saturation of the culture medium (/////). Additionally, increase of P450 content during oxygen limitation (3–5 % oxygen saturation, by nitrogen N_2 streaming, growth rate of 5–10 %) is shown. P450 content increased in *C. maltosa* faster than in *Y. lipolytica* (3- to 4-fold increased P450 content obtained after longer or by stronger oxygen limitation or by partial CO limitation; cf. Table 1 and **b**; Schunck et al. 1987a, b). (**b**) Enhanced P450 formation in alkane-growing *Y. lipolytica* EH52 and *D. formicarius* Y1555 caused by partial and selective CO inhibition of P450-catalysed alkane hydroxylation; constant P450 content reached during cultivation on 0.5 % C16 at pO_2 of 80–90 %; inhibition of growth rate to about 10–15 % by additional streaming with carbon monoxide (*CO*) at pO_2 of 65–75 % increased the P450 content 3- to 5-fold (effect was reversible). Under the same conditions, the small P450 content in glucose-grown *D. formicarius* was not influenced (Mauersberger and Sharyshev, unpublished results)

obviously distinguished by an extraordinarily high in vivo activity, which is significantly higher than the P450 activities determined in vitro using microsomal membrane fractions (2–8 nmol product/min × nmol P450, Table 3).

Influence of Oxygen on Alkane-Mediated Cytochrome P450 Expression. The cellular P450 content of *Y. lipolytica* was increased by transition to oxygen-limited growth during cultivation on *n*-alkanes as studied in detail with *C. maltosa, C. tropicalis* and *Debaryomyces formicarius* (for ref., see Schunck et al. 1987b; Mauersberger et al. 1996). Low oxygen levels (3–5 % of saturation) during cultivation on alkanes resulted in a moderately increased P450 content in *Y. lipolytica*, compared with the strong oxygen-mediated effect observed in *C. maltosa* (Fig. 4a; Schunck et al. 1987b), what was obviously connected with the above-mentioned different apparent K_{O_2}-values of alkane utilisation in the two yeast species (Fig. 3; Schunck et al. 1987a). However, prolonged strongly oxygen-limited growth of *Y. lipolytica* on hexadecane in bioreactors (reduced air supply from 3 to 0.3 vvm, apparent pO_2 <1 %) resulted in a P450-content increase from 50–60 to 120–140 pmol/mg cell protein (Table 1). Higher P450 content (220–240 pmol/mg protein) was reached when the growth rate on alkane was partially inhibited down to 10–15 % by streaming the culture medium with CO at 65–75 % oxygen saturation (Fig. 4b, Table 1). This CO effect was completely reversible, and comparable results were obtained with *Debaryomyces formicarius* (Fig. 4b, Sharyshev and Mauersberger, unpublished results) and *C. maltosa* (Table 1, cf. Schunck et al. 1987b). The enhanced P450 formation appears to depend on a decrease in the enzymatic activity of the alkane-hydroxylating P450 system (O_2-deficiency or CO inhibition) and the presence of alkane, which act both as substrate and as inducer of P450. The P450 formation is therefore mainly regulated by the intracellular inducer concentration which depends on the relative rates of alkane uptake and the actual P450-catalysed alkane-hydroxylating activity (AHA) as proposed by Schunck et al. (1987a, b; cf. Mauersberger et al. 1996).

3 Presence of Multiple Cytochrome P450 Genes in Fungi

First evidence on the presence of different alkane-inducible P450 genes in the yeasts accumulated in the late 1980s by molecular cloning and heterologous expression in *S. cerevisiae* of the first two P450s from both *C. maltosa* and *C. tropicalis* (Schunck et al. 1989, 1991; Sanglard and Fiechter 1989; Ohkuma et al. 1991a, b) and by purification of different P450s from one *C. maltosa* strain (P450Cm1 and Cm3; Mauersberger et al. 1992a, 1996).

3.1 Multiple Cytochrome P450 ALK *Genes in Yeasts*

The presence of multiple alkane- or fatty acid-inducible P450ALK forms in *C. maltosa* and *C. tropicalis* was finally shown by cloning of up to eight P450 *ALK* genes (excluding allelic variants), all belonging to the *CYP52* gene family (Seghezzi et al. 1992; Ohkuma et al. 1995a, 1998; Zimmer et al. 1996, 1998; Craft et al. 2003; Eschenfeldt et al. 2003). The yeast *Y. lipolytica* contains at least 17 P450 genes and several related electron transfer proteins (Table 4). Eight probably alkane- or fatty acid-inducible P450 isoforms encoding *ALK* genes (*ALK1* to *ALK8*, classified as *CYP52F1* to *CYP52F8* in the *CYP52F* gene subfamily) and one single *CPR* gene coding for the NADPH-P450 reductase were identified in *Y. lipolytica* strain CX161-1B (Iida et al. 1998, 2000; Table 4). Exploration of the genome sequence of strain E150 (CLIB122) revealed 12 P450 *ALK* genes including *ALK1* to *ALK8* (Fickers et al. 2005; Thevenieau 2006; Thevenieau et al. 2009, 2010).

Table 4 The cytochrome P450 genes of *Yarrowia lipolytica*, their putative functions and related genes encoding putative electron transfer proteins to cytochrome P450

Yarrowia lipolytica (*YALI*) gene name	Trivial gene names (% protein homology: identity/similarity)[a]	P450 subfamily or putative gene names[e]	Assumed alkane hydroxylase (AHA) specificity		Fatty acid ω-hydroxylase (FAHA)		Induction on alkanes	Effect on long-chain DCA production[g]
			C10	C16[f]	C12FA	LCFA[f,g]	C10/C16[h]	
YALI0E25982g	*YlALK1* (71/77 %)	*CYP52F1*	++	+/–	–	[–]	100/110	ns
YALI0B06248g	*YlALK9*	*CYP52F10*	n.d.	n.d.	n.d.	[–]	6/11	ns
YALI0F01320g	*YlALK2* (69/78 %)	*CYP52F2*	+/–	++/[+]	–	[+]	22/33	+
YALI0B20702g	*YlALK10*	*CYP52F11*	n.d.	n.d.	n.d.	[+]	0.03/0.04	+
YALI0E23474g	*YlALK3* (64/78 %)	*CYP52F3*	n.d.	n.d.	+	[–]	1.7/3.7	–
YALI0A20130g	*YlALK12*	*CYP52F9*	n.d.	n.d.	n.d.	n.d.	0.09/0.10	n.d.
YALI0B13816g	*YlALK4*	*CYP52F4*	–	+	–	n.d.	11/14	n.d.
YALI0B13838g	*YlALK5* (78/90 %)	*CYP52F5*	[+]	n.d.	+/[+]	[+]	0.4/1.6	–
YALI0A15488g	*YlALK7*	*CYP52F7*	n.d.	n.d.	+	[+]	0.01/0.02	–
YALI0B01848g	*YlALK6*	*CYP52F6*	–	+	–	n.d	14/12	n.d.
YALI0C12122g	*YlALK8*	*CYP52F8*	n.d.	n.d.	n.d.	n.d.	0.02/0.01	n.d.
YALI0C10054g	*YlALK11*	*CYP52S1*	[+]	n.d.	[+]	[–]	1.8/2.5	ns
			Protein size		% Identity to related proteins			
YALI0B05126g	P450 14DM, *YlERG11*	*CYP51F1*	523 aa		63 % to ScCYP51A1, 530 aa			
YALI0A18062g	P450 22DS, *YlERG5*	*CYP61A1*	507 aa		64 % to ScCYP61A1, 538 aa			
YALI0B21824g	$P450_1$[b]	*CYP548P1*	565 aa		41 % to NcCYP548A1, 539 aa			
YALI0E14509g	$P450_2$[c] (*YlDIT2*-like)	*CYP5223A1* (not *CYP56E*)	465 aa		36 % to CaCYP5217A1, 519 aa; 36 % CdDIT2, 31 % DhDIT2			
YALI0F03663g	$P450_3$[d] (*PAHA*)	*CYP504A17*	530 aa		49 % to NhCYP504A6, 515 aa			
YALI0D04422g	NADPH-P450 reductase	*YlCPR* (*NCP*)	722 aa		42 % ScCPR 691 aa, 49 % CmCPR			
YALI0D12122g	Cytochrome b_5	*YlCYB5*	110 aa		47 % CtCYB5 129aa, 38 % ScCYB5			
YALI0E24079g	Cytochrome b_5-like	*YlCYB5*	126 aa		45 % CtCYB5 128aa, 41 % CaCYB5			
YALI0A20394g	Cytochrome b_5-like	*YlCYB5*	149 aa		48 % CtCYB5 161aa, 46 % CaCYB5			
YALI0D11330g	NADH-b_5 reductase	*YlMCR1*	291 aa		55 % ScMCR1 302 aa, 58 % Ca 301 aa			
YALI0D04983g	NADH-b_5 reductase	*YlCBR1*	290 aa		47 % ScCBR1 284 aa, 57 % Ca 294 aa			
YALI0B14839g	*ARH1* (AdR-like)	*YlARH1*	464 aa		36 % ScARH1 493 aa, 40 % Sp 469 aa			

(continued)

Table 4 (continued)

Yarrowia lipolytica (*YALI*) gene name	Trivial gene names (% protein homology: identity/similarity)[a]	P450 subfamily or putative gene names[e]	Assumed alkane hydroxylase (AHA) specificity		Fatty acid ω-hydroxylase (FAHA)		Induction on alkanes C10/C16[h]	Effect on long-chain DCA production[g]
			C10	C16[f]	C12FA	LCFA[f,g]		
YALI0B02222g	*YAH1* (Adx-like)	*YlYAH1*	161 aa		66 % ScYAH1 172 aa (SpEtp1 631aa)			
YALI0F24651g	*COX15*	*YlCOX15*	515 aa		51 % ScCOX15 486 aa (SpEtp1 631aa)			

Abbreviations: P450 14DM, lanosterol 14α-demethylase, with homology to the *ScERG11* (*CYP51A1*), transforms lanosterol into 4,4′-dimethyl cholesta-8,14,24-triene-3β-ol; P450 22DS with homology to the *ScERG5* (*CYP61A1*), C22 sterol desaturase, step 4/5 from zymosterol to ergosterol (ergosta-5,7-dienol to ergosta-5,7,22,24-tetraene-3β-ol). Further explanations are given in the text

[a]Phylogenetic analysis of the deduced P450ALK protein sequences (513–583 aa) revealed that P450ALK1/ALK9, ALK2/ALK10, ALK3/ALK12, ALK5/ALK7 (64–78 % identity/77–90 % similarity) are pairwise closer related, respectively; the protein homology values of these pairs are shown in parenthesis; pair ALK1/ALK9 is closer related to pair ALK2/ALK10 (55–63/68–74 %), and pair ALK3/ALK12 is closer related to pair ALK5/ALK7 (51–52/67–68 %); other P450ALK show lower homology (37–50/57–69 %); *ALK4* and *ALK5* are tandemly arranged on chromosome B (2 kb distance) and exhibit a moderate sequence homology (55/71 %)

[b]$P450_1$ designated as CYP548P1, with best homology to *Neurospora crassa* benzoate-4-monooxygenase encoded by *CYP548A1* (NCU06895), according to Nelson's Cytochrome P450 homepage (http://drnelson.utmem.edu/cytochromeP450.html)

[c]$P450_2$ classified as CYP5223A1 with best homology to *C. albicans* CYP5217A1, highly similar to CAX42515.1 of *C. dubliniensis*, encoding, additionally to *CYP56E3* (CdDIT2, 477 aa), a second putative *N*-formyltyrosine oxidase (sporulation-specific microsomal P450s involved in the production of *N,N*-bisformyl dityrosine required for spore wall maturation, like P450 ScDIT2 *CYP56A1*), with homology to DIT2-like P450s of *D. hansenii* (DEHA2C00814g) and other yeasts

[d]$P450_3$ with best homology to *Nectria haematococca* CYP504A6 and *Aspergillus fumigatus* CYP504A7 with phenylacetate 2-hydroxylase activity (*PAHA* genes)

[e]*CYP52* subfamily numbers for P450 *ALK* genes (*CYP52F1* to *F11*, *CYP52S1*) are according to Nelson's Cytochrome P450 homepage

[f]Results on assumed or evidenced substrate specificities of *Y. lipolytica* P450s encoded by *ALK* genes were taken from Iida et al. (1998, 2000) and Takai et al. (2012) for *ALK* genes disruption results (see the text), from Hanley et al. (2003) for lauric acid (C12FA) ω-hydroxylase activity (FAHA) in vitro testing (both given as + or −) and from Thevenieau (2006) and Thevenieau et al. (2009, 2010) for the substrate preference of overexpressed P450ALK2, ALK5 or ALK11 during DCA production from C10/C16 alkane mixtures (given as [+], although in this case it was not distinguished between AHA and FAHA); n.d., not determined

[g]The effect of individual P450ALK overexpression on long-chain DCA production from sunflower oil (mixture of long-chain fatty acids, LCFA: C16:0, C18:0, C18:1, C18:2) was given according to Thevenieau (2006) and Thevenieau et al. (2010): ns, non significant; +, increase; −, decrease; n.d., not determined. The probable chain-length specificities for P450ALK with FAHA are proposed as [+] or [−] for LCFA (C16FA to C18FA) in addition to results obtained for FAHA with C12FA (Hanley et al. 2003)

[h]P450 induction by the alkanes C10 or C16 is given as 10^4 mRNA copies/ng total RNA determined by qRT-PCR (Hirakawa et al. 2009), and these results are in general in agreement (except slight differences in previous studies) with previous Northern blot results using C10 and C14 (Iida et al. 1998, 2000) for *ALK1* to *ALK8*, with C10 and C16 (Yamagami et al. 2004; Endoh-Yamagami et al. 2007; Kobayashi et al. 2008) for *ALK1*, and by testing the *ALK1* promoter activity using pALK1-*lacZ* with C10, C12, 1-dodecanol, dodecanal, lauric acid (C12FA), glucose or glycerol as substrates (Sumita et al. 2002a)

Three of these additionally detected CYP52F P450s, ALK9 (CYP52F10), ALK10 (52F11) and ALK12 (52F9) presented high protein sequence homology (64–78 %) to ALK1-ALK3, respectively (Table 4; Nelson's Cytochrome P450 homepage: http://drnelson.utmem.edu/cytochromeP450.html), whereas P450ALK11 with 44 % homology to ALK1 was classified as *CYP52S1* gene product, being more related to P450 52B and 52C of *C. tropicalis* and *C. maltosa*. Additionally to these 12 P450 *ALK*, genes for P450 14DM (subfamily *CYP51F, ERG11* related) and P450 22DS (subfamily *CYP61A*, *ERG5* related) involved in ergosterol biosynthesis, and moreover three putative P450 genes ($P450_1$ to $P450_3$ in Table 4, classified by sequence homology as *CYP548P1*, *CYP5223A1*, probably a *DIT2*-like P450, but different from *CYP56E*, and *CYP504A17*), were detected. The metabolic function of most of these classified by their sequence homology P450 genes remains to be elucidated experimentally (Table 4). Up till now no mitochondrial P450 forms were detected in yeasts.

Till now 3 to 12 P450 *ALK* genes (per haploid genome, not counting the allelic variants present in diploid yeasts) classified into the *CYP52* family were shown to be present in one yeast species, like *C. maltosa* (8), *C. tropicalis* (strain dependent 6–8), *C. albicans* (at least 5–7, but totally up to 19 P450 genes including allelic variants; Lah et al. 2008), *C. dubliniensis* (5), *C.* (*Clavispora*) *lusitaniae* (3), *C.* (*parakrusei*) *parapsilosis* (9), *C. (Pichia) guilliermondii* (4), *Debaryomyces* (*Torulaspora*) *hansenii* (5), *Lodderomyces elongisporus* (6), *Pichia stipitis* (5) and *Y. lipolytica* (12), for which the *CYP52* genes were almost completely discovered in particular due to genome sequencing. When detected allelic variants are included, up to 10–15 *CYP52* genes were found in different strains of *C. tropicalis* and *C. maltosa*. In *C.* (*Starmerella*) *bombicola* (3), *C. apicola* (2) and in other yeasts (*Pichia farinosa,* Iida et al. 2000; *C. digboiensis*, Sood and Lal 2008), less P450 *CYP52* genes were described as examples. Additionally, almost all fungi contain the housekeeping P450 genes *CYP51F*, *CYP61A* (both involved in ergosterol biosynthesis) and in some cases also *CYP56* or a related P450 gene (P450 DIT2, involved in spore wall formation). These two or three P450 are present in almost all yeasts, including all *Saccharomyces* spec. (only 3 P450), *C. glabrata* (3), *Eremothecium (Ashbya) gossypii* (3), *Kluyveromyces* spec. (3–5) and *Schizosaccharomyces* spec. (*S. pombe*, *S. japonicus*, *S. octosporus*; only two P450, no *CYP56*). Recent genome sequencing revealed the presence of one to four additional P450 genes (families *CYP53B*, *CYP501*, *CYP504*, *CYP537*; *CYP548*, *CYP557*, *CYP5217*, *CYP5139*, *CYP5215*, *CYP5216*, *CYP5223* and *CYP5251*) in different yeasts, including *Y. lipolytica* (Table 4; Nelson's Cytochrome P450 homepage: http://drnelson.utmem.edu/cytochromeP450.html and Fungal cytochrome P450 database: http://p450.riceblast.snu.ac.kr; Génolevures database: http://www.genolevures.org/). The function of these P450 genes is, if at all, only partially studied. Thus, among yeasts *Y. lipolytica* (17 P450 genes: 12 P450 *ALK* of *CYP52* family and 5 other *CYP* genes; Table 4) contains together with *C. albicans* (19, including allelic variants in this case) the highest number, followed by *C. parapsilosis* (14) and most other alkane-utilising yeasts (8–11, excluding the allelic variants present). Otherwise, the alkane non-utilising yeasts (no *CYP52* genes present) contain only two P450 genes for genus *Schizosaccharomyces*; three for *Saccharomyces* spec., *Kluyveromyces* spec., *Eremothecium gossypii* and *C. glabrata*; five for *Kluyveromyces lactis*; and, in case of the basidiomycetous yeasts *Sporobolomyces roseus* and *Cryptococcus neoformans*, from six to eight P450 genes, respectively.

The significant expansion of protein families containing paralogs of genes involved in hydrophobic substrate utilisation is a striking feature revealed by genome sequencing of *Y. lipolytica* and other alkane-utilising yeasts. Thus, the multiplicity in the *CYP52* family (up to 12 genes) probably reflects an adaptation to the utilisation of different hydrocarbons and fatty acids and is assumed to be a result of gene duplications and divergent evolution from an ancestral gene (Ohkuma et al. 1998; Zimmer et al. 1998; Iida et al. 2000; Craft et al. 2003). Besides the 12 P450 *ALK* genes of the CYP52 family of *Y. lipolytica*, this concerns also several other gene families, like acyl-CoA oxidases of peroxisomal β-oxidation (6 *POX* genes) and triacylglycerol lipases/carboxylesterases (up to 23 *LIP* genes: 4 *LIP1*-like, 19 *LIP2*-like), all containing more members than other yeasts, what correlates with the *Y. lipolytica* addiction for hydrophobic substrate utilisation (Fickers et al. 2005; Thevenieau et al. 2009, 2010; Beopoulos et al. 2010a, b, 2011; Petzsch 2012). In general, this multiplicity of alkane, fatty acid and triglyceride degradation genes is obviously the basis for the wide range of the substrate and chain-length spectrum and of the regio- and stereoselectivity of the enzymatic steps involved in these pathways in alkane-utilising yeasts.

3.2 Multiple Cytochrome P450 Genes in Other Fungi

Interestingly, in comparison to yeasts (up to 17–19 P450 genes in *Y. lipolytica* and *C. albicans*; cf. Sect. 3.1, Génolevures database), the filamentous and other fungi contain much larger numbers of P450 sequences in their genomes. Recent genome sequencing revealed the presence of 39–46 P450 genes described in *Neurospora crassa*, up to 206 in *Aspergillus* spp. (e.g. 172, 150–156, 111–119 and 155–206 P450 genes of *CYP51* to *CYP68*, *CYP501* to *CYP699* and *CYP5001* to *CYP5337* families detected in *A. fumigatus*, *A. niger*, *A. nidulans* and *A. oryzae*, respectively), 123–133 in *Magnaporthe grisea*, 107–110 in *Fusarium graminearum* and 150–160 in *Phanerochaete chrysosporium*. Among them are 2–5 *CYP52* genes and up to 10–12 P450 of *CYP53* (benzoate monooxygenases/hydroxylases), *CYP57* (pisatin demethylases), *CYP58* (trichodiene oxygenases), *CYP64* (P450 oxidoreductases), and CYP65 (trichothecene hydroxylases), respectively, and many P450 (25–49) with unknown functions (Machida et al. 2005; Doddapaneni et al. 2005; Lah et al. 2008; Wortman et al. 2009). Among the different phyla, plants have the highest number of P450 sequences, followed by fungi. Up to 8,170 (from 380 in 2005 to over 8,000 in 2010, most of them putative) fungal P450s from 156 fungal and oomycete genomes have been identified, and the number is increasing with the sequencing of new fungal genomes (Fungal cytochrome P450 database, FCPD:

http://p450.riceblast.snu.ac.kr or Nelson's Cytochrome P450 homepage: http://drnelson.utmem.edu/cytochromeP450.html; Nelson 2009; Crešnar and Petrič 2011).

4 Functional Analysis of Yeast Cytochrome P450 *ALK* Genes

The substrate specificity (chain-length, regio- and stereospecificities) of selected P450s ALK and regulation of most P450 *ALK* genes were studied in more detail up to now especially for *C. tropicalis* and *C. maltosa*, mainly by sequential gene disruption and after heterologous expression of single P450 genes in *S. cerevisiae* or in a homologous *C. maltosa* expression system (Ohkuma et al. 1991a, b, 1995a, 1998; Schunck et al. 1991; Seghezzi et al. 1992; Zimmer et al. 1998) as well as in the *Spodoptera frugiperda* Sf9 insect cells expression system (Eschenfeldt et al. 2003). As discussed in Sect. 2, the CYP52 P450s catalyse predominantly the terminal hydroxylation of *n*-alkanes (AHA), the first and rate-limiting step in alkane degradation, and the ω-hydroxylation of fatty acids (FAHA; Mauersberger et al. 1996; Ohkuma et al. 1998; Zimmer et al. 1998; Fickers et al. 2005). For six out of eight P450ALK of *C. maltosa*, it was demonstrated that individual P450s catalyse predominantly the terminal hydroxylation of short- (C12) and long-chain (C16) alkanes (AHA, CYP52A3 or P450Cm1) or the fatty acid ω-hydroxylation (FAHA) of either short-chain (C12FA, CYP52A10 and A11) or both short- and long-chain (C16FA) fatty acids (CYP52A9). For other P450s the substrate and chain-length specificities were found to be broader. These P450 52A5 (P450Cm2) and 52A4 (P450Cm3) catalyse the hydroxylation of both alkanes and fatty acids, although with P450, 52A5 activity towards C12FA is prevailing (Schunck et al. 1991; Mauersberger et al. 1992a; Ohkuma et al. 1995a, 1998; Zimmer et al. 1996, 1998). Comparable results with slight differences were presented for five P450 52A of *C. tropicalis* ATCC750 (Alk1 to Alk5) and selected allelic variants Alk13 and Alk17 of strain ATCC20336 (Seghezzi et al. 1992; Eschenfeldt et al. 2003). Thus, for all 11 P450s of CYP52A subfamily from *C. maltosa* and *C. tropicalis*, the substrate specificity was studied in detail, whereas no data were not yet reported for the other subfamilies CYP52B1 (CtAlk6), CYP52C1 (CtAlk7), CYP52C2 (CmAlk6), CYP52D1 (CmAlk4) and CYP52D2 (CtAlk8), except results on their induction by alkanes and their oxidation products 1-alkanols, alkanals and fatty acids.

4.1 Analysis of Yarrowia lipolytica *Cytochrome P450ALK Function and Regulation*

The metabolic function of the 12 P450ALK1 to ALK12 (*CYP52F1* to *F11*, *CYP52S1*, Table 4) of *Y. lipolytica* was not so comprehensively studied experimentally compared to P450 of the CYP52A subfamily of *C. maltosa* and *C. tropicalis* (cf. Sect. 4). The key role of these P450ALK in the degradation pathway of alkanes and fatty acids has been mentioned above (cf. Sect. 2) and in previous reviews (Fickers et al. 2005; Thevenieau et al. 2009, 2010). The enzymatic functions attributed so far to individual *Y. lipolytica* P450 forms are the primary hydroxylation of alkanes (AHA) and the ω-hydroxylation of fatty acids (FAHA, Fig. 2, Table 3; Fickers et al. 2005; Fukuda and Ohta 2013).

The P450 forms mainly involved in alkane degradation are 52F1 (ALK1), 52F2 (ALK2) and to a lesser extent also 52F4 (ALK4) and 52F6 (ALK6), as demonstrated, (1) by their strong alkane inducibility in Northern blot, quantitative real-time PCR (qRT-PCR) and differential expression studies; (2) by P450 *ALK* expression regulation and *ALK*-promoter studies; and (3) in particular by gene disruption experiments, an independent approach applied to reveal the in vivo function of P450 forms (Iida et al. 1998, 2000; Sumita et al. 2002a, b; Yamagami et al. 2004; Endoh-Yamagami et al. 2007; Hirakawa et al. 2009; Takai et al. 2012; Table 4; cf. sections below), although participation of other *ALK* genes in alkane metabolism either as AHA or FAHA cannot be excluded. Till now no direct evidence for the function of individual P450ALK of *Y. lipolytica* as alkane hydroxylases (AHA) with chain-length specificities by in vitro experiments was provided, as earlier demonstrated for related P450ALK from *Candida* yeasts (Ohkuma et al. 1998; Zimmer et al. 1998). The P450ALK1 to ALK7 of *Y. lipolytica* were tested only for lauric acid ω-hydroxylase activity (C12FAHA) after expression in plant leaves (Hanley et al. 2003; cf. section "Cytochrome P450ALK Substrate Specificity", Table 4).

Inducibility of *ALK* Genes. The transcription regulation of *ALK* genes by alkanes C10, C12, C14 and C16 can be regarded as indirect evidence for their participation in alkane assimilation, but will not differentiate between their enzymatic function as chain-length-specific AHA or FAHA. First Northern blot (with transcript quantification) and *ALK1*-promoter (pALK1-*lacZ*) studies (Iida et al. 1998, 2000; Sumita et al. 2002a, b; Hanley et al. 2003) demonstrated the strong (25–100 times) induction of *ALK1* and *ALK2* by alkanes (C10 to C16) compared to repressing glycerol and no or weak expression on fatty alcohols (C12-1-ol), fatty aldehydes (C12-al) or fatty acids (lauric, C12FA, myristic, C14FA and oleic, C18:1FA, acids). In accordance with the proposed role of *ALK1* (Iida et al. 2000; cf. section "*ALK* Gene Disruption"), its induction of transcription by C10 was stronger than by C16 (Yamagami et al. 2004; Endoh-Yamagami et al. 2007). On glucose several P450 *ALK* genes were not completely repressed, and, in contrast to glycerol, presence of glucose did not repress the *ALK* induction by alkanes (C10 or C14, in accordance with the effect of glucose on P450 induction by C10 and C16; Table 2, Mauersberger 1991), as demonstrated also for other hydrophobic substrate utilisation genes (*POX*, *POT1*, *PAT1*; for ref., cf. Fickers et al. 2005), indicating a pronounced general glycerol repression system compared to glucose (Iida et al. 1998, 2000). Furthermore, a weak or very weak induction by C14 and no induction by myristic acid (C14FA) were observed for *ALK3* and *ALK4*, or *ALK5* and *ALK6*, respectively, whereas no induction occurred for *ALK7* and *ALK8* (Iida et al. 2000).

Comparable results were obtained by Drennan (2000) on differential expression (RT-PCR and Northern hybridisation) of the *Y. lipolytica* genes *ALK1* to *ALK8* on longer-chain alkanes (C14–C28) and fatty acids (C18FA, C22FA).

Quantitative real-time PCR analysis (qRT-PCR) of transcriptional regulation of *ALK* genes revealed that relative high transcript levels of several *ALK* genes were detectable on both alkanes C10 and C16, in comparison to glucose or glycerol (Hirakawa et al. 2009; Table 4). Besides strongly induced *ALK1* (maximal level) and *ALK2*, moderate transcript levels for *ALK4*, *ALK6*, *ALK9* and a low level for *ALK11* were observed. The transcription of these *ALK* genes is activated on alkanes by the transcription factors Yas1p and Yas2p heterocomplex and repressed by the Opi1 family transcription factor Yas3p (Hirakawa et al. 2009; cf. section "Cytochrome P450 *ALK* Expression Regulation" below). A low transcription level was also detected for *ALK3* and *ALK5*, whereas very low levels were found for *ALK7*, *ALK8* (in agreement with Northern blot results; Iida et al. 2000) and surprisingly for *ALK10* and *ALK12*. For the weakly expressed *ALK3* and *ALK5* and the very weakly expressed *ALK12*, transcription was differently regulated (decreased in Δyas3 cells, no strong effect in Δyas1 or Δyas2 cells) compared to the above-mentioned *ALK* genes. Interestingly, for *ALK3* and *ALK5* the transcript level was 2.2–4 times higher on C16 than on C10, never observed for the other *ALK* genes (Table 4).

In contrast to these studies on P450 induction by different alkane chain lengths, only few data are available on the inducibility of individual P450 *ALK* genes of *Y. lipolytica* by alkane oxidation products or other hydrophobic compounds, which were shown to induce P450 in yeasts (Mauersberger et al. 1980, 1981). A distinctly lower total P450 content (7–17 % of C16 cells) was observed during growth of *Y. lipolytica* on 1-hexadecanol and palmitic acid in earlier studies (Mauersberger and Matyashova 1980; Ilchenko et al. 1980; Schunck et al. 1987b; cf. Sect. 2.1, Table 1). In Northern blot experiments on individual P450 genes of *Y. lipolytica*, only weak or very weak expression of *ALK1* was detected on oleic acid (C18:1FA) or glucose (Endoh-Yamagami et al. 2007) and of *ALK2* on myristic acid (C14FA; no expression of *ALK1*, *ALK3* to *ALK8*), comparable with expression on glucose (Iida et al. 1998, 2000). Additionally, a slight induction of pALK1-*lacZ* (β-galactosidase expression) was observed in glucose-grown cells with 0.1–1.0 % of the alkane oxidation products 1-dodecanol (14–22 %), dodecanal (5–6 %) and C12FA (11–18 %, in comparison with 100 and 84 % for alkanes C10 and C12), whereas controls with glucose (18 %) and glycerol (4 %) exhibited low but striking different expression levels (Sumita et al. 2002a). The inducibility of individual P450 *ALK* genes by alkane oxidation products (fatty alcohols, aldehydes and fatty acids) of different chain lengths in addition to alkanes therefore remains to be studied more in detail.

Recently, first differential gene expression studies in *Y. lipolytica* have been presented comparing cells grown on fatty acids (long-chain oleic acid, C18:1FA and medium-chain pentadecanoic acid, C15FA) and glucose (YPD or minimal glucose medium) by sequencing cDNA libraries to obtain the cDNA copy number for each gene (Mekouar et al. 2010) and by microarray analysis (Petzsch 2012). Whereas P450 *ALK* genes were found only little represented in the cDNA libraries

(*ALK1*, *ALK3* and *ALK5* out of 12 *ALK* genes found only among 3818 detected genes; *ALK1* with slightly increased cDNA copy number in C18:1FA-grown cells), other genes, like *POX2*, *POT1* and *ICL1*, were expectedly highly expressed on fatty acid (Mekouar et al. 2010). Otherwise, comparison of microarray data obtained with glucose and C15FA cells revealed that several P450 *ALK* genes were expressed during growth on fatty acid, indicated by fold change values from glucose to C15FA of 6 (*ALK2*), 10 (*ALK5*), 11 (*ALK6*), 12 (*ALK1*), 18 (*ALK4*), 20 (*ALK9*) and 40 (*ALK11*), whereas other *ALK* and the housekeeping *CYP51* and *CYP61* genes exhibited FC-values below 5, and control genes, like *POT1* (13), *ICL1* (12) and *MLS1* (18), were found comparably induced on fatty acid (Petzsch 2012; Petzsch and Mauersberger, unpublished results). These somewhat contradictory results together with the available data on substrate specificities of P450ALK (cf. sections "*ALK* Gene Disruption" and "Cytochrome P450ALK Substrate Specificity"; Table 4) indicate the necessity of further studies to elucidate the induction of individual P450 genes by fatty acids and compounds other than alkanes. The results on alkane induction indicate that besides *ALK1* and *ALK2*, the genes *ALK4*, *ALK6*, *ALK9* and *ALK11* are probably involved in *n*-alkane assimilation in *Y. lipolytica* (Hirakawa et al. 2009). Whether the encoded P450 function as alkane hydroxylases (AHA), catalysing the primary alkane monooxygenation step, or as fatty acid ω-hydroxylases (FAHA) of different chain lengths (Fickers et al. 2005) has to be studied, as it was more thoroughly elucidated for 6 alkane-inducible out of 8 P450 ALK genes (*CYP52*) of *C. maltosa* (Ohkuma et al. 1998; Zimmer et al. 1998).

Cytochrome P450 *ALK* Expression Regulation. Detailed studies on the transcriptional regulation mechanism of *ALK1* expression by alkanes (C10, C16) identified an alkane-responsive region (ARR1) in the *ALK1*-promoter, consisting of two alkane-responsive *cis*-acting elements ARE1 and ARE2 (Sumita et al. 2002b; Yamagami et al. 2004). ARE1-like sequences are present in other genes encoding alkane degradation enzymes in *Y. lipolytica*, including *ALK2*, *POX1*, *POX3*, *POT1*, *PAT1* and *YAS1*, and similar sequences (conserved motif: TGTG or CACA) were detected in alkane-inducible genes (P450 *ALK* of CYP52 family, *CPR* or *POT1* of *C. tropicalis* and *C. maltosa*) in other yeasts. Therefore, the alkane response mediated by Yas1p–Yas2p/Yas3p-like bHLH family proteins is conserved among alkane-assimilating yeasts (Yamagami et al. 2004). The two *YAS1* and *YAS2* (yeast alkane signalling) genes essential for transcriptional induction of *ALK1* by C10 and C16 were identified, encoding the two basic helix-loop-helix (bHLH family) transcription factors Yas1p and Yas2p, activating the transcription of *ALK1* in an alkane-dependent manner by forming a heteroduplex and binding to ARE1 for alkane signalling (Yamagami et al. 2004; Endoh-Yamagami et al. 2007). Both *YAS2* and *YAS1* are essential for growth of *Y. lipolytica* on both alkanes C10 and C16, but not for oleic acid, glucose or glycerol utilisation. A model for the transcription activation in response to alkanes by the Yas1p–Yas2p heterocomplex was postulated, including a positive autoregulatory feedback loop of Yas1p, what makes it possible to induce massive changes in gene expression in response to alkanes and to adapt quickly to exposure to alkanes. However, how alkanes are recognised in the yeast is still an open question (Endoh-Yamagami et al. 2007).

More recently, the *YAS3* encoding the repressing transcription factor Yas3p (ortholog of the negative regulator Opi1 family) was described in *Y. lipolytica*, which is involved as a repressor in transcriptional regulation of *ALK1* and other *ALK* genes in response to alkanes through interaction with Yas2p, but not with Yas1p (Hirakawa et al. 2009). According to a proposed model, Yas3p is transported into the nucleus and represses the expression of *ALK1* and other target genes by binding to Yas2p in absence of alkanes, whereas Yas3p is trapped on the ER membrane, and consequently the transcription of target genes can be activated by Yas1p–Yas2p complex in the presence of alkanes. Thus, Yas3p functions as a master regulator of transcriptional response by changing its localisation between the nucleus (on glucose) and ER (upon transfer to alkanes, like C10) in response to carbon sources (Hirakawa et al. 2009). It is of interest that not all *ALK* genes are regulated according to this model in the same way as *ALK1*, *ALK2*, *ALK4*, *ALK6*, *ALK9* and *ALK11*. In contrast, for *ALK3*, *ALK12* and *ALK5* (not for related *ALK7*), encoding P450 with probable substrate preference towards fatty acids (FAHA, cf. Table 4), the inducing/repressing effects of Yas1p–Yas2p/Yas3p were not observed, indicating the presence of different regulation mechanisms for the *ALK* genes in *Y. lipolytica* (Hirakawa et al. 2009). Additionally, it was shown that peroxisome deficiency repressed alkane-mediated *ALK1* induction in *Y. lipolytica*, involving *PEX10*, *PEX5* and *PEX6* gene products (Sumita et al. 2002a), and that disruption of the *SCS2* ortholog impaired its growth on decane, although the transcript level of *ALK1* was not much affected (Kobayashi et al. 2008).

***ALK* Gene Disruption.** Gene disruptions of *ALK* were performed to study the in vivo function of P450ALK forms in *Y. lipolytica* (Iida et al. 1998, 2000; Takai et al. 2012). A strain disrupted in *ALK1* showed almost no growth on C10 and significantly reduced growth on C11 to C15, while it grew well on C16 (Takai et al. 2012). Single deletions of *ALK2*, *ALK3*, *ALK4* or *ALK6* did not affect growth on both alkanes C10 and C16 (Iida et al. 2000). The phenotype AlkAb ($C10^-$ $C16^+$, C12-1-ol^+, $C12FA^+$, C18:$1FA^+$) of *ALK1* deletion was beside AlkAa ($C10^-$ $C16^-$) and AlkAc ($C10^+$ $C16^-$) frequently found among chemical and insertion mutants of *Y. lipolytica* (Mauersberger 1991; Mauersberger et al. 2001; Mauersberger and Nicaud 2003; Takagi, personal communication) and was shown to be connected also with mutations in different non-related genes, e.g. acetoacetyl-CoA thiolase gene *PAT1* (Yamagami et al. 2001, 2004), *PEX10* encoding peroxisomal integral membrane protein involved in peroxisomes biogenesis (Sumita et al. 2002a, b) and *ANT1* encoding a peroxisomal membrane localised adenine nucleotide transporter protein, providing ATP for activation of short-chain fatty acids by acyl-CoA synthetase II in peroxisomes (Thevenieau et al. 2007). The phenotype AlkAc ($C10^+$ $C16^-$) was not detected among P450 *ALK* gene deletions, whereas two insertion mutants of AlkAc phenotype were disrupted in ABC transporter encoding *ABC1*, thus providing first evidence for its participation in chain-length-dependent alkane transport processes (Mauersberger et al. 2001; Thevenieau et al. 2007).

Furthermore, a double disruptant Δ*alk1* Δ*alk2* was not growing on C10 to C14 and very poorly growing (delayed slow growth) on C15 and C16 alkanes, whereas a double mutant Δ*alk1* Δ*alk3* grew like Δ*alk1*. Finally, the *Y. lipolytica* Δ*alk1-12* mutant deleted in all 12 genes encoding P450ALKs was constructed (Takai et al. 2012; Fukuda and Ohta 2013). The mutant Δ*alk1-12* completely lost the ability to grow on alkanes from C10 to C16, whereas its growth on the C12 alkane oxidation products 1-dodecanol, dodecanal, dodecanoic acid (C12FA) and on glucose was comparable with the wild-type strain. The same phenotype AlkAa ($C10^-$ $C16^-$) was observed for the mutant Δ*alk1* Δ*alk2* Δ*alk4* Δ*alk6* and all subsequent mutants with increasing *ALK* gene deletions, introduced in the order *ALK11*, *ALK9*, *ALK3*, *ALK12*, *ALK5*, *ALK7*, *ALK8* and *ALK10*, indicating that these genes are not very much involved in alkane assimilation (Takai et al. 2012). Thus, individual expression of each P450ALK in mutant Δ*alk1-12* will greatly contribute to elucidate its function and substrate specificity in vivo and in vitro. In this respect it was of interest that expression of *ALK1* from its native promoter in mutant Δ*alk1-12* restored the Alk^- phenotype completely to a wild-type strain comparable growth of the transformant Δ*alk1-12 ALK1* on all alkanes from C10 to C16, indicating the major role of P450ALK1 in the metabolism of alkanes (Takai et al. 2012). Therefore, it will be of interest to test the expression and also overexpression of each *ALK* gene in the same way in mutant Δ*alk1-12*.

Cytochrome P450ALK Substrate Specificity. The role of single P450ALK of *Y. lipolytica* as alkane hydroxylases (AHA) was concluded from their alkane induction and mainly from gene disruption experiments, because in contrast to testing different P450ALK for lauric acid ω-hydroxylase activity (C12FAHA, Hanley et al. 2003), they were not yet studied in vitro for AHA after individual expression in an appropriate host not containing any *CYP52* genes (*S. cerevisiae* or *Y. lipolytica* mutant strain Δ*alk1-12*). Iida et al. (1998, 2000) and Takai et al. (2012) concluded from the alkane induction and gene disruption results that among the 8 alkane-inducible P450 genes *ALK1* to *ALK6*, *ALK9* and *ALK11* of *Y. lipolytica* (Table 4), P450ALK1 is involved predominantly in shorter-chain alkane hydroxylation (C10, although it is probably not very specific with a broad substrate spectrum from C10 to C15), i.e. it plays a major role in short- to middle-chain alkane assimilation, and perhaps it is involved in detoxification of short-chain alkanes like C10. Furthermore, P450ALK2 functions coordinately with P450ALK1 in longer-chain alkane (C14-C16) degradation (Iida et al. 2000), thus indicating the prominent roles of P450ALK1 and ALK2 in terminal alkane hydroxylation (AHA). Accordingly, *ALK1* and *ALK2* are the most abundantly expressed among all 12 *ALK* genes during growth on alkanes (Table 4). P450ALK4 and ALK6 are not significantly involved in C10–C14 assimilation, but these P450s might be active towards C15–C16 or longer alkanes >C16 (Takai et al. 2012, Table 4).

At present one can assume overlapping chain-length specificities. For P450ALK1 a chain-length range of its AHA from C10 (high) to C15 (low activity) is likely, whereas P450ALK2 might be more specific for longer alkanes (C12 to C16 with increasing activity, Table 4), as previously demonstrated for selected P450ALK (52A3 to A5, 52A9) of *C. maltosa* (Scheller et al. 1996; Ohkuma et al.

1998; Zimmer et al. 1998). The AHA of P450ALK6 and ALK4 might be lower than for P450ALK1 and ALK2 and restricted more to long-chain alkanes (C15–C16 or even longer), and one cannot exclude additional FAHA towards long-chain fatty acids for these P450ALK, as previously shown for selected *C. maltosa* P450ALK. The function of moderately or weakly alkane-induced *ALK9* and *ALK11*, and the very weakly alkane-induced *ALK8*, *ALK10* and *ALK12* remains to be clarified. On one hand, the expression level of these genes on alkanes might be too low to support growth on alkanes in respective deletion mutants, and on the other hand, several of these P450ALK might be not involved in alkane hydroxylation but rather in the oxidation of unknown substrates.

The *ALK1* to *ALK7* genes from *Y. lipolytica* have been expressed in leaves of *Nicotiana benthamiana*, revealing that *ALK3*, *ALK5* and *ALK7* code for lauric acid ω-hydroxylases (C12FAHA), whereas no C12FAHA was detected with *ALK1*, *ALK2*, *ALK4* and *ALK6* (Hanley et al. 2003; Table 4), although the chain-length specificity of these P450s with FAHA was not studied. The P450ALK3 and ALK5 with C12FAHA were shown to be weakly induced by alkanes, whereas P450ALK7 was not induced (Table 4). Interestingly, the C12FAHA-positive P450s pairs ALK3/ALK12 and ALK5/ALK7 are phylogenetically more similar to each other than to the rest of the *YlALK* gene *CYP52F/S* subfamilies (Table 4). Furthermore, P450ALK2 and ALK10 exhibit 44 and 50 % protein sequence identity to P450 52A3 (P450Cm1, *CmALK1*) of *C. maltosa*, which is the major P450 in alkane-assimilating cells of this species, and it hydroxylates preferably longer-chain alkanes (C16-C18) and shows a moderate ω-hydroxylase activity with long-chain fatty acids (C16FAHA-C18FAHA), in contrast to P450 52A4 (P450Cm2, *CmALK3*) with preferences to shorter alkane chain lengths (C12AHA) and especially to lauric acid (C12FAHA, Scheller et al. 1996; Zimmer et al. 1998). Whether these P450ALK of *Y. lipolytica* themselves are able to catalyse efficiently a cascade of sequential mono- and diterminal monooxygenation reactions, thus oxidising alkanes directly to fatty acids and dicarboxylic acids (DCA), as shown in vitro for *C. maltosa* P450 52A3 (Scheller et al. 1998), and whether this alternative P450 function plays a role in vivo in diterminal alkane oxidation remain to be clarified.

A different functional study of eight P450 *ALK* genes of *Y. lipolytica* indicated that selected P450ALK may be specific for the alkane or fatty acid substrate chain lengths and their overexpression made it possible to increase the production of dicarboxylic acid (DCA) from long-chain fatty acid substrates (sunflower oil: C16:0, C18:0, C18:1, C18:2FA) or C10/C16-alkane mixtures (Thevenieau 2006; Thevenieau et al. 2009, 2010; Table 4). The weakly alkane-induced genes *ALK3* and *ALK5* might be involved in alkane degradation, perhaps more probably as fatty acid ω-hydroxylases (FAHA) involved in the diterminal or ω-oxidation part. The P450ALK3, ALK5 and ALK7 with C12FAHA in vitro exhibited negative (ALK3) or positive (ALK5/ALK7) effects on long-chain DCA formation from sunflower C16 to C18 fatty acids due to their overexpression, what might be connected with their FA chain-length specificities (Thevenieau et al. 2009, 2010; Table 4). In case of P450ALK1, ALK2, ALK9, ALK10 and ALK11 (probably without C12FAHA), also different effects on long-chain DCA formation from C16FA to C18FA were observed (Thevenieau et al. 2010; cf. [+]/[−] in Table 4 for LCFA). Furthermore, Thevenieau et al. (2010) proposed that in addition to the function of P450ALK1 (short-chain alkane specific) and ALK2 (long-chain alkane specific), also ALK5 and ALK11 might be specific for short-chain alkanes (indicated by [+] in Table 4 for AHA), based on results obtained for DCA production from a C10/C16 1:1 mixture with strains overexpressing P450ALK2, ALK5 or ALK11, respectively.

Despite the progress made in the last decade, the metabolic function (substrate and chain-length specificities towards alkanes and fatty acids, including regio- and stereoselectivity of ω- or ω-n hydroxylase activities) of the P450ALK1 to ALK12 in *Y. lipolytica* and the regulation of their gene expression remain to be elucidated more precisely, as it was made for eight *CYP52* genes of *C. maltosa* or *C. tropicalis* mainly after gene disruption and heterologous expression in *S. cerevisiae* or insect cells to study the enzyme activity of individual P450 (Seghezzi et al. 1992; Ohkuma et al. 1995a, 1998; Mauersberger et al. 1996; Zimmer et al. 1998; Eschenfeldt et al. 2003; Fukuda and Ohta 2013). These studies demonstrated the presence of P450ALK with restricted substrate chain-length specificities for fatty acids (like P450 52A4, 52A10 and 52A11) or alkanes (P450 52A3, also YlALK1 might be an example) as well as P450 with overlapping functions, exhibiting bride chain-length and substrate specificities towards alkanes and fatty acids in one P450 enzyme, like P450 52A5 (P450Cm3). Besides the 8 alkane-induced P450ALK, the functions remain to be studied experimentally also for the four not or very weakly alkane-induced *ALK* genes (*ALK7*, *ALK8*, *ALK10*, *ALK12*) and the other three P450 genes ($P450_1$, $P450_2$, $P450_3$) classified so far only by sequence homology (Table 4), perhaps except for CYP51F1 and CYP61A1, which functions are almost well defined in these two families.

4.2 Electron Transport Proteins to Cytochrome P450 in Yeasts

Besides the 17 P450s several related to P450 electron transport proteins are present in *Y. lipolytica* as revealed by genome sequencing (Table 4; Fickers et al. 2005). The main electron transfer protein to the host-own, ER-resident P450s is the microsomal (ER-localised) NADPH-P450 reductase (CPR) encoded by the single *YlCPR* gene (YALI0D04422g, homolog to *NCP1* or *CPR* of *S. cerevisiae* and fungi). Whereas a single *CPR* gene was found in *Y. lipolytica*, *S. cerevisiae* (genome sequence data), *C. maltosa* (Ohkuma et al. 1995c) and other yeasts (Lah et al. 2008), two *CPRA* and *B* genes were detected in *C. tropicalis* (Craft et al. 2003) and from one to three *CPR* genes in filamentous fungi (Lah et al. 2008). Additionally to the NADPH-P450 reductase gene *YlCPR*, three putative cytochrome b_5 genes (*CYB5*, YALI0D12122g, YALI0E24079g, YALI0A20394g, with slightly different b_5-like gene products) and two NADH-cytochrome b_5 reductase genes (*CBR1*, YALI0D04983g, and *MCR1*, YALI0D11330g, Table 4) are probably involved in electron transport to class II P450s in the yeast ER. The yeast CPR was repeatedly demonstrated as functional partner for heterologously expressed microsomal P450 in different yeasts, including *S. cerevisiae*, *S. pombe*, *P. pastoris* and *Y. lipolytica* (Oeda et al. 1985; Urban et al. 1994; Sakaki et al. 1996; Pompon et al. 1996; Juretzek et al. 2000; Shkumatov et al. 2002, 2006; Mauersberger et al. 2013). Partially purified YlCPR and purified CmCPR were shown to be considerably active in heterologous in vitro reconstitution experiments with class II microsomal P450c21 or P450Coh, but very weakly active with class I mitochondrial P450scc or P450c11 (Shkumatov and Smettan 1991; Novikova et al. 2009).

The NADPH-P450 reductase and most P450 forms in fungi are integral membrane proteins co-located in the ER. Interestingly, gene-dose-dependent high-level expression of homologous CPR and heterologous bovine P45017α (CYP17) resulted in a strong proliferation of different ER types in *Y. lipolytica* (Förster and Mauersberger, unpublished), as it was repeatedly demonstrated by overexpression of P450, P450 reductase or other membrane proteins in *S. cerevisiae* or *C. maltosa*, a phenomenon called inducible membranes (Schunck et al. 1991; Wright 1993; Ohkuma et al. 1995b; Menzel et al. 1997). Like *S. cerevisiae*, *Y. lipolytica* also exhibited different types of proliferated ER membranes depending on expressed protein. Overexpression of only CYP17 alone induced the formation of a mostly tubular network of ER membranes in various parts of cytoplasm and of plasma membrane-associated ER. Contrarily, overexpression of the homologous Yl*CPR* was leading to a special type of ER proliferation, forming karmellae-like stacks of up to 25 membrane layers, mainly in close vicinity to the nucleus, but partially also extending into cytoplasm. In diploid cells coexpressing both genes, both forms of proliferated ER were evident. Furthermore, whereas mostly a tubular network of ER membranes proliferated during CYP17 overexpression and in the early phase of coexpression, karmellae-like membrane stacks were observed almost in the later phase of high-level Yl*CPR* expression or in coexpressing cells (Förster and Mauersberger, unpublished results; Mauersberger et al. 2013).

Furthermore, other internal electron transfer proteins (ferredoxin and NADPH-ferredoxin reductase of mitochondrial matrix) were described in yeasts (although no class I type mitochondrial P450 forms have been detected in yeasts), which might substitute the mammalian proteins NADPH-adrenodoxin reductase (AdR) and adrenodoxin (Adx), involved in reduction of class I P450s. In *S. cerevisiae*, the genes *ARH1* (Adrenodoxin Reductase Homologues *1*; Lacour et al. 1998; Manzella et al. 1998) and *YAH1* (Yeast Adrenodoxin Homologues *1*; Barros and Nobrega 1999) encode proteins that are highly homologous to mammalian AdR and Adx, respectively. Orthologous proteins to ScArh1p with high similarity to human mitochondrial AdR were found in *Y. lipolytica* (Table 4; YALI0B14839g; YlArh1p) and in *S. pombe* (SpArh1p). In cooperation with Adx, ScArh1p can support the 11β-hydroxylase activity of P450c11 (Dumas et al. 1996; Manzella et al. 1998; Lacour et al. 1998). In fission yeast both heterologous and homologous redox chains, SpArh1p-etp1^{fd}, SpArh1p-Adx, AdR-etp1^{fd} and AdR-Adx, can function with the heterologous P450c18, P450c11 or P450scc (Bureik et al. 2002;

Schiffler et al. 2004; Ewen et al. 2008). The in vivo function of the orthologous gene products of *ARH1* (highly homologous to human mitochondrial AdR) probably located in the mitochondrial matrix remains to be elucidated. Furthermore, it should be studied whether YlArh1p can function in *Y. lipolytica* also as electron donor protein to heterologously expressed mitochondrial class I P450 forms (like P450c11, P450c18 or P450scc), as it was demonstrated for *S. cerevisiae* and *S. pombe* (Manzella et al. 1998; Lacour et al. 1998; Bureik et al. 2002; Schiffler et al. 2004; Ewen et al. 2008).

The adrenodoxin homologous yeast gene *YAH1*, encoding a mitochondrial matrix iron–sulphur protein (*S. cerevisiae* ferredoxin Yah1p; Barros and Nobrega 1999), involved in the biogenesis of iron–sulphur proteins and heme A synthesis (for ref., see Schiffler et al. 2004; Ewen et al. 2008), is highly conserved in fungi, plant and animals. This Adx homolog is also contained in *Y. lipolytica* (YALI0B02222p, Table 4), *P. pastoris* and in most yeast and filamentous fungi. However, ScYah1p could not substitute Adx in reconstitution of steroid hydroxylation systems in vivo (Dumas et al. 1996). Whether YlYah1p can support the electron transfer (substituting Adx) from heterologous AdR or homologous Arh1p to mitochondrial P450, heterologously expressed in *Y. lipolytica*, has to be tested.

A single ferredoxin ScYah1p-like encoding gene was not found in fission yeasts *Schizosaccharomyces pombe* or *S. japonicus*, which in contrast contain ETP1, encoding the adrenodoxin-like mitochondrial electron transfer protein 1 (etp1). These fusion proteins consist of the N-terminal COX15 etp1^{cd} (COX15-CtaA family, from aa 85 to 473 in SpEtp1p, which functions in the assembly of the cytochrome oxidase COX complex) and the carboxy terminal ferredoxin etp1^{fd} (Fer2, ferredoxin-like [2Fe-2S]-cluster, with high homology to the ferredoxin family) domains (Bureik et al. 2002; Schiffler et al. 2004). Evidence was given that etp1^{fd} is cleaved off from the COX15 domain during mitochondrial import. Moreover, the bacterially expressed ferredoxin domain etp1^{fd} can replace adrenodoxin in steroid hydroxylase assays and transfer electrons from bovine NADPH-adrenodoxin reductase (AdR) bovine or human P450scc, bovine P450c11, human P450c18 or bacterial CYP106A2 in vivo and in vitro (Bureik et al. 2002; Schiffler et al. 2004). Thus, in contrast to the *S. cerevisiae* ferredoxin Yah1, another closely related iron–sulphur protein etp1^{fd} can replace Adx in the interaction with its redox partners AdR and P450. Therefore, etp1^{fd} resembles Adx more than yeast ferredoxin Yah1 in its structural and functional features. Whereas the ScYah1p homolog was found in *Y. lipolytica*, SpEtp1p-like fusion proteins were not detected in *Y. lipolytica* (Table 4), *S. cerevisiae*, other yeast and filamentous fungi. When using the complete SpEtp1 sequence for BLAST, several homologous proteins were found in the databases, including YlCox15p (YALI0F24651g, Table 4) and ScCox15p (YER141W), although their homology is restricted to the N-terminal COX15 domain of etp1, involved in cytochrome c oxidase assembly (hydroxylation of heme O to heme A). Thus, YALI0F24651g contains probably no Adx-like etp1^{fd} domain which might be functional towards mammalian P450. Otherwise, when using the etp1^{fd} (502–606 aa) for BLAST, homologous Adx-like ferredoxin proteins were found, including YlYah1p and ScYah1p. Thus, the appearance of etp1 fusion proteins containing the adrenodoxin-like etp1^{fd} domain, which is after cleavage from the COX15 etp1^{cd} domain in mitochondria functional in electron transfer, is obviously restricted to fission yeasts.

5 Conclusions

In the genome of *Y. lipolytica*, 17 P450 genes and related genes for electron transfer proteins (*CPR*, *CYB5*, *MCR1*, *CBR1*) to these ER-located class II P450 were detected by molecular cloning and genome sequencing, the highest number of P450 genes among yeasts, when excluding the allelic variants present in diploid *Candida* yeasts. Two of these P450 genes encode the housekeeping P450 14DM (*CYP51F1*) and P450 22DS (*CYP61A1*), both involved in ergosterol biosynthesis, which are present in all tested fungi. Additionally, three P450$_1$ to P450$_3$ genes are contained, including *CYP5223A1*, a *DIT2* (*CYP56*)-like P450$_2$, probably involved in spore wall formation, as well as P450$_1$ (*CYP548P1*) with assumed benzoate-4-hydroxylase activity and P450$_3$ (*CYP504A17*) with assumed phenylacetate 2-hydroxyalse activity (PAHA). The majority of the *Y. lipolytica* P450 genes (12 *ALK* genes) belongs to the CYP52 family of the CYP supergene family (Nelson 2009), consisting of fungal alkane- or fatty acid-inducible genes, involved as alkane or fatty acid hydroxylase enzyme systems in the terminal hydroxylation of alkanes (AHA) and fatty acid ω-hydroxylation (FAHA).

The induction of P450 by the growth of *Y. lipolytica* on alkanes, fatty alcohols or fatty acids was shown in first studies in earlier 1980s, and a high, interfering cytochrome oxidase content required a modified CO-difference spectral detection method for P450 in this yeast. As demonstrated by in vivo and in vitro studies, alkane-grown *Y. lipolytica* cells contained both P450 activities, AHA (terminal alkane hydroxylation step in primary alkane oxidation pathway) and FAHA (fatty acid ω-oxidation), however no discrimination between individual P450 isoforms was made at that time. Although the main conclusions about the function and regulation of alkane-induced P450s of *Candida* yeasts also apply to *Y. lipolytica*, the in vivo function of individual P450 forms remains to be elucidated in detail. The multiple paralog P450 *CYP52* genes of *Candida* spp., *Debaryomyces* spp. and *Y. lipolytica* exhibit functional diversity in their inducibility and regulation, as well as in the substrate and chain-length specificities, the regio- and probable stereoselectivity of the encoded P450 proteins (Ohkuma et al. 1995a, 1998; Zimmer et al. 1996, 1998; Iida et al. 2000; Craft et al. 2003; Eschenfeldt et al. 2003; Takai et al. 2012). However, functional analyses have been performed so far only for a subset of P450s of *C. tropicalis*, *C. maltosa*, *C. albicans*, *C. apicola*, *D. hansenii* and *Y. lipolytica*, and the substrate specificity of most P450 of CYP52 family remains to be studied. For the most thoroughly studied eight P450ALK encoded by *CYP52* genes of *C. maltosa* or *C. tropicalis*, restricted substrate chain-length specificities for fatty acids (like P450Cm2) or alkanes (P450Cm1), respectively, were described at one hand; as well as at the other hand, P450 isoforms with overlapping, bride chain-length and substrate specificities for alkanes and fatty acids in one P450 enzyme (P450Cm3) have been demonstrated (Ohkuma et al. 1995a, 1998; Zimmer et al. 1998; Eschenfeldt et al. 2003). Thus, despite the progress made in phenotype analysis of P450 *ALK* deletion mutants (especially for *ALK1*, *ALK2*), *ALK* gene expression studies and first testing of C12FAHA (whereas comparable data on AHA are still missing) for individual P450ALK forms after heterologous expression, the in vivo function, the physiological substrates and chain-length specificities of P450ALK of *Y. lipolytica*, in particular those of P450ALK3 to ALK12, remain to be studied. This is difficult to prove by gene disruption due to the existence of multiple *ALK* gene paralogs in the yeast genome (Takai et al. 2012). Therefore, individual P450ALK genes should be expressed in the P450ALK, AHA and FAHA free yeast *S. cerevisiae* or insect cells to study the substrate specificity in vitro (Schunck et al. 1991; Zimmer et al. 1998; Eschenfeldt et al. 2003). Finally, the recently constructed deletion mutant *Y. lipolytica* Δ*alk1-12* in all 12 P450 *ALK* genes enables first functional and regulatory studies of individual *ALK* genes in *Y. lipolytica* itself (Takai et al. 2012).

One characteristic feature of alkane-utilising yeasts is the presence of several multigene families (*LIP*, P450 *ALK*, *FALD*, *POX* genes) involved in the degradation pathways of hydrophobic substrates, which seems to be most pronounced in *Y. lipolytica* (Fickers et al. 2005). The observed multiplicity of alkane, fatty acid or triglyceride degradation genes is obviously the basis for the wide substrate and chain-length spectrum and for the regio- and stereoselectivity of enzymatic steps involved in these pathways in alkane-utilising yeasts. In particular, the multiplicity of paralog P450 *ALK* genes in the CYP52 family probably reflects an adaptation to the utilisation of different alkanes and fatty acids and is assumed to be a result of gene duplications and divergent evolution starting from an ancestral gene.

One of the most striking features of alkane-utilising yeasts (e.g. *Y. lipolytica*, *Candida* spp.) is that they exhibit in vivo a high catalytic activity of their alkane-inducible P450 systems of the CYP52 family involved in primary alkane and fatty acid oxidation (substrate turnover number for alkanes in the range of 1–3 μmol product/nmol P450 × min), calculated from the growth rate on alkanes, the alkane substrate utilisation rate and the total P450 content of whole cells (Schunck et al. 1987a; Blasig et al. 1988; Mauersberger et al. 1996). This high in vivo turnover number of host-own P450ALK is obviously supported by metabolic adaptations and an effective subcellular organisation of the hydrophobic substrate and product transport processes in the alkane-utilising yeast cells (inducible systems of facilitated uptake and intracellular translocation of alkanes and fatty acids: excretion of surfactants, adjustment of cell surface hydrophobicity to interact with water-immiscible substrates, like alkanes, fatty acids and triglycerides; plasma membrane-associated ER and association of ER and peroxisomes; selective mechanisms of oxidised products secretion) and of the efficiency of electron transfer systems (increased expression of the inherent electron transport components NADPH-P450 reductase CPR, cytochrome b_5 and NADH-b_5 reductase for class II P450 in alkane-cells) used by the host-own P450 (Mauersberger et al. 1987, 1996; Barth and Gaillardin 1996; Mlícková et al. 2004; Fickers et al. 2005;

Thevenieau et al. 2010). Therefore, it was of special interest to address the question whether these unique properties of *Y. lipolytica* will support also the function of heterologously expressed P450s in this yeast and will give this organism an advantage in P450-catalysed bioconversion of hydrophobic substrates (Juretzek 1999; Juretzek et al. 2000; Shkumatov et al. 2006; Novikova et al. 2009; Braun et al. 2012; Mauersberger et al. 2013).

Acknowledgements I would like to thank my collaborators Masamichi Takagi (Tokyo University), Wolf-Hagen Schunck (MDC Berlin), Andrey Sharyshev (IBPM Pushchino), Thomas Juretzek, André Förster and Patrick Petzsch (TU Dresden) for sharing unpublished results and for fruitful discussions and permanent interest in this research field. Part of this work was supported in part by grant 0310257A from the Bundesministerium für Forschung und Technologie (BMFT) of Germany.

References

Baroncelli V, Boccalon G, Giannini I, Renzi P (1979) An inducible n-alkane hydroxylase system containing cytochrome "o" from *Candida lipolytica*. Mol Cell Biochem 28:3–6

Barros MH, Nobrega FG (1999) *YAH1* of *Saccharomyces cerevisiae*: a new essential gene that codes for a protein homologous to human adrenodoxin. Gene 233:197–203

Barth G, Gaillardin C (1996) Yarrowia lipolytica. In: Wolf K (ed) Nonconventional yeasts in biotechnology. A handbook. Springer, Berlin, pp 313–388

Beopoulos A, Desfougéres T, Sabirova J, Zinjarde S, Neuveglise C, Nicaud J-M (2010a) The hydrocarbon-degrading oleaginous yeast *Yarrowia lipolytica*. In: Timmis KN (ed) Handbook of hydrocarbon and lipid microbiology, volume 3, part 1, chapter 35. Springer, Berlin, pp 2111–2121

Beopoulos A, Desfougéres T, Sabirova J, Nicaud J-M (2010b) *Yarrowia lipolytica* as a cell factory for oleochemical biotechnology. In: Timmis KN (ed) Handbook of hydrocarbon and lipid microbiology, volume 4, part 5, chapter 45. Springer, Berlin, pp 3003–3010

Beopoulos A, Nicaud JM, Gaillardin C (2011) An overview of lipid metabolism in yeasts and its impact on biotechnological processes. Appl Microbiol Biotechnol 90:1193–1206

Blasig R, Mauersberger S, Riege P, Schunck W-H, Jockisch W, Franke P, Müller H-G (1988) Degradation of long-chain *n*-alkanes by the yeast *Candida maltosa* II. In vitro oxidation of n-alkanes and intermediates using microsomal membrane fractions. Appl Microbiol Biotechnol 28:589–597

Braun A, Geier M, Buehler B, Schmid A, Mauersberger S, Glieder A (2012) Steroid biotransformation in biphasic systems with *Yarrowia lipolytica* expressing human liver cytochrome P450 genes. Microb Cell Fact 11:106. doi:10.1186/1475-2859-11-106

Bureik M, Schiffler B, Hiraoka Y, Vogel F, Bernhardt R (2002) Functional expression of human mitochondrial CYP11B2 in fission yeast and identification of a new internal electron transfer protein, etp1. Biochemistry 41:2311–2321

Coelho MAZ, Amaral PFF, Belo I (2010) *Yarrowia lipolytica*: an industrial workhorse. In: Méndez-Vilas A (ed) Current research, technology and education topics in applied microbiology and microbial biotechnology, vol 2. Formatex, Badajoz, pp 930–944

Craft DL, Madduri KM, Eshoo M, Wilson CR (2003) Identification and characterization of the CYP52 family of Candida tropicalis ATCC 20336, important for the conversion of fatty acids and alkanes to α, ω-dicarboxylic acids. Appl Environ Microbiol 69:5983–5991

Crešnar B, Petrič S (2011) Cytochrome P450 enzymes in the fungal kingdom. Biochim Biophys Acta 1814(1):29–35

Delaissé JM, Martin P, Verkeyen-Bouvy MF, Nyns EJ (1981) Subcellular distribution of enzymes in the yeast *Saccharomycopsis lipolytica*, grown on *n*-hexadecane, with special reference to the ω-hydroxylase. Biochim Biophys Acta 676:77–90

Doddapaneni H, Chakraborty R, Yadav JS (2005) Genome-wide structural and evolutionary analysis of the P450 monooxygenase genes (P450omr) in the white rot fungus *Phanerochaete chrysosporium*: evidence for gene duplications and extensive gene clustering. BMC Genomics 6:92

Drennan MB (2000) Differential transcription of *CYP52* genes of *Yarrowia lipolytica* during growth on hydrocarbons. Master thesis, Department Microbiology and Biochemistry, The University of the Orange Free State, Bloemfontein, Republic of South Africa

Dujon B, Sherman D, Fischer G et al (2004) Genome evolution in yeasts. Nature 430:35–44

Dumas B, Cauet G, Lacour T, Degryse E, Laruelle L, Ledoux C, Spagnoli R, Achstetter T (1996) 11β-hydroxylase activity in recombinant yeast mitochondria. In vivo conversion of 11-deoxycortisol to hydrocortisone. Eur J Biochem 238:495–504

Endoh-Yamagami S, Hirakawa K, Morioka D, Fukuda R, Ohta A (2007) Basic helix-loop-helix transcription factor heterocomplex of Yas1p and Yas2p regulates cytochrome P450 expression in response to alkanes in the yeast *Yarrowia lipolytica*. Eukaryot Cell 6:734–743

Eschenfeldt WH, Zhang Y, Samaha H, Stols L, Eirich LD, Wilson CR, Donnelly MI (2003) Transformation of fatty acids catalyzed by cytochrome P450 monooxygenase enzymes of *Candida tropicalis*. Appl Environ Microbiol 69:5992–5999

Ewen KM, Schiffler B, Uhlmann-Schiffler H, Bernhardt R, Hannemann F (2008) The endogenous adrenodoxin reductase-like flavoprotein arh1 supports heterologous cytochrome P450-dependent substrate conversions in Schizosaccharomyces pombe. FEMS Yeast Res 8:432–441

Fickers P, Benetti PH, Wache Y, Marty A, Mauersberger S, Smit MS, Nicaud JM (2005) Hydrophobic substrate utilisation by the yeast *Yarrowia lipolytica*, and its potential applications. FEMS Yeast Res 5:527–543

Finogenova TV, Morgunov IG, Kamzolova SV, Chernyavskaya OG (2005) Organic acid production by the yeast *Yarrowia lipolytica*: a review of prospects. Appl Biochem Microbiol 41:418–425

Fukuda R, Ohta A (2013) Utilization of hydrophobic substrates by *Yarrowia lipolytica*. In: Barth G (ed) *Yarrowia lipolytica*. Springer, Heidelberg

Hanley K, Nguyen LV, Khan F, Pogue GP, Vojdani F, Panda S, Pinot F, Oriedo VP, Rasochova L, Subramanian M, Miller B, White EL (2003) Development of a plant viral-vector-based gene expression assay for the screening of yeast cytochrome P450 monooxygenases. Assay Drug Dev Technol 1:147–160

Hirakawa K, Kobayashi S, Inoue T, Endoh-Yamagami S, Fukuda R, Ohta A (2009) Yas3p, an Opi1 family transcription factor, regulates cytochrome P450 expression in response to *n*-alkanes in *Yarrowia lipolytica*. J Biol Chem 284:7126–7137

Iida T, Ohta A, Takagi M (1998) Cloning and characterization of an *n*-alkane-inducible cytochrome P450 gene essential for *n*-decane assimilation by *Yarrowia lipolytica*. Yeast 14:1387–1397

Iida T, Sumita T, Ohta A, Takagi M (2000) The cytochrome P450ALK multigene family of an *n*-alkane-assimilating yeast, *Yarrowia lipolytica*: cloning and characterization of genes coding for new CYP52 family members. Yeast 16:1077–1087

Ilchenko AP, Mauersberger S, Matyashova RN, Losinov AB (1980) Induction of cytochrome P-450 in the course of yeast growth on different substrates (Russian). Mikrobiologiya 49:452–458

Ilchenko AP, Morgunov IG, Honeck H, Mauersberger S, Vasilkova NN, Müller H-G (1994) Purification and some properties of alcohol oxidase from the yeast *Yarrowia lipolytica* H222 (Russian). Biokhimiya (Moscow) 59:1312–1319

Ilchenko AP, Chernyavskaya OG, Shishkanova NV, Finogenova TV (2003) Induction of cytochrome P-450 and ethanol oxidation in *Yarrowia lipolytica* cells (Russian). Mikrobiologiya 72:168–173

Juretzek T (1999) Entwicklung von Wirts-Vektor-Systemen zur heterologen Expression von Proteinen in der nichtkonventionellen Hefe *Yarrowia lipolytica* und ihre Anwendung für die Cytochrom P450-katalysierte Stoffumwandlung. PhD thesis, Technische Universität Dresden

Juretzek T, Mauersberger S, Barth G (2000) Recombinant haploid or diploid *Yarrowia lipolytica* cells for the functional heterologous expression of cytochrome P450 systems. Patent WO0003008, DE19932811

Kemp GD, Dickinson FM, Ratledge C (1990) Light sensitivity of the n-alkane-induced fatty alcohol oxidase from *Candida tropicalis* and *Yarrowia lipolytica*. Appl Microbiol Biotechnol 32:461–464

Kobayashi S, Hirakawa K, Fukuda R, Ohta A (2008) Disruption of the *SCS2* ortholog in the alkane-assimilating yeast *Yarrowia lipolytica* impairs its growth on *n*-decane, but does not impair inositol prototrophy. Biosci Biotechnol Biochem 72:2219–2223

Lacour T, Achstetter T, Dumas B (1998) Characterization of recombinant adrenodoxin reductase homologue (Arh1p) from yeast. Implication in *in vitro* cytochrome P45011beta monooxygenase system. J Biol Chem 273:23984–23992

Lah L, Krasevec N, Trontelj P, Komel R (2008) High diversity and complex evolution of fungal cytochrome P450 reductase: cytochrome P450 systems. Fungal Genet Biol 45:446–458

Lamb DC, Lei L, Warrilow AG, Lepesheva GI, Mullins JG, Waterman MR, Kelly SL (2009) The first virally encoded cytochrome p450. J Virol 83(16):8266–8269

Lottermoser K, Schunck W-H, Asperger O (1996) Cytochrome P450 of the sophorose lipid-producing yeast *Candida apicola*: heterogeneity and polymerase chain reaction mediated cloning of two genes. Yeast 12:565–575

Machida M, Asai K, Sano M, Tanaka T et al (2005) Genome sequencing and analysis of *Aspergillus oryzae*. Nature 438:1157–1161

Manzella L, Barros MH, Nobrega FG (1998) *ARH1* of *Saccharomyces cerevisiae*: a new essential gene that codes for a protein homologous to the human adrenodoxin reductase. Yeast 14:839–846

Marchal R, Metche M, Vandecasteele J-P (1982) Hydroxylation of alkanes and fatty acids in *Saccharomycopsis lipolytica*. Evidence for the involvement of cytochrome P-450. J Gen Microbiol 128:1125–1134

Mauersberger S (1991) Mutants of alkane oxidation in the yeasts *Yarrowia lipolytica* and *Candida maltosa*. In: Finogenova TV, Sharyshev AA (eds) Alkane metabolism and oversynthesis of metabolites by microorganisms. Center for Biological Research, USSR Academy of Sciences, Pushchino, Russia, pp 59–78

Mauersberger S, Matyashova RN (1980) The content of cytochrome P-450 in yeast cells growing on hexadecane (Russian). Mikrobiologiya (Moscow) 49:571–577

Mauersberger S, Nicaud J-M (2003) Chapter 56. Tagging of genes by insertional mutagenesis in the yeast *Yarrowia lipolytica*. In: Wolf K, Breunig K, Barth G (eds) Non-conventional yeasts in genetics, biochemistry and biotechnology. Practical protocols. Springer, Berlin, pp 343–356

Mauersberger S, Matyashova RN, Müller H-G, Losinov AB (1980) Influence of the growth substrate and the oxygen concentration in the medium on the cytochrome P-450 content in *Candida guilliermondii*. Eur J Appl Microbiol Biotechnol 9:285–294

Mauersberger S, Schunck W-H, Müller H-G (1981) The induction of cytochrome P-450 in *Lodderomyces elongisporus*. Z Allg Mikrobiol 21:313–321

Mauersberger S, Schunck W-H, Müller H-G (1984) The induction of cytochrome P-450 in the alkane-utilizing yeast *Lodderomyces elongisporus*: alterations in the microsomal membrane fraction. Appl Microbiol Biotechnol 19:29–35

Mauersberger S, Kärgel E, Matyashova RN, Müller H-G (1987) Subcellular organization of alkane oxidation in the yeast *Candida maltosa*. J Basic Microbiol 27:565–582

Mauersberger S, Böhmer A, Schunck W-H, Müller H-G (1991) Cytochrome P-450 of the yeast *Yarrowia lipolytica*. In: Abstract international symposium cytochrome P-450 of microorganisms, Berlin, p 63; and Abstracts international conference on biochemistry, biophysics of cytochrome P-450: structure, function, biotechnological and ecological aspects, Moscow

Mauersberger S, Persiyanova TB, Avetisova SM, Sokolov YI, Kärgel E, Kraft R, Schunck W-H, Davidov ER, Müller H-G (1992a) Characterization of two cytochrome P-450 forms purified from the yeast *Candida maltosa*. In: Archakov AI, Bachmanova GI (eds) Cytochrome P-450: biochemistry and biophysics. INCO - TNC, Joint Stock Company, Moscow, pp 651–653

Mauersberger S, Drechsler H, Oehme G, Müller H-G (1992b) Substrate specificity and stereoselectivity of the fatty alcohol oxidase from the yeast *Candida maltosa*. Appl Microbiol Biotechnol 37:66–73

Mauersberger S, Ohkuma M, Schunck W-H, Takagi M (1996) Chapter 12. *Candida maltosa*. In: Wolf K (ed) Non-conventional yeasts in biotechnology. Springer, Berlin, pp 411–580

Mauersberger S, Wang H-J, Gaillardin C, Barth G, Nicaud J-M (2001) Insertional mutagenesis in the *n*-alkane-assimilating yeast *Yarrowia lipolytica*: generation of tagged mutants in genes involved in hydrophobic substrates utilization. J Bacteriol 183:5102–5109

Mauersberger S, Novikova LA, Shkumatov VM (2013) Cytochrome P450 expression in *Yarrowia lipolytica* and its use in steroid biotransformation. In: Barth G (ed) *Yarrowia lipolytica*. Springer, Heidelberg

Mekouar M, Blanc-Lenfle I, Ozanne C, Da Silva C, Cruaud C, Wincker P, Gaillardin C, Neuvéglise C (2010) Detection and analysis of alternative splicing in *Yarrowia lipolytica* reveal structural constraints facilitating nonsense-mediated decay of intron-retaining transcripts. Genome Biol 11:R65

Menzel R, Vogel F, Kärgel E, Schunck W-H (1997) Inducible membranes in yeast: relation to the unfolded-protein-response pathway. Yeast 13:1211–1229

Mlícková K, Roux E, Athenstaedt K, d'Andrea S, Daum G, Chardot T, Nicaud J-M (2004) Lipid accumulation, lipid body formation, and acyl coenzyme A oxidases of the yeast *Yarrowia lipolytica*. Appl Environ Microbiol 70:3918–3924

Nelson DR (2009) The cytochrome P450 homepage. Hum Genomics 4:59–65

Novikova LA, Faletrov YV, Kovaleva IE, Mauersberger S, Luzikov VN, Shkumatov VM (2009) From structure and functions of steroidogenic enzymes to new technologies of gene engineering. Review. Biochemistry (Moscow) 74:1482–1504

Oeda K, Sakaki T, Ohkawa H (1985) Expression of rat liver cytochrome P-450MC cDNA in *Saccharomyces cerevisiae*. DNA 4:203–210

Ohkuma M, Hikiji T, Tanimoto T, Schunck W-H, Müller H-G, Yano K, Takagi M (1991a) Evidence that more than one gene encodes *n*-alkane-inducible cytochrome P-450s in *Candida maltosa*, found by two-step gene disruption. Agric Biol Chem 55:1757–1764

Ohkuma M, Tanimoto T, Yano K, Takagi M (1991b) *CYP52* (cytochrome P450alk) multigene family in *Candida maltosa*: molecular cloning and nucleotide sequence of the two tandemly arranged genes. DNA Cell Biol 10:271–282

Ohkuma M, Muraoka S-I, Tanimoto T, Fujii M, Ohta A, Takagi M (1995a) *CYP52* (cytochrome P450alk) multigene family in *Candida maltosa*: identification and characterization of eight members. DNA Cell Biol 14:163–173

Ohkuma M, Park S-M, Zimmer T, Menzel R, Vogel F, Schunck W-H, Ohta A, Takagi M (1995b) Proliferation of intracellular membrane structures upon homologous overproduction of cytochrome P-450 in *Candida maltosa*. Biochim Biophys Acta (Biomembranes) 1236:163–169

Ohkuma M, Masuda Y, Park S-M, Ohtomo R, Ohta A, Takagi M (1995c) Evidence that the expression of the gene for NADPH-cytochrome P-450 reductase is *n*-alkane-inducible in *Candida maltosa*. Biosci Biotechnol Biochem 59:1328–1330

Ohkuma M, Zimmer T, Iida T, Schunck W-H, Ohta A, Takagi M (1998) Isozyme function of *n*-alkane-inducible cytochromes P450 in *Candida maltosa* revealed by sequential gene disruption. J Biol Chem 273:3948–3953

Ortiz de Montellano PR (ed) (2005) Cytochrome P450: structure, mechanism and biochemistry, 3rd edn. Plenum, New York

Peterson JA (1970) Cytochrome content of two Pseudomonads containing mixed function oxidase systems. J Bacteriol 103:714–721

Petzsch P (2012) Generierung eines funktionellen Kataloges für das Genom der Hefe *Yarrowia lipolytica* und dessen Anwendung bei Microarray-Untersuchungen zur Genexpression. PhD thesis, Technische Universität Dresden

Pompon D, Louerat B, Bronine A, Urban P (1996) Yeast expression of animal and plant P450s in optimized redox environments. Methods Enzymol 272:51–64

Sakaki T, Kominami S, Hayashi K, Akiyoshi-Shibata M, Yabusaki Y (1996) Molecular engineering study on electron transfer from NADPH-P450 reductase to rat mitochondrial P450c27 in yeast microsomes. J Biol Chem 271:2609–26213

Sanglard D, Fiechter A (1989) Heterogeneity within the alkane-inducible cytochrome P450 gene family of the yeast *Candida tropicalis*. FEBS Lett 256:128–133

Scheller U, Zimmer T, Kärgel E, Schunk W-H (1996) Characterization of the *n*-alkane and fatty acid hydroxylating cytochrome P450 forms 52A3 and 52A4. Arch Biochem Biophys 328:245–254

Scheller U, Zimmer T, Becher D, Schauer F, Schunck W-H (1998) Oxygenation cascade in conversion of *n*-alkanes to α, ω-dioic acids catalyzed by cytochrome P450 52A3. J Biol Chem 273:32528–32534

Schiffler B, Bureik M, Reinle W, Müller EC, Hannemann F, Bernhardt R (2004) The adrenodoxin-like ferredoxin of *Schizosaccharomyces pombe* mitochondria. J Inorg Biochem 98:1229–1237

Schunck W-H, Mauersberger S, Huth J, Riege P, Müller H-G (1987a) Function and regulation of cytochrome P-450 in alkane-assimilating yeast I. Selective inhibition with carbon monoxide in growing cells. Arch Microbiol 147:240–244

Schunck W-H, Mauersberger S, Kärgel E, Huth J, Müller H-G (1987b) Function and regulation of cytochrome P-450 in alkane-assimilating yeast II. Effect of oxygen-limitation. Arch Microbiol 147:245–248

Schunck W-H, Kärgel E, Gross B, Wiedmann B, Mauersberger S, Köpke K, Kießling U, Strauss M, Gaestel M, Müller H-G (1989) Molecular cloning and characterization of the primary structure of the alkane hydroxylating cytochrome P-450 from the yeast Candida *maltosa*. Biochem Biophys Res Commun 181:843–850

Schunck W-H, Vogel F, Gross B, Kärgel E, Mauersberger S, Köpke K, Gengnagel C, Müller HG (1991) Comparison of two cytochromes P-450 from *Candida maltosa*: primary structures, substrate specificities and effects of their expression in *Saccharomyces cerevisiae* on the proliferation of the endoplasmic reticulum. Eur J Cell Biol 55:336–345

Seghezzi W, Meili C, Ruffiner R, Kuenzi R, Sanglard D, Fiechter A (1992) Identification and characterization of additional members of the cytochrome P450 multigene family CYP52 of *Candida tropicalis*. DNA Cell Biol 11:767–780

Shkumatov VM, Smettan G (1991) Heterologous reconstruction of monooxygenase systems: substrate specific cytochrome P-450-microorganism (Russian). In: Finogenova TV, Sharyshev AA (eds) Alkane metabolism and oversynthesis of metabolites by microorganisms. Center for Biological Research, USSR Academy of Sciences, Pushchino, Russia, pp 78–86

Shkumatov VM, Usova EV, Poljakov YS, Frolova NS, Radyuk VG, Mauersberger S, Chernogolov AA, Honeck H, Schunck W-H (2002) Biotransformation of steroids by a recombinant yeast strain expressing bovine cytochrome P-45017α. Biochemistry (Moscow) 67:456–467

Shkumatov VM, Frolova NS, Rudaya EV, Faletrov YV, Mauersberger S, Barth G (2006) Range of substrates and steroid bioconversion reactions performed by recombinant microorganisms *Saccharomyces cerevisiae* and *Yarrowia lipolytica* expressing cytochrome P450c17. Prikl Biokhim Mikrobiol 42:539–546

Sood N, Lal B (2008) Isolation of a novel yeast strain *Candida digboiensis* TERI ASN6 capable of degrading petroleum hydrocarbons in acidic conditions. J Environ Manage 90:1728–1736

Sumita T, Iida T, Hirata A, Horiuchi H, Takagi M, Ohta A (2002a) Peroxisome deficiency represses the expression of *n*-alkane-inducible *YlALK1* encoding cytochrome P450ALK1 in *Yarrowia lipolytica*. FEMS Microbiol Lett 214:31–38

Sumita T, Iida T, Yamagami S, Horiuchi H, Takagi M, Ohta A (2002b) YlALK1 encoding the cytochrome P450ALK1 in *Yarrowia lipolytica* is transcriptionally induced by *n*-alkane through two distinct *cis*-elements on its promoter. Biochem Biophys Res Commun 294:1071–1078

Takagi M, Moriya K, Yano K (1980) Induction of cytochrome P450 in petroleum-assimilating yeast. I. Selection of a strain and basic characterization of cytochrome P450 induction in the strain. Cell Mol Biol 25:363–369

Takai H, Iwama R, Kobayashi S, Horiuchi H, Fukuda R, Ohta A (2012) Construction and characterization of a *Yarrowia lipolytica* mutant lacking genes encoding cytochromes P450 subfamily 52. Fungal Genet Biol 49(1):58–64

Tanaka A, Fukui S (1989) Metabolism of *n*-alkanes. In: Rose AH, Harrison JS (eds) The yeast. Metabolism and physiology of yeasts, vol 3, 2nd edn. Academic, London, pp 261–287

Thevenieau (2006) Ingenierie metabolique de la levure *Yarrowia lipolytica* pour la production d'acides dicarboxyliques a partier d'huiles vegetales. PhD thesis, INRA, Paris-Grignon

Thevenieau F, Le Dall MT, Nthangeni B, Mauersberger S, Marchal R, Nicaud JM (2007) Characterization of Yarrowia lipolytica mutants affected in hydrophobic substrate utilization. Fungal Genet Biol 44(6):531–542 (Epub 2006 Oct 30)

Thevenieau F, Nicaud J-M, Gaillardin C (2009) Chapter 26. Applications of the non-conventional yeast *Yarrowia lipolytica*. In: Satyanarayana T, Kunze G (eds) Yeast biotechnology: diversity and applications. Springer, Dordrecht, pp 589–613. doi:10.1007/978-1-4020-8292-4_26

Thevenieau F, Beopoulos F, Desfougeres T, Sabirova J, Albertin K, Zinjarde S, Nicaud J-M (2010) Uptake and assimilation of hydrophobic substrates by the oleaginous yeast *Yarrowia lipolytica*. In: Timmis KN (ed) Handbook of hydrocarbon and lipid microbiology, vol 2, part 7, chapter 48. Springer, Berlin, pp 1513–1527

Urban P, Truan G, Bellamine A, Laine R, Gautier JC, Pompon D (1994) Engineered yeasts simulating P450-dependent metabolisms: tricks, myths and reality. Review. Drug Metabol Drug Interact 11:169–200

van den Brink HM, van Gorcom RF, van den Hondel CA, Punt PJ (1998) Cytochrome P450 enzyme systems in fungi. Fungal Genet Biol 23:1–17

Vogel F, Gengnagel C, Kärgel E, Müller H-G, Schunck W-H (1992) Immunocytochemical localization of alkane-inducible cytochrome P450 and its NADPH-dependent reductase in the yeast *Candida maltosa*. Eur J Cell Biol 57:285–291

Wortman JR, Gilsenan JM, Joardar V et al (2009) The 2008 update of the *Aspergillus nidulans* genome annotation: a community effort. Fungal Genet Biol 46(Suppl 1):S2–S13

Wright R (1993) Insights from inducible membranes. Curr Biol 3:870–873

Yamagami S, Iida T, Nagata Y, Ohta A, Takagi M (2001) Isolation and characterization of acetoacetyl-CoA thiolase gene essential for *n*-decane assimilation in yeast *Yarrowia lipolytica*. Biochem Biophys Res Commun 282:832–838

Yamagami S, Morioka D, Fukuda R, Ohta A (2004) A basic helix-loop-helix transcription factor essential for cytochrome P450 induction in response to alkanes in yeast *Yarrowia lipolytica*. J Biol Chem 279:22183–22189

Zimmer T, Ohkuma M, Ohta A, Takagi M, Schunck W-H (1996) The *CYP52* multigene family of *Candida maltosa* encodes functionally diverse *n*-alkane-inducible cytochromes P450. Biochem Biophys Res Commun 224:784–789

Zimmer T, Iida T, Schunck W-H, Yoshida Y, Ohta A, Takagi M (1998) Relation between evolutionary distance and enzymatic properties among the members of the CYP52A subfamily of *Candida maltosa*. Biochem Biophys Res Commun 251:244–247

Index

A
Acetic acid, 139
Acetyl-CoA synthetase, 143
Aconitate hydratase, 139
Acyl-CoA
  oxidase, 158, 163, 238, 244
  synthetase, 229
Adrenodoxin, 174, 179, 207, 210
Adrenodoxin homologous yeast gene *YAH1*, 187, 254
*A*drenodoxin *r*eductase *h*omologues 1 (*ARH1*), 184, 186, 253
Aerobic sterol exclusion phenomenon, 209
Alcoholysis, 130
Aldo-keto reductase superfamily AKR, 203
Alkaline extracellular protease (AEP), 78. *See also* XPR2
  induction, 92
  processing, 79–82
    dipeptidyl aminopeptidase, 79
    Xpr6p processing, 80
  signal peptide cleavage
    prediction, 79
    radiosequencing, 79
  translocation
    co-translational, 82
    SRP-dependent, 82
Alkane(s), 153, 228, 234
  assimilating yeasts, 228
  induced P450s, 190
  inducibility, 246
  inducible P450 genes, 231, 240, 256
  oxidation, 229
  responsive *cis*-acting elements, 248
  responsive elements, 160
  utilizing yeast, 181
Alkane hydroxylase (AHA), 229, 236
Alkane monooxygenase system (AMOS), 158
ALK gene(s), 229
ALK gene disruption, 249
Alk mutants, 234
ALK1 promoter activity, 242
Amino acid sensing
  OPT1, 85
  SSY5, 85
Amylase, 47, 48, 52, 64
Androstenedione, 186
Application(s)
  environmental, 4, 37, 56, 62, 63
  industrial, 4, 12, 56, 59, 60, 62
  therapeutic, 58, 60, 61, 62, 65
  of two-phase systems, 218
Aroma precursors, 131
ARS/CEN low copy expression plasmid, 190
Association of ER and peroxisomes, 256
ATP-citrate lyase, 143
AXP processing
  pH dependent autoprocessing, 82
  predicted signal peptide cleavage, 82
AXP1 regulation
  maximum at acidic pH, 86
  Rim pathway involved, 86
AXP1 sequence
  CLIB122, 82
  Yarrowia lipolytica 148, 82
AXP1/XPR2 regulatory mutations
  MED4, 85
  NUP85, 85
  OPT1, 85
  Rim pH signaling pathway, 85
  SSY5, 85
  VPS28, 85

G. Barth (ed.), *Yarrowia lipolytica*, Microbiology Monographs 25,
DOI 10.1007/978-3-642-38583-4, © Springer-Verlag Berlin Heidelberg 2013

B
Baker's yeast, 183
Benzoate para-hydroxylase, 182
Biocatalysis, 152
Biochemical oxygen demand (BOD), 123
Bioconversion, 57, 59, 61
  of cholesterol, 215, 218
  of hydrophobic substrates, 182
Bioplastics, 61
Biosensor, 63
Biotransformation
  capacity, 218
  of hydrophobic substrates, 218
  reactions, 181

C
Campesterol, 183, 208
Candida spp., 174
  C. maltosa, 158, 179, 229
  C. tropicalis, 158, 179
Carbon monoxide, 236, 238
Cell density, 52
Cell surface hydrophobicity, 256
Chain-lengths specificities, 242, 245, 250, 251
Cheese ripening, 131–132
Chemical oxygen demand (COD), 123
Chiral auxiliaries, 128
Cholesterol, 174, 183, 208
Cholesterol side-chain cleavage P450scc, 182
Citrate synthase, 141
Citric acid (CA), 133–134, 138–140
Class II P450, 256
CO-difference spectra (COD), 190, 230, 233, 235
Codon bias, 33, 45
Coexpression of P450c17, 185
CO-inhibition, 237
Comparison with
  Aspergillus niger, 45, 49, 53, 55
  Hansenula polymorpha, 6, 65
  Kluyveromyces lactis, 3, 53
  mammals, 5, 6, 7, 45, 65
  Pichia pastoris, 3, 6, 49, 53, 54, 65
  Saccharomyces cerevisiae, 4, 5, 6, 12, 36, 39, 42, 45, 54
Copy number(s), 192
Copy number ratios, 212
Corticosterone, 184
CO-sensitivity, 238
Cosmetics, 153
Crossover
  double, 10
  single, 10, 37
C22 sterol desaturase, 242
Culture medium, 14, 50, 52
CYP11A1, 182
CYP52A subfamily, 245
CYP74B, 181
CYP2D6, 182
CYP52 family, 174, 229, 244
Cytochrome b_5, 178, 252
Cytochrome oxidase, 192, 232, 237
Cytochrome P450, 229, 231–240
Cytochrome P450 reductase (CPR), 158, 173, 178
Cytokinin oxidase, 6, 11, 43, 47, 48, 49, 50, 53, 56

D
Dairy products, 130
DCA production, 242
Debaryomyces formicarius, 239
γ-Decalactone, 162
Deletion, 10, 12, 37, 38, 65
Deoxycortisol, 184
α,ω-Dicarboxylic acids (DCA), 158–160, 230
α,ω-Dicarboxylic compounds, 153–155
Differential gene expression, 247
Dimorphic, 4
Dimorphism, 50, 64
Diploidization, 218
Diploid *Yarrowia lipolytica* strain lines, 192, 197
Disruption. *See* Deletion
Docking platform, 12, 37, 41

E
Electron transfer components, 175, 178, 208
Electron transfer protein, 179, 240, 252, 255
Electron-transfer protein 1 (ETP1), 187, 254
Enantiomer, 128
Enantioselectivity, 127
Endoplasmic reticulum (ER), 195, 235
Epoxide hydrolase, 52, 53, 57
Ergosterol, 166
Ergosterol biosynthesis, 243, 255
ER-proliferation, 197
Escherichia coli, 175
Ethanol, 139, 190, 206, 234
7-Ethoxyresorufin O-deethylase, 181
Ethyl oleate, 182, 218
E value, 128
Extracellular acid proteases
  AXP processing, 82
  one *versus* three, 88

F
Fatty acid(s), 4, 32, 41, 43, 57, 60, 153, 228
Fatty acid hydroperoxide lyase, 181
Fatty acid ω-hydroxylase (FAHA), 229, 248
Fatty acid ω-oxidation, 255
Fermented sausages, 132–133
Filamentous fungi, 244, 252
Fine chemistry, 127–130
Fish waste, 126
Flavor compounds, 131
Fluoroacetate, 139
Formation of steryl esters, 209
Fumarase (FUM), 143
Fumarate (FA), 143
Functional heterologous expression, 181
Fusarium verticillioidis, 175
Fusion protein mAdx-mP450scc, 210

G
β-Galactosidase expression, 32, 39, 42, 44, 51, 247
Gene disruption, 245, 246, 256
Gene-dose-effect, 193
Gene duplications, 256
Generally Regarded As Safe (GRAS), 3, 4, 62
Genetic engineering, 37, 56, 60, 64
Genome, 4, 12
GHSLGAA motif, 103
Glucose, 234
Glucose repression, 235
Glycerol, 139, 234
Glycosylation, 6, 49, 65
GO Term Finder, 94
Green-notes aroma, 181
Gum arabic, 111
GX type lipase, 110

H
Heterologous expression, 256
Heterologous P450, 179
Heterologous protein production
  axp1 Äxpr2 host, 87
  UAS1B hybrid promoters, 87
  XPR2 prepro-region, 87
  XPR2 promoter, 87
Hexadecane, 190, 206, 234
Hexadecane-1-ol, 233, 235
Hexokinase, 105
High copy replicative vector, 187
High-throughput screening, 41
Homologs to *XPR2*, 90
hp4d promoter, 13, 14, 32, 37, 41, 42, 43, 50
HPO lyase, 181
20α,β-HSD, 186
Hydrocarbon-assimilating yeast, 175
Hydrocortisone, 184
α/β-Hydrolases, 100
Hydrophobic compounds, 174
Hydrophobic substrate, 103
Hydrophobic substrate and product transport, 256
ω-Hydroxyacids, 153–155
3-Hydroxy-γ-decalactone, 163
Hydroxy-fatty acids, 155
17α-Hydroxylase, 200
ω-Hydroxylase activity, 246
ω-Hydroxylation of fatty acids, 245
17α-Hydroxypregnenolone, 203, 207
17α-Hydroxyprogesterone, 184, 207
Hydroxysteroid dehydrogenase, 185, 206
20α-, 20β-, and 17β-Hydroxysteroid dehydrogenases (HSDs), 203
Hyperglycosylation, 6, 49

I
Inducible membranes, 195
Induction of P450, 230
Integration
  multiple, 13, 37, 38, 40, 51, 52
  random, 39
  targeting, 10, 37, 38, 39, 41, 63
Integrative multicopy transformants, 192
Integrative multicopy vectors, 191
Intron, 32, 45
Inulinase, 59, 60, 61
Invertase, 12, 13, 46, 47, 49, 139
In vitro reconstitution experiments, 207
Isocitrate dehydrogenases, 141
Isocitrate lyase (ICL), 139
Isocitrate lyase promoter pICL1, 181, 188
Isocitric acid (ICA), 138, 139

K
Karmellae-like membrane stacks, 195–196, 253
α-Ketoglutarate dehydrogenase (KGDH), 143
α-Ketoglutaric acid (KGA), 138, 141–144
20-Keto reduction, 184
Kit, 3, 37, 54, 56
Kluyveromyces lactis, 175

L
Laccase, 47, 48, 49, 50, 51, 53, 56
Lactones, 155
Lanosterol 14α-demethylase, 242
Lauric acid ω-hydroxylase, 251
LIP2, 101
Lipase, 6, 39, 41, 46, 49, 50, 51, 57, 65, 100
  immobilization, 115
  isoenzyme, 114
  production, 104
  purification, 107
Liquid two-phase systems, 174
Long-chain fatty alcohol, 234
Long-chain triglycerides, 113
Low copy transformants, 192
17,20-Lyase, 200

M
Macrocyclic ketones, 154
Macrocyclic lactones, 154
Malate (MA) synthase, 141, 143
Mammalian steroidogenesis, 183
Mammalian steroidogenic D450, 185
Mature AEP activity
  pro-mAEP inhibition, 81
Mature (m) protein forms, 183
Membrane fluidity, 166
MEROPS Peptidase Database, 87
Methyl-β-cyclodextrin, 166
Methyloleate, 106
Microarray analysis, 247
Microsomal membrane fraction, 232, 235
Monopropylene glycol, 112
Multicopy integration vectors, 218
Multicopy selection marker, 191
Multicopy transformants, 182, 212
Multigene families, 231, 256
Multiple expression cassettes, 192
Multiple paralog P450 *CYP52* genes, 255
Musk, 153

N
NADH-b_5 reductases, 186
NADH-cytochrome b_5 reductase, 252
NADPH-adrenodoxin reductase, 174, 179, 207, 210
NADPH cytochrome c reductase, 193, 236
NADPH-dependent P450 reductase, 229, 240, 252
NADPH-P450 reductase, 179, 193, 207
n-alkanes, 140
Natural, 152
N-formyltyrosine oxidase, 242
Nicotiana benthamiana, 251
Nitrogen limitiation, 139
N-methyl-*N*'-nitro-*N*-nitrosoguanidine, 106
n-paraffin, 141
Nucleophilic elbow, 100

O
O-acetyltransferase, 184, 203
Oleaginous yeast, 181
Oleic acid, 105
Olive mill, 123
Oxidation, 153
β-Oxidation, 160, 229
ω-Oxidation, 158, 229
Oxyanion hole, 100
Oxygen
  concentration, 141
  limitation, 233, 237, 239

P
P4501A1, 181
P450 activity, 174
P450 *ALK* genes, 240–244
P450ALK substrate specificity, 250
Palmitic acid, 235
P45053B1, 182
P450c17, 175, 178
P450 catalyzed biotransformations, 181
P450c17-catalyzed progesterone biotransformation, 190
P450 14DM, 242, 243, 255
P450 22DS, 242, 243, 255
Peptidase family lineage specific expansion
  A1 (pepsin family), 88
  S8 (subtilisin family), 88
Peroxisome, 35
pH values, 141
Phytosterols, 208
Pichia pastoris, 175
Plasma membrane associated ER, 196, 253, 256
P17M translocation
  post-translational, 82
  SRP-independent, 82
Polymer, 153
Polymerisation reactions, 129–130
Pregnenolone, 183, 208, 215
Primary alkane oxidation, 255
Pristane (2,6,10,14-tetramethylpentadecane), 233

Progesterone, 182, 185
17α-hydroxylase, 190
biotransformation, 205, 206
biotransformation activity, 192, 197
Proliferation
of different ER types, 253
of ER, 195
Pro-mAEP forms dimers, 81
Promoter pGAL10, 187
P450scc, 178, 210, 215
Pyruvate carboxylase (PYC), 143
Pyruvic acid (PA), 138

R
Racemic mixture, 127–129
Rapeseed oil, 142
Recombinant diploid strains, 213
Reconstituted P450scc-system, 215
Retrotransposon. *See* Zeta
Rhodotorula minuta, 182
Ribosomal DNA (rDNA), 38, 63
Ricinoleic acid, 155, 161

S
Saccharomyces cerevisiae, 175
Schizosaccharomyces pombe, 175
Secretion, 3, 5
Secretory pathway induced genes
SSA4, 94
SSO1, 94
VPS13, 94
Secretory pathway regulation
genomic transcription, 91
secretory demand, 91
Secretory pathway repressed genes
SEC61, 95
SRP14, 95
SRP21, 95
Selective CO-inhibition, 239
Self-sufficient steroid biosynthesis, 183
Sequence. *See* Genome
Sewage sludge, 126
Side-chain cleavage P450scc-system, 185
Side products
formation, 200
reduction, 202
Simultaneous integration in clusters, 213
Simultaneous integrative transformation, 210
Single cell
oil, 60
protein, 4, 59, 60, 61
β-Sitosterol, 208, 215
Spectral interference, 234
Spore wall maturation, 242
Square zones
skim milk (SKM) plates, 80
xpr6 segregants, 80
Steroid
bioconversion capacity, 197
biotransformation, 181, 195
hormones, 173
hydroxylation, 186
Steroidogenic enzymes, 174
Sterol
bioconversion, 216
D7-reductase, 183
fraction of rape oil distillate, 208
homeostasis, 166
Steryl ester
hydrolases, 209
synthases, 184, 209
Strains, 12, 37, 38, 39, 43, 60
Substrate induction, 239
Substrate specificity, 245
of P450c17, 200
Substrate turnover number, 238, 256
Succinate (SA), 143
Succinate dehydrogenase (SDH), 145
Succinic acid, 138, 145–146
Sucrose, 4, 12, 13, 37, 50
Sunflower oil, 140, 242, 251
Surface display, 36, 54, 61

T
Tag, 36, 42, 61
Terminal hydroxylation of alkanes, 255
Testosterone, 182, 185
Tetrahedral intermediate, 110
Thiamine, 138
Transcriptional regulation, 248
Transcriptional response
AEP induction, 93
DNA microarray, 93
secretion stress, 93
Transcription factors Yas1p and Yas2p, 247, 248
Transesterification, 130
Transformation, 10, 12, 37, 40
Translocation pathway choice
signal peptide hydrophobicity, 83
signal peptide kinkiness, 83
Tricarboxylic-acid-cycle, 138
Two-liquid biphasic culture systems, 182, 218
Type I spectral change, 208

U
UAS1
  cis-acting element, 83
  insensitivity to repressing conditions, 84
UAS2
  cis-acting element, 83
  C-, N-and pH regulation, 84
Unfolded protein response, 95

W
Waste treatment, 122–126
Whole cell biotransformation, 181

X
XPR2, 5, 12, 14, 34, 38, 43, 46, 64
XPR2 multicopy strains, 92
  AEP production, 93
XPR2 pH regulation
  UAS2A and UAS2B, 84
  YlRIM101, 85
Xpr6p protease, 48, 64
XPR2 promoter
  deletion analysis, 83
  trans-acting factors, 84
XPR2 sequence
  CLIB122, 78
  CX161-1B, 78
  NRRL Y-1094, 78
  W29, 78

Y
Yarrowia lipolytica, 174, 228
Yeast cell factory, 184, 218

Z
Zeta, 11, 12, 38, 39, 40, 41